T0289929

The Fractional Laplacian

The
Fractional
Laplacian

Wenxiong Chen
Yeshiva University, USA

Yan Li
Yeshiva University, USA

Pei Ma
Nanjing Forestry University, China

World Scientific

NEW JERSEY · LONDON · SINGAPORE · BEIJING · SHANGHAI · HONG KONG · TAIPEI · CHENNAI · TOKYO

Published by

World Scientific Publishing Co. Pte. Ltd.

5 Toh Tuck Link, Singapore 596224

USA office: 27 Warren Street, Suite 401-402, Hackensack, NJ 07601

UK office: 57 Shelton Street, Covent Garden, London WC2H 9HE

Library of Congress Cataloging-in-Publication Data

Names: Chen, Wenxiong (Mathematician), author.

Title: The fractional Laplacian / Wenxiong Chen, Yeshiva University, USA,
    Yan Li, Yeshiva University, USA, Pei Ma, Nanjing Forestry University, China.

Description: New Jersey : World Scientific Publishing Co. Pte. Ltd., [2020] | Includes bibliographical references.

Identifiers: LCCN 2020013220 | ISBN 9789813223998 (hardcover) | ISBN 9789813224001 (ebook) |
    ISBN 9789813224018 (ebook other)

Subjects: LCSH: Laplacian operator. | Fractional differential equations.

Classification: LCC QA374 .C457 2020 | DDC 515/.3533--dc23

LC record available at https://lccn.loc.gov/2020013220

British Library Cataloguing-in-Publication Data

A catalogue record for this book is available from the British Library.

For any available supplementary material, please visit
https://www.worldscientific.com/worldscibooks/10.1142/10550#t=suppl

Contents

viii   Contents

Preface

Recently, fractional Laplacian has seen more and more applications in various branches of sciences and has attracted great attention from the mathematical community. Our research experience on this subject during the last few years led us to a point where we sensed a need to write a book that deals with such nonlocal operators comprehensively and thoroughly, to address the necessity in providing a good reading material for graduate students and young researchers to help them step into this interesting field quickly, and also to offer a good handbook for those who come across nonlocal operators in their research and want to know more.

We try to make this book unique in providing a comprehensive understanding of nonlinear equations involving the fractional Laplacian as well as other nonlocal operators. Beginning from the definition of the fractional Laplacian, and after the necessary preparations for basic knowledge, it gradually guides the readers to the frontier of current research. The explanations and illustrations are clear and elementary enough so that the first-year graduate students can follow rather easily, while it is advanced enough to include many new ideas, methods, and developments that just appeared recently in research literature so that even the professionals in this field would find them helpful in doing their research. It focuses on introducing direct methods on the nonlocal problems without going through extensions, such as the direct methods of moving planes, direct method of moving spheres, and direct blowing up and rescaling arguments and so on. Different from most other books, it emphasizes on illuminating the ideas behind the formal concepts and proofs by using simple examples, so that readers can quickly grasp the essence and will be able to apply them to solve various problems in this area.

Chapters 1-4 are devoted to the preparations for the needed basic material.

In Chapter 1, we introduce three different definitions of the fractional Laplacian via the singular integral, the Fourier transform, and the extension. Then we show the equivalences among them. Chapter 2 shows how to obtain the explicit formula for the Green's functions for both Laplacian and fractional Laplacian on some special domains, such as the unit ball, the half space, and the whole space. In Chapter 3, a simple maximum principle is established via two different approaches. In Chapter 4, we establish the Poisson representation for fractional harmonic functions. In Chapter 5, based on the Poisson representation, we derive two Liouville theorems for fractional harmonic functions, one on the whole space and the other on a half space. Then we show how to use these Liouville theorems to establish the equivalence between pseudo-differential equations and integral equations in both spaces.

Chapter 6 is the highlight of the book. It summarizes the recent developments in the direct method of moving planes for fractional Laplacians, a new idea introduced in [CLL] and its applications to various equations and systems involving fractional Laplacians as well as other nonlocal pseudo-differential operators in different domains. We believe that, after learning the techniques in this chapter, readers will be able to apply them to solve problems in the current research front.

In Chapter 7, we introduce another method to study fractional equations, the method of moving planes in integral forms. Since its appearance in [CLO] a decade ago, this method has been applied widely to solve a variety of problems involving fractional equations and higher order equations.

In Chapter 8, another direct method, the method of moving spheres on fractional equations is illustrated. Although this method does not enjoy such broad applications as the method of moving planes, it is simpler in the cases where it works.

Chapter 9 deals with the direct method of blowing up and rescaling and its applications in obtaining the a priori estimate on the solutions of fractional equations. Combining this with the Leray-Schauder degree theory, one will be able to derive the existence of solutions.

Chapter 10 is devoted to the variational methods and Pohozaev identities. The former is a powerful tool in obtaining the existence of weak solutions, and the latter can be adapted to show the non-existence of solutions.

In Chapter 11, we study higher order fractional equations via two different approaches. One is to reduce them to systems of lower order ones, the other is to consider an equivalent integral equation.

Chapter 12 focuses on the regularity of solutions to fractional equations in bounded domains with Dirichlet conditions. An interior Schauder type estimate is established and a Hölder continuity is obtained near the boundary.

1

Introduction to Fractional Laplacian

1.1 Overview

In recent years, there has been a great deal of interest in using the fractional Laplacian to model diverse physical phenomena, such as anomalous diffusion and quasi-geostrophic flows, turbulence and water waves, molecular dynamics, and relativistic quantum mechanics of stars (see [BoG] [CaV] [Co] [TZ] and the references therein). It also has various applications in probability and finance (see [A] [Be] [CT]). In particular, the fractional Laplacian can be understood as the infinitesimal generator of a stable Lévy diffusion process (see [Be]) and appears in anomalous diffusions in plasmas, flames propagation and chemical reactions in liquids, population dynamics, geographical fluid dynamics, and American options in finance. We refer the readers to Di Nezza, Palatucci, and Valdinoci's survey paper [NPV] for a detailed exposition of the function spaces involved in the analysis of the operator and a long list of relevant references.

1.2 Definitions

The fractional Laplacian in $\mathbb{R}^n$ is a nonlocal operator, taking the form

$$(-\triangle)^{\frac{\alpha}{2}} u(x) = C_{n,\alpha} \, PV \int_{\mathbb{R}^n} \frac{u(x) - u(z)}{|x - z|^{n+\alpha}} dz, \qquad (1.1)$$

where α is any real number between 0 and 2, PV stands for the Cauchy principal value, and

$$C_{n,\alpha} = \frac{2^\alpha \Gamma(\frac{n+\alpha}{2})}{\pi^{n/2} \Gamma(-\frac{\alpha}{2})}.$$

This singular integral can be evaluated as

$$\lim_{\epsilon \to 0^+} \int_{\mathbb{R}^n \backslash B_\epsilon(x)} \frac{u(x) - u(z)}{|x - z|^{n+\alpha}} dz. \qquad (1.2)$$

1

In order that the integral on the right-hand side of (1.1) is well defined, we require that $u \in C_{loc}^{1,1} \cap \mathcal{L}_\alpha$, with

$$\mathcal{L}_\alpha = \{u \in L_{loc}^1 \mid \int_{\mathbb{R}^n} \frac{|u(x)|}{1 + |x|^{n+\alpha}} dx < \infty\}.$$

In the case $\alpha \geq 1$, it is necessary to take the principal value of the integral. The following simple example in $\mathbb{R}^1$ shows the difference between a singular integral defined in the usual sense and in the PV sense.

We say that the integral

$$\int_{-1}^1 f(x)dx$$

converges in the usual sense if and only if both limits

$$\lim_{\delta \to 0} \int_{-1}^\delta f(x)dx \text{ and } \lim_{\epsilon \to 0} \int_\epsilon^1 f(x)dx$$

exist. Obviously, for $f(x) = \frac{1}{x}$, none of the above limits exist and hence the integral diverges.

However, if we consider it in a principal value sense, then

$$PV \int_{-1}^1 \frac{1}{x}dx = \lim_{\epsilon \to 0} \left(\int_{-1}^{-\epsilon} \frac{1}{x}dx + \int_\epsilon^1 \frac{1}{x}dx \right)$$
$$= 0.$$

Here the positive one cancels with the negative one, and thus the integral converges in the PV sense.

In (1.1), $\frac{1}{|x-z|^{n+\alpha}}$ is singular at $z = x$, while the numerator

$$u(x) - u(z) = -\nabla u(x) \cdot (z - x) + O(|z - x|^2)$$

can produce at most one power of $|x - z|$ to cancel out this singular effect. Hence in the case $\alpha \geq 1$, the integral diverges in the usual sense.

Yet the PV definition of the integral manages to overcome this "singularity", as can be seen from below. Assume that $u \in C_{loc}^{1,1}(\mathbb{R}^n)$. By Taylor expansion, we have

$$PV \int_{B_1(x)} \frac{u(x) - u(z)}{|x - z|^{n+\alpha}} dz$$

$$= \lim_{\epsilon \to 0} \int_{B_1(x) \backslash B_\epsilon(x)} \frac{u(x) - u(z)}{|x - z|^{n+\alpha}} dz$$

$$= \lim_{\epsilon \to 0} \int_{B_1(x) \backslash B_\epsilon(x)} \frac{-\nabla u(x) \cdot (z - x) + O(|z - x|^2)}{|z - x|^{n+\alpha}} dz$$

$$= \lim_{\epsilon \to 0} \int_{B_1(x) \backslash B_\epsilon(x)} \frac{O(|z - x|^2)}{|z - x|^{n+\alpha}} dz$$

$$\leq C \lim_{\epsilon \to 0} \int_{B_1(x) \backslash B_\epsilon(x)} \frac{1}{|z - x|^{n+\alpha-2}} dz$$

$$< \infty.$$

The third equality is a consequence of the symmetry of the domain of integration. In fact, since $\nabla u(x)$ is fixed, for any $z_1 \in \partial B_r(x)$, there exists a corresponding $z_2 \in \partial B_r(x)$ (see Figure 1.1) such that

$$\nabla u(x) \cdot (z_1 - x) = -\nabla u(x) \cdot (z_2 - x).$$

Therefore,

$$\int_{B_1(x)\backslash B_\epsilon(x)} \frac{-\nabla u(x) \cdot (z - x)}{|z - x|^{n+\alpha}} dz = 0,$$

and the third equality holds.

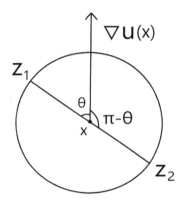

Fig. 1.1

For usual differential operators, such as the Laplacian, if a function u is identically zero near a point x, then $\triangle u(x) = 0$. However, this is not the case for the fractional Laplacian. For instance, let Ω be a bounded domain in $\mathbb{R}^n$; and $u > 0$ in Ω, $u = 0$ in $\mathbb{R}^n \backslash \Omega$. Let x be a point in $\mathbb{R}^n \backslash \Omega$. Then

$$(-\triangle)^{\alpha/2} u(x) = C_{n,\alpha} PV \int_\Omega \frac{-u(y)}{|x - y|^{n+\alpha}} dy < 0.$$

Here the value of $(-\triangle)^{\alpha/2} u(x)$ depends not only on the value of u near x, but also on the value of u on the whole $\mathbb{R}^n$. Hence, we say that $(-\triangle)^{\alpha/2}$ is a nonlocal operator.

For u in $\mathcal{S}$, the Schwartz space of rapidly decreasing C^∞ functions in $\mathbb{R}^n$, the fractional Laplacian in (1.1) can be equivalently defined by the Fourier transform:

$$\widehat{(-\triangle)^{\frac{\alpha}{2}} u}(\xi) = |\xi|^\alpha \hat{u}(\xi), \tag{1.3}$$

where $\hat{u} = \mathcal{F}(u)$ is the Fourier transform of u, i.e.,

$$\mathcal{F}(u)(\xi) = \int_{\mathbb{R}^n} e^{-2\pi i x \cdot \xi} u(x) \, dx.$$

We will show the equivalence in the next section.

Another equivalent definition of the fractional Laplacian is via extension [CS]. For a function $u : \mathbb{R}^n \to \mathbb{R}$, consider its extension to one higher dimension

$$U : \mathbb{R}^n \times [0, \infty) \to \mathbb{R}$$

that satisfies equations

$$\begin{cases} \triangle_x U + \frac{1-\alpha}{y} U_y + U_{yy} = 0, & (x, y) \in \mathbb{R}^n \times [0, \infty), \\ U(x, 0) = u(x), & x \in \mathbb{R}^n. \end{cases} \tag{1.4}$$

Later we'll show that, for $u \in C^{1,1}_{loc} \cap \mathcal{L}_\alpha$,

$$C_{n,\alpha}(-\triangle)^{\alpha/2} u = \lim_{y \to 0^+} -y^{1-\alpha} U_y = \alpha \lim_{y \to 0^+} \frac{U(x, y) - U(x, 0)}{y^\alpha},$$

which reduces to the regular normal derivative in the case $\alpha = 1$.

Remark 1.2.1 *In order the solution U of (1.4) be unique, one needs to impose some decay condition on U near infinity.*

One can extend the definition of the fractional Laplacian to the distributions in the space $\mathcal{L}_\alpha$ by

$$< (-\triangle)^{\frac{\alpha}{2}} u, \phi > = \int_{\mathbb{R}^n} u (-\triangle)^{\frac{\alpha}{2}} \phi \, dx, \quad \text{for all } \phi \in C_0^\infty(\mathbb{R}^n).$$

Given any $f \in L^1_{loc}(\mathbb{R}^n)$, we say that $u \in \mathcal{L}_\alpha$ solves the problem

$$(-\triangle)^{\frac{\alpha}{2}} u = f(x), \quad x \in \mathbb{R}^n$$

in the sense of distribution if and only if

$$\int_{\mathbb{R}^n} u (-\triangle)^{\frac{\alpha}{2}} \phi \, dx = \int_{\mathbb{R}^n} f(x) \phi(x) dx, \quad \text{for all } \phi \in C_0^\infty(\mathbb{R}^n).$$

One may notice that, for an integer order differential operator, say for $\triangle$, in order the integral

$$\int_{\mathbb{R}^n} u (-\triangle \phi) \, dx$$

to be valid, we only require u to be locally integrable, i.e. $u \in L^1_{loc}$, because $\triangle \phi$ has a compact support. Then why do we need $u \in \mathcal{L}_\alpha$ here? This is actually the essential difference between a local and nonlocal operator. For the latter, even though x is beyond the support of ϕ, $(-\triangle)^{\alpha/2} \phi(x)$ may still not vanish as opposed to the case of $\triangle \phi(x)$.

However, for $|x|$ large, we have

$$|(-\triangle)^{\frac{\alpha}{2}} \phi(x)| = \left| C_{n,\alpha} PV \int_{\mathbb{R}^n} \frac{\phi(x) - \phi(y)}{|x - y|^{n+\alpha}} dy \right|$$

$$= \left| C_{n,\alpha} \int_{supp\,\phi} \frac{-\phi(y)}{|x - y|^{n+\alpha}} dy \right|$$

$$\leq \int_{supp\,\phi} \frac{C}{|x - y|^{n+\alpha}} dy$$

$$\sim \frac{C}{|x|^{n+\alpha}}.$$

Together with $u \in \mathcal{L}_\alpha$, it implies that

$$\int_{\mathbb{R}^n} u\,(-\triangle)^{\frac{\alpha}{2}}\phi\,dx < \infty.$$

Besides the one defined by (1.1), there are other kinds of fractional Laplacians, among which one is called the spectral fractional Laplacian defined by the eigenvalues and eigenfunctions of the Laplacian operator. Let $\{\lambda_k\}_{k=1}^\infty$ be the eigenvalues of $-\triangle$ in Ω, whose corresponding eigenfunctions $\{\varphi_k\}_{k=1}^\infty$ satisfy

$$\begin{cases} -\triangle\varphi_k = \lambda_k\varphi_k, & \text{in } \Omega, \\ \varphi_k = 0, & \text{on } \partial\Omega, \end{cases} \tag{1.5}$$

and are normalized by $\|\varphi_k\|_{L^2(\Omega)} = 1$. Let

$$H_0^{\alpha/2}(\Omega) = \{u = \sum_{k=1}^\infty c_k\varphi_k \mid \sum_{k=1}^\infty c_k^2\lambda_k^{\alpha/2} < \infty\}.$$

Then for $u \in H_0^{\alpha/2}$, the $A^{\alpha/2}$ operator is given by

$$u \mapsto A^{\alpha/2}u = \sum_{k=1}^\infty c_k\lambda_k^{\alpha/2}\varphi_k.$$

Another one is known as the regional fractional Laplacian, and is obtained by restricting the integration to the domain:

$$(-\triangle)_\Omega^{\alpha/2}u(x) = C_{n,\alpha}\,PV\int_\Omega \frac{u(x) - u(z)}{|x - z|^{n+\alpha}}dz.$$

In this book, we will study the fractional Laplacian defined by (1.1).

1.2.1 The Equivalence of the Fourier Transform Definition

We prove that, for the functions in the Schwartz space $\mathcal{S}$, the integral definition of $(-\triangle)^{\frac{\alpha}{2}}$ in (1.1) is equivalent to that via the Fourier transform in (1.3).

Proposition 1.2.1 ([Si]) *For $\alpha \in (0,2)$, let $(-\triangle)^{\frac{\alpha}{2}} : \mathcal{S} \to L^2(\mathbb{R}^n)$ be the fractional Laplacian defined by*

$$(-\triangle)^{\frac{\alpha}{2}}u(x) = C_{n,\alpha}PV\int_{\mathbb{R}^n} \frac{u(x) - u(y)}{|x - y|^{n+\alpha}}dy, \tag{1.6}$$

where

$$C_{n,\alpha} = \left(\int_{\mathbb{R}^n} \frac{1 - cos(2\pi\zeta_1)}{|\zeta|^{n+\alpha}}d\zeta\right)^{-1}. \tag{1.7}$$

Then, for any $u \in \mathcal{S}$,

$$\widehat{(-\triangle)^{\frac{\alpha}{2}}u}(\xi) = |\xi|^\alpha\hat{u}(\xi), \quad \forall \xi \in \mathbb{R}^n. \tag{1.8}$$

Proof. Let $y = x + z$ and $y = x - z$ respectively in the defining integral in (1.6), then

$$C_{n,\alpha} PV \int_{\mathbb{R}^n} \frac{u(x) - u(x + z)}{|z|^{n+\alpha}} dz = C_{n,\alpha} PV \int_{\mathbb{R}^n} \frac{u(x) - u(x - z)}{|z|^{n+\alpha}} dz.$$

Therefore,

$$(-\triangle)^{\frac{\alpha}{2}} u(x) = \frac{C_{n,\alpha}}{2} \int_{\mathbb{R}^n} \frac{2u(x) - u(x + y) - u(x - y)}{|y|^{n+\alpha}} dy. \qquad (1.9)$$

Applying Taylor expansion to $u(x + y)$ and $u(x - y)$ near x and from the assumption that $u \in \mathcal{S}$, we compute

$$\int_{\mathbb{R}^n} \int_{\mathbb{R}^n} |e^{-2\pi i x \cdot \xi} \frac{u(x + y) + u(x - y) - 2u(x)}{|y|^{n+\alpha}}| \, dx \, dy$$

$$\leq \int_{|y| \leq 1} \int_{\mathbb{R}^n} \frac{\sup_{B_1(x)} |D^2 u| |y|^2}{|y|^{n+\alpha}} dx \, dy + \int_{|y| \geq 1} \int_{\mathbb{R}^n} \frac{4|u(x)|}{|y|^{n+\alpha}} dx \, dy$$

$$\leq C \int_{|y| \leq 1} \frac{1}{|y|^{n+\alpha-2}} dy + C \int_{|y| \geq 1} \frac{1}{|y|^{n+\alpha}} dy$$

$$< \infty.$$

This enables us to apply the Fubini-Tonelli's Theorem and arrive at

$$\mathcal{F}((-\triangle)^{\frac{\alpha}{2}} u) = -\frac{1}{2} C_{n,\alpha} \int_{\mathbb{R}^n} \frac{\mathcal{F}(u(x + y) + u(x - y) - 2u(x))}{|y|^{n+\alpha}} dy$$

$$= -\frac{1}{2} C_{n,\alpha} \int_{\mathbb{R}^n} \frac{e^{2\pi i \xi \cdot y} + e^{-2\pi i \xi \cdot y} - 2}{|y|^{n+\alpha}} (\mathcal{F}u)(\xi) \, dy$$

$$= C_{n,\alpha} (\mathcal{F}u)(\xi) \int_{\mathbb{R}^n} \frac{1 - \cos(2\pi \xi \cdot y)}{|y|^{n+\alpha}} dy.$$

Now, to obtain (1.8), it suffices to show that

$$\int_{\mathbb{R}^n} \frac{1 - \cos(2\pi \xi \cdot y)}{|y|^{n+\alpha}} dy = C_{n,\alpha}^{-1} |\xi|^\alpha. \qquad (1.10)$$

In fact, let $y = \frac{z}{|\xi|}$ in (1.10), then

$$\int_{\mathbb{R}^n} \frac{|\xi|^{n+\alpha} (1 - \cos(\frac{2\pi z \cdot \xi}{|\xi|}))}{|\xi|^n |z|^{n+\alpha}} dz$$

$$= |\xi|^\alpha \int_{\mathbb{R}^n} \frac{1 - \cos(2\pi z \cdot \frac{\xi}{|\xi|})}{|z|^{n+\alpha}} dz$$

$$= |\xi|^\alpha \int_{\mathbb{R}^n} \frac{1 - \cos(2\pi z_1)}{|z|^{n+\alpha}} dz$$

$$= |\xi|^\alpha C_{n,\alpha}^{-1}.$$

This completes the proof.

1.2.2 The Equivalence of the Extension Definition

Let
$$U : \mathbb{R}^n \times [0, \infty) \rightarrow \mathbb{R}$$
be the extension of u satisfying

$$\begin{cases} \triangle_x U + \frac{b}{y} U_y + U_{yy} = 0, & (x, y) \in \mathbb{R}^n \times [0, \infty), \\ U(x, 0) = u(x), & x \in \mathbb{R}^n. \end{cases} \quad (1.11)$$

We will show that for $\alpha = 1 - b$,

$$C(-\triangle)^{\alpha/2} u = \lim_{y \to 0+} -y^{1-\alpha} U_y = \alpha \lim_{y \to 0+} \frac{U(x, y) - U(x, 0)}{y^\alpha},$$

under two different conditions:
  (i) $u \in C_{loc}^{1,1} \cap \mathcal{L}_\alpha$, or
  (ii) $u \in H_0^{\alpha/2}(\mathbb{R}^n)$ with

$$\int_{\mathbb{R}^n} |\xi|^\alpha |\hat{u}(\xi)|^2 d\xi < \infty.$$

Under condition (i), we present the solution of (1.11) in terms of the Poisson kernel $P(x, y)$ that satisfies

$$\begin{cases} \triangle_x P + \frac{b}{y} P_y + P_{yy} = 0, & (x, y) \in \mathbb{R}_+^{n+1}, \\ P(x, 0) = \delta_0(x), & x \in \mathbb{R}^n = \partial \mathbb{R}_+^{n+1}, \end{cases}$$

where $\mathbb{R}_+^{n+1} = \{(x, y) \mid (x, y) \in \mathbb{R}^n \times [0, \infty)\}$.
  Then one can easily verify that

$$U(x, y) = \int_{\mathbb{R}^n} P(x - \xi, y) u(\xi) d\xi \quad (1.12)$$

is a solution of (1.11). Indeed, let

$$LU \equiv \triangle_x U + \frac{b}{y} U_y + U_{yy}, \quad (x, y) \in \mathbb{R}_+^{n+1}.$$

Then
$$LU = \int_{\mathbb{R}^n} LP(x - \xi, y) u(\xi) \, d\xi = \int_{\mathbb{R}^n} 0 \cdot u(\xi) \, d\xi = 0$$
and
$$U(x, 0) = \int_{\mathbb{R}^n} \delta(x - \xi) u(\xi) d\xi = u(x).$$

In the following, we indicate how to seek such a $P(x, y)$.
  In the special case when $b = 0$, this Poisson kernel $P(x, y)$ can be derived from the *Green's function representation*. Consider the Dirichlet problem

$$\begin{cases} \triangle U(x,y) = 0, & (x,\,y) \in \mathbb{R}^{n+1}_+, \\ U(x,0) = u(x), & x \in \mathbb{R}^n. \end{cases} \tag{1.13}$$

It is well-known that the *Green's function* in $\mathbb{R}^{n+1}_+$ is

$$G(X,Z) = C\left(\frac{1}{|X-Z|^{n-1}} - \frac{1}{|X-\bar{Z}|^{n-1}}\right),$$

where $X = (x,y)$, $Z = (\xi,\eta)$, and $\bar{Z} = (\xi,-\eta)$ is the reflection of Z about the boundary $\partial \mathbb{R}^{n+1}_+ = \{Z \in \mathbb{R}^{n+1} \mid \eta = 0\}$. Obviously,

$$\begin{cases} -\triangle G(X,Z) = \delta(X-Z), & X \in \mathbb{R}^{n+1}_+, \\ G(X,Z) = 0, & X \text{ or } Z \in \partial \mathbb{R}^{n+1}_+. \end{cases} \tag{1.14}$$

Assume that $U \in H^1_0(\mathbb{R}^{n+1}_+)$ is a solution for (1.13) and ν is the unit outward normal vector on $\partial \mathbb{R}^{n+1}_+$. By the *Green's second identity*, we have

$$\begin{aligned} 0 &= \int_{\mathbb{R}^{n+1}_+} [-\triangle U(Z)\, G(X,Z)]\, dZ \\ &= \int_{\mathbb{R}^{n+1}_+} U(Z)\,(-\triangle G(X,Z))\, dZ + \int_{\mathbb{R}^n} \left(U\frac{\partial G}{\partial \nu} - G\frac{\partial U}{\partial \nu}\right) d\xi \\ &= U(X) - \int_{\mathbb{R}^n} u(\xi)\frac{\partial G}{\partial \nu}\, d\xi. \end{aligned} \tag{1.15}$$

By a direct calculation (as an exercise), one can derive that

$$-\frac{\partial G}{\partial \nu}\bigg|_{\eta=0} = \frac{\partial G}{\partial \eta}\bigg|_{\eta=0} = C\frac{y}{[(x-\xi)^2 + y^2]^{\frac{n+1}{2}}}, \tag{1.16}$$

and this is the Poisson kernel $P(x - \xi, y)$ in the case $b = 0$. Combining (1.15) and (1.16), we arrive at

$$U(x,y) = \int_{\mathbb{R}^n} P(x - \xi, y)u(\xi)\, d\xi.$$

For b taking more general values, without knowing the explicit expression of the *Green's function* associated with the operator L, we can approach as follows.

Suppose b is a nonnegative integer and $U(x,y) : \mathbb{R}^n \times \mathbb{R}^{1+b} \to \mathbb{R}$ is radially symmetric in the y variable. In other words, if $|y| = |y'| = r$, then $U(x,y) = U(x,y')$. Then the Laplacian of $U(x,r)$ takes the form:

$$\triangle U = \triangle_x U + \frac{b}{r}U_r + U_{rr}. \tag{1.17}$$

Notice that (1.17) is exactly the expression in (1.11) with y replaced by r. Even for non-integer b, (1.17) still bears interesting information. In fact, (1.11) can be seen as the harmonic extension of u in $1+b$ more dimensions. Although when b is not an integer, we can not describe the meaning for $\mathbb{R}^{n+1+b}$, the solutions of (1.11) have many properties common to harmonic functions.

To seek the fundamental solution for (1.11) at the origin, we only need to consider the fundamental solution of the Laplacian in $n+1+b$ dimensions:

$$\Gamma(X) = C_{n+1+b}\frac{1}{|X|^{n-1+b}}. \tag{1.18}$$

Through a straightforward calculation, one can see that Γ is a solution of (1.11) when $y \neq 0$; and for some constant C,

$$\lim_{y\to 0+} y^b \Gamma_y = -C\delta_0(x). \tag{1.19}$$

It is quite interesting to note that $\Gamma(x,0) = \frac{C}{|x|^{n-1+b}}$ is also the fundamental solution of the fractional Laplacian $(-\triangle)^{\alpha/2}$ when $\alpha = 1 - b$, that is

$$(-\triangle)^{\alpha/2}\Gamma(x,0) = C\delta_0(x).$$

Property (1.19) suggests that $y^b \Gamma_y$ is a promising candidate for $P(x,y)$. We check if it satisfies the required equation (1.11):

$$L(y^b \Gamma_y) = 0.$$

By (1.17),

$$0 = \triangle_x \Gamma + \frac{b}{y}\Gamma_y + \Gamma_{yy}$$
$$= y^b \triangle_x \Gamma + (y^b \Gamma_y)_y. \tag{1.20}$$

Differentiating (1.20) with respect to y,

$$0 = by^{b-1}\triangle_x\Gamma + y^b\triangle_x\Gamma_y + (y^b\Gamma_y)_{yy}$$
$$= by^{b-1}(-y^{-b}(y^b\Gamma_y)_y) + y^b\triangle_x\Gamma_y + (y^b\Gamma_y)_{yy} \tag{1.21}$$
$$= \triangle_x(y^b\Gamma_y) - \frac{b}{y}(y^b\Gamma_y)_y + (y^b\Gamma_y)_{yy}. \tag{1.22}$$

This is almost the equation we need. Unfortunately, the middle term bears a " $-$ " sign instead of a " $+$ " sign as appeared in $L(y^b\Gamma_y)$. An easy remedy is to replace "b" in $y^b\Gamma_y$ by "$-b$" (note: also replace "b" in Γ by "$-b$"). More precisely, choose

$$P(x,y) = -Cy^{-b}\frac{\partial}{\partial y}\left(\frac{1}{(|x|^2+y^2)^{\frac{n-1-b}{2}}}\right)$$
$$= C_1\frac{y^{1-b}}{(|x|^2+y^2)^{\frac{n+1-b}{2}}}.$$

Then P satisfies

$$LP \equiv \triangle_x P + \frac{b}{y}P_y + P_{yy} = 0.$$

Here the constant C_1 is properly chosen so that

$$\lim_{y \to 0} P(x, y) = \delta_0(x).$$

$P(x, 0)$ is indeed a "Delta" function. To see this, for all $\varphi(x) \in C_0^\infty(\mathbb{R}^n)$,

$$\lim_{y \to 0} \int_{\mathbb{R}^n} \frac{y^{1-b}}{(|x|^2 + |y|^2)^{\frac{n+1-b}{2}}} \varphi(x) dx$$

$$= \lim_{y \to 0} \int_{\mathbb{R}^n} \frac{y^{1-b}}{(|y|^2|z|^2 + |y|^2)^{\frac{n+1-b}{2}}} \varphi(yz) y^n dz$$

$$= \lim_{y \to 0} \int_{\mathbb{R}^n} \frac{1}{(|z|^2 + 1)^{\frac{n+1-b}{2}}} \varphi(yz) dz$$

$$= \varphi(0) \int_{\mathbb{R}^n} \frac{1}{(|z|^2 + 1)^{\frac{n+1-b}{2}}} dz$$

$$= C_b \varphi(0).$$

From the above calculation, one can also find another interesting property of the Poisson kernel:

$$\int_{\mathbb{R}^n} P(x - \xi, y) \, d\xi = 1, \quad \forall \, y > 0. \tag{1.23}$$

Now we show how to relate (1.11) to the operator $(-\triangle)^{\alpha/2}$. In other words, we prove that

$$C \lim_{y \to 0+} y^b U_y(x, y) = -(-\triangle)^{\alpha/2} u(x) = CPV \int_{\mathbb{R}^n} \frac{u(\xi) - u(x)}{|\xi - x|^{n+\alpha}} d\xi. \tag{1.24}$$

Our proof of (1.24) is based on the Poisson representation of a solution of (1.11):

$$U(x, y) = \int_{\mathbb{R}^n} P(x - \xi, y) u(\xi) d\xi,$$

with

$$U(x, 0) = u(x).$$

It follows from these and (1.23) that

$$\frac{1}{1-b} \lim_{y \to 0+} y^b U_y(x, y) = \lim_{y \to 0+} \frac{U(x, y) - U(x, 0)}{y^{1-b}} \tag{1.25}$$

$$= \lim_{y \to 0+} \frac{1}{y^{1-b}} \int_{\mathbb{R}^n} P(x - \xi, y)[u(\xi) - u(x)] d\xi$$

$$= C_1 \lim_{y \to 0+} \int_{\mathbb{R}^n} \frac{u(\xi) - u(x)}{(|x - \xi|^2 + y^2)^{\frac{n+1-b}{2}}} d\xi$$

$$= C_1 PV \int_{\mathbb{R}^n} \frac{u(\xi) - u(x)}{|x - \xi|^{n+\alpha}} d\xi$$

$$= -C(-\triangle)^{\alpha/2} u(x).$$

Here we explain (1.25). Through a change of variable $z = y^{1-b}$,

$$
\begin{aligned}
\lim_{y \to 0+} \frac{U(x,y) - U(x,0)}{y^{1-b}} &= \lim_{z \to 0+} \frac{U(x, z^{\frac{1}{1-b}}) - U(x,0)}{z} \\
&= \lim_{z \to 0+} U_y(x, z^{\frac{1}{1-b}}) \frac{d}{dz}(z^{\frac{1}{1-b}}) \\
&= \lim_{z \to 0+} U_y(x, z^{\frac{1}{1-b}}) \frac{1}{1-b} z^{\frac{b}{1-b}} \\
&= \frac{1}{1-b} \lim_{y \to 0+} U_y(x,y) y^b.
\end{aligned}
$$

Under condition (ii), we prove that (1.24) can also be obtained by showing that the corresponding energy functionals coincide, i.e.,

$$
\int_{y>0} |\nabla U|^2 y^b dX = J(\phi) \int_{\mathbb{R}^n} |\xi|^\alpha |\hat{u}(\xi)|^2 d\xi. \tag{1.26}
$$

Taking the Fourier transform in x and (1.11) becomes

$$
-|\xi|^2 \hat{U}(\xi, y) + \frac{b}{y} \hat{U}_y(\xi, y) + \hat{U}_{yy}(\xi, y) = 0. \tag{1.27}
$$

We try to seek a solution of the form

$$
\hat{U}(\xi, y) = \hat{u}(\xi)\phi(|\xi|y) \tag{1.28}
$$

with $\phi(0) = 1$, so that $\hat{U}(\xi, 0) = \hat{u}(\xi)$. Substituting (1.28) into (1.27) yields

$$
\begin{cases}
-\phi(y) + \frac{b}{y}\phi'(y) + \phi''(y) = 0 \\
\phi(0) = 1 \\
\lim_{y \to \infty} \phi(y) = 0.
\end{cases} \tag{1.29}
$$

The last condition is to ensure that $\lim_{y \to \infty} \hat{U}(\xi, y) = 0$.

Such a function $\phi : [0, \infty) \to \mathbb{R}$ is actually a minimizer (or a critical point) of the functional

$$
J(\phi) := \int_0^\infty (|\phi'|^2 + |\phi|^2) y^b dy,
$$

and the energy of U is

$$
\begin{aligned}
\int_{y>0} |\nabla U|^2 y^b dX &= \int_{\mathbb{R}^n} \int_0^\infty (|\xi|^2 |\hat{U}|^2 + |\hat{U}_y|^2) y^b \, dy \, d\xi \\
&= \int_{\mathbb{R}^n} \int_0^\infty |\hat{u}(\xi)|^2 |\xi|^2 \left(|\phi(|\xi|y)|^2 + |\phi'(|\xi|y)|^2 \right) y^b \, dy \, d\xi \\
&= \int_{\mathbb{R}^n} |\hat{u}(\xi)|^2 |\xi|^{1-b} \int_0^\infty \left(|\phi(|\bar{y}|)|^2 + |\phi'(|\bar{y}|)|^2 \right) \bar{y}^b \, d\bar{y} \, d\xi \\
&= \int_{\mathbb{R}^n} |\hat{u}(\xi)|^2 |\xi|^{1-b} J(\phi) d\xi \\
&= J(\phi) \int_{\mathbb{R}^n} u(-\triangle)^{\alpha/2} u \, dx. \tag{1.30}
\end{aligned}
$$

This proves (1.26).

For $u \in H_0^{\alpha/2}(\mathbb{R}^n)$, let

$$F(U) := \int_{y>0} |\nabla U|^2 y^b dX, \tag{1.31}$$

and

$$H(u) := \int_{\mathbb{R}^n} (-\triangle)^{\alpha/2} u \, u \, dx. \tag{1.32}$$

To see how (1.30) implies (1.24), we perform the integration by parts on both integrals.

For any $V(x, y)$ of the form

$$\hat{V}(\xi, y) = \hat{v}(\xi)\phi(|\xi|y)$$

with $v(x) \in C_0^\infty(\mathbb{R}^n)$, similar to (1.30), one can derive

$$F(U + tV) = CJ(\phi) \cdot H(u + tv).$$

Hence we must have

$$\frac{d}{dt}F(U+tV)\bigg|_{t=0} = CJ(\phi)\frac{d}{dt}H(u+tv)\bigg|_{t=0}. \tag{1.33}$$

Through integration by parts and via equation (1.11), we have

$$\text{LHS of (1.33)} = 2\int_{y>0} \nabla U \cdot \nabla V y^b dX$$

$$= -2\int_{y>0} div(y^b \nabla U) \, V dX - 2\int_{\mathbb{R}^n} \lim_{y \to 0^+} y^b U_y V(x,0) dx$$

$$= -2\int_{\mathbb{R}^n} \lim_{y \to 0^+} y^b U_y v(x) dx.$$

$$\text{RHS of (1.33)} = 2CJ(\phi)\int_{\mathbb{R}^n} (-\triangle)^{\alpha/2} u \, v \, dx.$$

Hence $\forall v \in C_0^\infty(\mathbb{R}^n)$,

$$\int_{\mathbb{R}^n} - \lim_{y \to 0^+} y^b U_y v(x) dx = CJ(\phi)\int_{\mathbb{R}^n} (-\triangle)^{\alpha/2} u \, v \, dx. \tag{1.34}$$

Therefore, we arrive at

$$- \lim_{y \to 0^+} y^b U_y = C(-\triangle)^{\alpha/2} u.$$

This proves (1.24).

2

The Green's Function

2.1 The Green's Function of $-\triangle$

2.1.1 The Representation Formula

Let Ω be a bounded domain in n-dimensional Euclidean space $\mathbb{R}^n$ with smooth boundary $\partial\Omega$. We study the boundary value problem

$$\begin{cases} -\triangle u(x) = f(x), & x \in \Omega, \\ u(x) = 0, & x \in \partial\Omega. \end{cases} \tag{2.1}$$

When Ω is a ball or a half space, we will obtain an integral representation formula for the solution. The idea is to use the *Green's function* $G(x, y)$ which solves

$$\begin{cases} -\triangle G(x, y) = \delta(x - y), & x, y \in \Omega, \\ G(x, y) = 0, & x \text{ or } y \in \partial\Omega, \end{cases} \tag{2.2}$$

where $\delta(x, y)$ is the *Dirac Delta function*

$$\delta(x, y) = \begin{cases} 0, & \text{if } x \neq y \\ \infty, & \text{if } x = y; \end{cases}$$

and

$$\int_{\mathbb{R}^n} \delta(x, y) dx = 1.$$

Before constructing the *Green's function*, let's first see how it can be used to represent the solutions of the boundary value problems. To this end, we need the *Green's Second Identity* which can be derived from the *Green's First Identity*

$$\int_\Omega (\triangledown u \cdot \triangledown v + u \triangle v) dx = \int_{\partial\Omega} u \frac{\partial v}{\partial \nu} d\sigma. \tag{2.3}$$

In (2.3), interchanging u and v, we have

$$\int_\Omega (\triangledown v \cdot \triangledown u + v \triangle u) dx = \int_{\partial\Omega} v \frac{\partial u}{\partial \nu} d\sigma. \tag{2.4}$$

Subtracting (2.4) from (2.3), we arrive at the *Green's Second Identity*

$$\int_\Omega (u\triangle v - v\triangle u)dx = \int_{\partial\Omega} \left(u\frac{\partial v}{\partial\nu} - v\frac{\partial u}{\partial\nu}\right)d\sigma. \tag{2.5}$$

In the above identity, if we let u be the solution of boundary value problem (2.1) and $v = G$, then

$$\int_\Omega [u\left(-\delta(x-y)\right) - G(x,y)\triangle u]dx = \int_{\partial\Omega} \left(u\frac{\partial G}{\partial\nu} - G\frac{\partial u}{\partial\nu}\right)d\sigma = 0. \tag{2.6}$$

Due to the property of the *Delta* function

$$\int_\Omega \delta(x,y)u(x)dx = u(y), \quad \forall\, y \in \Omega,$$

Eq. (2.6) becomes

$$u(y) = \int_\Omega G(x,y)f(x)dx. \tag{2.7}$$

This is known as the *Green's function representation* of the solution.

2.1.2 The Green's Functions Associated with Various Domains

The *Green's function* usually consists of two parts:

$$G(x,y) = \Gamma(x,y) - h(x,y).$$

The first part $\Gamma(x,y)$ is called the *fundamental solution* of the *Laplace's equation* which satisfies

$$-\triangle\Gamma(x,y) = \delta(x-y). \tag{2.8}$$

While the second part $h(x,y)$ is introduced to accommodate the zero boundary condition:

$$h(x,y) = \Gamma(x,y), \quad \text{if } x \text{ or } y \in \partial\Omega, \tag{2.9}$$

and we also want $\triangle h = 0$.

Let's first seek $\Gamma(x,y)$. We may try the form $\Gamma(x,y) = |x-y|^a$; and to determine the exponent, we may consider $\Gamma(x,0) = |x|^a$. Since this function is rotationally symmetric, it is more convenient to apply the polar form of the Laplace's operator:

$$\triangle = \frac{d^2}{dr^2} + \frac{n-1}{r}\frac{d}{dr}, \tag{2.10}$$

where $r = |x|$ is the distance from the point x to the origin. By (2.10), we have

$$\triangle r^a = [a(a-1) + a(n-1)]r^{a-2}.$$

To make this "Delta like" function, we choose $a = 2-n$, and consequently

$$\Gamma(x,y) = \frac{c}{|x-y|^{n-2}}, \tag{2.11}$$

where c is a constant to be determined so that

$$-\int_{\mathbb{R}^n} \triangle \Gamma(x,y)dx = 1.$$

Notice that, when the dimension $n = 2$, the right-hand side of (2.11) gives us a constant, which is obvious not the desired "Delta like" function. In this case, we can approach as the following, which actually works for all dimensions.

Let $v(r)$ be a function of r only. Express

$$\triangle v(r) = \frac{d^2v}{dr^2} + \frac{n-1}{r}\frac{dv}{dr} = \frac{1}{r^{n-1}}(r^{n-1}v'(r))'.$$

Solve

$$\frac{1}{r^{n-1}}(r^{n-1}v'(r))' = 0.$$

Integrating both sides once, we have

$$r^{n-1}v'(r) = C_1,$$

hence

$$v'(r) = \frac{C_1}{r^{n-1}}.$$

Integrating a second time, we arrive at

$$v(r) = \begin{cases} \frac{C_1}{(2-n)}\frac{1}{r^{n-2}} + C_2, & \text{if } n \neq 2 \\ C_1 \ln r + C_2, & \text{if } n = 2. \end{cases} \tag{2.12}$$

Therefore, we can take

$$\Gamma(x,y) = \begin{cases} \frac{c}{|x-y|^{n-2}}, & \text{if } n \neq 2, \\ c\ln|x-y|, & \text{if } n = 2. \end{cases} \tag{2.13}$$

Due to the symmetric nature of $\Gamma(x,y)$ with respect to x and y, no matter whether we apply $\triangle_x$ or $\triangle_y$ to Γ, we will arrive at the same result, where

$$\triangle_x = \frac{\partial^2}{\partial x_1^2} + \cdots + \frac{\partial^2}{\partial x_n^2},$$

and $\triangle_y$ is defined similarly.

Now we consider the second part $h(x,y)$. Obviously, for different domains, it should be different. We first start from the simplest case when Ω is the upper half Euclidean space

$$\mathbb{R}_+^n = \{x = (x_1, \cdots, x_n) \mid x_n > 0\}.$$

Here $n \geq 3$. Let $\bar{x} = (x_1, \cdots, -x_n)$, the mirror image of the point x about the boundary $\partial \mathbb{R}_+^n = \{x \mid x_n = 0\}$.

From the above calculation, we can see that

$$\triangle \frac{c}{|\bar{x} - y|^{n-2}} = 0, \text{ for any } x, y \in \mathbb{R}^n_+,$$

because $\bar{x}$ can never equal y. Moreover, by symmetry, one can easily verify that

$$\frac{c}{|x - y|^{n-2}} - \frac{c}{|\bar{x} - y|^{n-2}} = 0, \text{ for either } x \text{ or } y \in \partial\mathbb{R}^n_+.$$

Therefore, our *Green's function* for the upper half space is

$$G(x, y) = \frac{c}{|x - y|^{n-2}} - \frac{c}{|\bar{x} - y|^{n-2}}. \tag{2.14}$$

And for this *Green's function*, we can express the solution of the boundary value problem

$$\begin{cases} -\triangle u = f(x), x \in \mathbb{R}^n_+, \\ u(x) = 0, \qquad x \in \partial\mathbb{R}^n_+, \end{cases} \tag{2.15}$$

as

$$u(x) = \int_{\mathbb{R}^n_+} G(x, y)f(y)dy.$$

Here, because $\mathbb{R}^n_+$ is an unbounded domain, we need to require that u and its derivatives go to zero at infinity at proper rates.

Now if v is a solution of a different boundary value problem

$$\begin{cases} \triangle v = 0, \qquad x \in \mathbb{R}^n_+, \\ v(x) = g(x), x \in \partial\mathbb{R}^n_+, \end{cases} \tag{2.16}$$

we can use the *Green's Second Identity* in a similar way to obtain the representation

$$v(x) = -\int_{\partial\mathbb{R}^n_+} g(y)\frac{\partial G}{\partial \nu}(x, y)d\sigma.$$

To solve the more general problem

$$\begin{cases} -\triangle w = f(x), x \in \mathbb{R}^n_+, \\ w(x) = g(x), \quad x \in \partial\mathbb{R}^n_+, \end{cases} \tag{2.17}$$

we can just add the previous two solutions u and v to derive

$$w(x) = \int_{\mathbb{R}^n_+} G(x, y)f(y)dy - \int_{\partial\mathbb{R}^n_+} g(y)\frac{\partial G}{\partial \nu}(x, y)d\sigma.$$

Next, we determine $h(x, y)$ in the case when $\Omega = B_1(0)$, the unit ball centered at the origin. The idea is to use the inversion point $\frac{x}{|x|^2}$ of the point x in $B_1(0)$. It's easy to see that if either x or y is on the boundary $\partial B_1(0)$, then

$$\frac{1}{|x - y|^{n-2}} = \frac{1}{(|x||\frac{x}{|x|^2} - y|)^{n-2}} = \frac{1}{(|y||\frac{y}{|y|^2} - x|)^{n-2}}.$$

Hence, we choose

$$h(x,y) = \frac{c}{(|x|\mid \frac{x}{|x|^2} - y|)^{n-2}}.$$

To see that $h(x,y)$ is harmonic, i.e. $\triangle h = 0$, we differentiate h with respect to x and y respectively,

$$\triangle_y h = \triangle_y \frac{c}{(|x|\mid \frac{x}{|x|^2} - y|)^{n-2}} = \frac{c}{|x|^{n-2}} \triangle_y \frac{1}{\mid \frac{x}{|x|^2} - y|^{n-2}} = 0,$$

because for x and y in $B_1(0)$, $\frac{x}{|x|^2}$ is a point outside $B_1(0)$, and y can never be equal to $\frac{x}{|x|^2}$.

Similarly, employing the second expression of h, we have

$$\triangle_x h = \triangle_x \frac{c}{(|y|\mid \frac{y}{|y|^2} - x|)^{n-2}} = \frac{c}{|y|^{n-2}} \triangle_x \frac{1}{\mid \frac{y}{|y|^2} - x|^{n-2}} = 0.$$

Finally, our *Green's function* for $B_1(0)$ is

$$G_1(x,y) = \frac{c}{|x-y|^{n-2}} - \frac{c}{(|x|\mid \frac{x}{|x|^2} - y|)^{n-2}}.$$

The *Green's function* $G_R(x,y)$ for $B_R(0)$ follows effortlessly from a rescaling of $G_1(x,y)$.

$$G_R(x,y) = \frac{1}{R^{n-2}} G_1(\frac{x}{R}, \frac{y}{R})$$

$$= \frac{c}{|x-y|^{n-2}} - \frac{c}{(|x| \cdot \mid \frac{Rx}{|x|^2} - \frac{y}{R}|)^{n-2}}.$$

2.2 The Green's Function of $(-\triangle)^{\alpha/2}$

Now we consider the Dirichlet problem for the fractional Laplacian

$$\begin{cases} (-\triangle)^{\alpha/2} u(x) = f(x), & x \in \Omega, \\ u(x) \equiv 0, & x \notin \Omega. \end{cases} \tag{2.18}$$

Notice that $(-\triangle)^{\alpha/2}$ is a nonlocal operator, hence it is more appropriate to assume that $u(x) \equiv 0$ in the whole complement of Ω instead of just on $\partial\Omega$. Even so, $(-\triangle)^{\alpha/2} u(x)$ may still not vanish for $x \notin \Omega$. For example, if we assume that $u > 0$ in Ω, then $\forall x^o \in \mathbb{R}^n \backslash \Omega$,

$$(-\triangle)^{\alpha/2} u(x^o) = C_{n,\alpha} PV \int_{\mathbb{R}^n} \frac{-u(y)}{|x^o - y|^{n+\alpha}} dy < 0.$$

The corresponding Green's function $G_\alpha(x,y)$ satisfies

$$\begin{cases} (-\triangle)^{\alpha/2} G_\alpha(x,y) = \delta(x-y), & x,y \in \Omega, \\ G_\alpha(x,y) = 0, & x \text{ or } y \notin \Omega. \end{cases} \tag{2.19}$$

If we can find such a Green's function, then the solutions of (2.18) can be represented as

$$u(x) = \int_\Omega G_\alpha(x,y)\,f(y)\,dy,\ x \in \mathbb{R}^n. \tag{2.20}$$

In particular, when $\Omega = \mathbb{R}^n$, it's well known that the associated Green's function is

$$G_\alpha(x,y) = \frac{C_{n,\alpha}}{|x-y|^{n-\alpha}},$$

which is also called the fundamental solution for the fractional Laplacian.

In this case, (2.20) becomes

$$u(x) = C_{n,\alpha} \int_{\mathbb{R}^n} \frac{f(y)}{|x-y|^{n-\alpha}}\,dy. \tag{2.21}$$

One of the interesting application of the representation formula (2.20) is that we can express solutions of a nonlinear problem

$$\begin{cases} (-\triangle)^{\frac{\alpha}{2}} u(x) = f(u),\ x \in \Omega, \\ u(x) \equiv 0, \qquad\qquad x \notin \Omega. \end{cases} \tag{2.22}$$

as

$$u(x) = \int_\Omega G_\alpha(x,y)\,f(u(y))\,dy, \quad x \in \mathbb{R}^n. \tag{2.23}$$

Then using the methods for integral equations, in particular, the *method of moving planes in integral forms*, we are able to investigate the properties of the solutions, such as symmetry and regularity.

Similar to $-\triangle$, the Green's function of $(-\triangle)^{\alpha/2}$ also consists of two parts: the fundamental solution $\Gamma(x,y)$ and the α-harmonic part $h(x,y)$.

2.2.1 The Fundamental Solution

It's well-known that

$$(-\triangle)^{\alpha/2}\Big(\frac{C_{n,\alpha}}{|x-y|^{n-\alpha}}\Big) = \delta(x-y). \tag{2.24}$$

Thus we choose the fundamental solution in $G_\alpha(x,y)$ to be

$$\Gamma(x,y) = \frac{C_{n,\alpha}}{|x-y|^{n-\alpha}}.$$

Note the value of $C_{n,\alpha}$ in this section may slightly differ from the one in the integral definition of $(-\triangle)^{\alpha/2}$.

In order to show (2.24), it suffices to derive that for all $\varphi \in C_0^\infty(\mathbb{R}^n)$,

$$< (-\triangle)^{\alpha/2}\Gamma, \varphi > := \int_{\mathbb{R}^n} \Gamma(x,y)(-\triangle)^{\alpha/2}\varphi(x)dx = \varphi(y). \tag{2.25}$$

One of the approaches is to prove that the Fourier transforms of both sides of the equation are equal. To this end, we need

Proposition 2.2.1 ([L]) *Assume that $f \in C_0^\infty(\mathbb{R}^n)$. Then*

$$(|\xi|^{-\alpha}\hat{f}(\xi))^{\vee}(x) = C_{n,\alpha} \int_{\mathbb{R}^n} \frac{f(y)}{|x-y|^{n-\alpha}}dy, \qquad (2.26)$$

where $\vee$ is the symbol for the inverse Fourier transform.

Proof. Let $\lambda = \frac{t}{\pi|\xi|^2}$, then

$$\begin{aligned}
\int_0^{+\infty} e^{-\pi|\xi|^2\lambda}\lambda^{\frac{\alpha}{2}-1}d\lambda &= \int_0^{+\infty} e^{-t}\frac{t^{\frac{\alpha}{2}-1}}{(\pi|\xi|^2)^{\frac{\alpha}{2}}}dt \\
&= \frac{1}{(\pi|\xi|^2)^{\frac{\alpha}{2}}}\int_0^{+\infty} e^{-t}t^{\frac{\alpha}{2}-1}dt \\
&= \pi^{-\frac{\alpha}{2}}\Gamma(\frac{\alpha}{2})|\xi|^{-\alpha} \\
&:= c_\alpha|\xi|^{-\alpha}. \qquad (2.27)
\end{aligned}$$

Then we have

$$\begin{aligned}
&(|\xi|^{-\alpha}\hat{f}(\xi))^{\vee}(x) \\
&= c_\alpha^{-1}\int_{\mathbb{R}^n} e^{2\pi i x\cdot\xi}c_\alpha|\xi|^{-\alpha}\hat{f}(\xi)d\xi \\
&= c_\alpha^{-1}\int_{\mathbb{R}^n} e^{2\pi i x\cdot\xi}\{\int_0^{+\infty} e^{-\pi|\xi|^2\lambda}\lambda^{\frac{\alpha}{2}-1}d\lambda\}\hat{f}(\xi)d\xi \\
&= c_\alpha^{-1}\int_0^{+\infty}\{\int_{\mathbb{R}^n} e^{2\pi i x\cdot\xi}e^{-\pi|\xi|^2\lambda}\hat{f}(\xi)d\xi\}\lambda^{\frac{\alpha}{2}-1}d\lambda \\
&= c_\alpha^{-1}\int_0^{+\infty}\{\int_{\mathbb{R}^n} e^{2\pi i x\cdot\xi}e^{-\pi|\xi|^2\lambda}\int_{\mathbb{R}^n} e^{-2\pi i\xi\cdot y}f(y)dyd\xi\}\lambda^{\frac{\alpha}{2}-1}d\lambda \\
&= c_\alpha^{-1}\int_0^{+\infty}\{\int_{\mathbb{R}^n}(\int_{\mathbb{R}^n} e^{2\pi i\xi\cdot(x-y)}e^{-\pi|\xi|^2\lambda}d\xi)f(y)dy\}\lambda^{\frac{\alpha}{2}-1}d\lambda. \qquad (2.28)
\end{aligned}$$

To calculate

$$\int_{\mathbb{R}_n} e^{2\pi i\xi\cdot(x-y)-\pi|\xi|^2\lambda}d\xi,$$

we integrate with respect to ξ_j, $j = 1, 2, \cdots, n$, respectively.

$$\begin{aligned}
&\int_{\mathbb{R}} e^{2\pi i\xi_1(x_1-y_1)-\pi\xi_1^2\lambda}d\xi_1 \\
&= \int_{\mathbb{R}} e^{-\pi\lambda(\xi_1-\frac{i(x_1-y_1)}{\lambda})^2-\frac{\pi}{\lambda}(x_1-y_1)^2}d\xi_1 \\
&= e^{-\frac{\pi}{\lambda}(x_1-y_1)^2}\int_{\mathbb{R}} e^{-\pi\lambda(\xi_1-\frac{i(x_1-y_1)}{\lambda})^2}d\xi_1.
\end{aligned}$$

To continue, we need the following well-known lemma on contour integrals from complex analysis.

Lemma 2.2.1 *Assume that C is a simple closed curve and $f(z)$ is analytic. Then*

$$\oint_C f(z)dz = 0.$$

Let $z = \xi_1 - \frac{i(x_1-y_1)}{\lambda} = a+bi$ and $C = \bigcup_{i=1}^{4} C_i$ (see Fig. 2.1), then the above lemma implies

$$\oint_C e^{-\pi\lambda z^2} dz = 0. \tag{2.29}$$

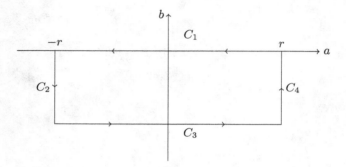

Fig. 2.1: Contour

Let

$$I_1 = \int_{C_1} e^{-\pi\lambda z^2} dz = \int_r^{-r} e^{-\pi\lambda \xi_1^2} d\xi_1,$$

$$I_2 = \int_{C_2} e^{-\pi\lambda z^2} dz = \int_0^{-\frac{x_1-y_1}{\lambda}} e^{-\pi\lambda(-r+ib)^2} db,$$

$$I_3 = \int_{C_3} e^{-\pi\lambda z^2} dz = \int_{-r}^{r} e^{-\pi\lambda(\xi_1 - \frac{i(x_1-y_1)}{\lambda})^2} d\xi_1,$$

$$I_4 = \int_{C_4} e^{-\pi\lambda z^2} dz = \int_{-\frac{x_1-y_1}{\lambda}}^{0} e^{-\pi\lambda(r+ib)^2} db.$$

As $r \to \infty$,

$$I_1 \to \int_\infty^{-\infty} e^{-\pi\lambda \xi_1^2} d\xi_1 = -\lambda^{-\frac{1}{2}},$$

$$I_2 \to 0, \qquad I_4 \to 0,$$

$$I_3 \to \int_{-\infty}^{\infty} e^{-\pi\lambda(\xi_1 - \frac{i(x_1-y_1)}{\lambda})^2} d\xi_1.$$

Together with (2.29), it yields

$$\int_{-\infty}^{\infty} e^{-\pi\lambda(\xi_1 - \frac{i(x_1-y_1)}{\lambda})^2} d\xi_1 = \lambda^{-\frac{1}{2}}.$$

Similarly, we can obtain the same result for the integrals with respect to the other $n-1$ variables and hence arrive at

$$\int_{\mathbb{R}^n} e^{2\pi i \xi \cdot (x-y) - \pi |\xi|^2 \lambda} d\xi = \lambda^{-\frac{n}{2}}.$$

Together with (2.28), it gives

$$\begin{aligned}
(|\xi|^{-\alpha} \hat{f}(\xi))^\vee (x) &= c_\alpha^{-1} \int_0^{+\infty} \{ \int_{\mathbb{R}^n} \lambda^{-\frac{n}{2}} e^{-\frac{\pi}{\lambda}|x-y|^2} f(y) dy \} \lambda^{\frac{\alpha}{2}-1} d\lambda \\
&= c_\alpha^{-1} \int_{\mathbb{R}^n} \{ \int_0^{+\infty} \lambda^{\frac{-n+\alpha}{2}-1} e^{-\frac{\pi |x-y|^2}{\lambda}} d\lambda \} f(y) dy \\
&= c_\alpha^{-1} \int_{\mathbb{R}^n} \{ \int_0^{+\infty} \lambda^{\frac{n-\alpha}{2}-1} e^{-\lambda \pi |x-y|^2} d\lambda \} f(y) dy \\
&= \int_{\mathbb{R}^n} c_{n,\alpha} |x-y|^{\alpha-n} f(y) dy \\
&= c_{n,\alpha} \int_{\mathbb{R}^n} |x-y|^{\alpha-n} f(y) dy.
\end{aligned}$$

Here the second to last identity is obtained by using (2.27) again with α replaced by $n-\alpha$ and

$$c_{n,\alpha} = \frac{c_{n-\alpha}}{c_\alpha}.$$

Remark 2.2.1 *One can see from the proof of Proposition 2.2.1 that actually the condition on f can be at least weakened to $f \in L^1(\mathbb{R}^n)$.*

Now, using (2.26), we are able to continue the proof of (2.25). Indeed,

$$\begin{aligned}
\mathcal{F} \left(\int_{\mathbb{R}^n} \Gamma(x,y)(-\triangle)^{\alpha/2} \varphi(x) \, dx \right) &= |\xi|^{-\alpha} \mathcal{F}((-\triangle)^{\alpha/2} \varphi)(\xi) \\
&= |\xi|^{-\alpha} |\xi|^\alpha \mathcal{F}(\varphi)(\xi) \\
&= \mathcal{F}(\varphi)(\xi).
\end{aligned}$$

Therefore

$$\Gamma(x,y) := \frac{C_{n,\alpha}}{|x-y|^{n-\alpha}}$$

is the fundamental solution for the fractional Laplacian $(-\triangle)^{\alpha/2}$.

2.2.2 The α-Harmonic Part

It then follows from (2.19) that we wish $h(x,y)$ to satisfy

$$\begin{cases} (-\triangle)^{\alpha/2} h(x,y) = 0, & x,y \in \Omega, \\ h(x,y) = \frac{C_{n,\alpha}}{|x-y|^{n-\alpha}}, & x \text{ or } y \notin \Omega. \end{cases} \tag{2.30}$$

When Ω is a special domain, a ball or a half space for example, we will find the explicit expression of $h(x, y)$.

When $\Omega = B_r(0)$, it has been proved in [L] that

$$h(x, y) = \int_{|z| > r} \frac{P_r(z, x) dz}{|y - z|^{n-\alpha}}, \qquad (2.31)$$

where $P_r(z, x)$ is the Poisson kernel:

$$P_r(z, x) = \frac{c(r^2 - |x|^2)^{\alpha/2}}{(|z|^2 - r^2)^{\alpha/2} |z - x|^n}.$$

Using the potential theory, we will show that for $|y| < r$, $|x| \geq r$,

$$h(x, y) = \frac{c_{n,\alpha}}{|x - y|^{n-\alpha}}, \qquad c_{n,\alpha} = \frac{\pi^{n/2+1}}{\Gamma(n/2) \sin(\pi\alpha/2)}. \qquad (2.32)$$

Definition 2.2.1 *For a measure $\lambda \in L^1(\mathbb{R}^n)$, its related potential is defined as*

$$U_\alpha^\lambda(x) := \int_{\mathbb{R}^n} \frac{d\lambda(y)}{|x - y|^{n-\alpha}} = \int_{\mathbb{R}^n} \frac{\lambda(y)}{|x - y|^{n-\alpha}} dy.$$

Before carrying out the detailed proof, we first introduce a useful change of coordinates (an inversion of a sphere):

$$x \to x^* = x_o - \frac{1 - |x_o|^2}{|x - x_o|^2}(x - x_o). \qquad (2.33)$$

Through elementary calculation, one can derive the following identities:

$$(a) \ |x - x_o||x^* - x_o| = |1 - |x_o|^2|,$$

$$(b) \ dx = \left(\frac{|1 - |x_o|^2|}{|x^* - x_o|^2}\right)^n dx^*,$$

$$(c) \ 1 - |x|^2 = \frac{(|x^*|^2 - 1)|1 - |x_o|^2|}{|x^* - x_o|^2},$$

$$(d) \ |x^* - y^*| = \frac{|y^* - x_o||x - y|}{|x - x_o|}.$$

The proof is based on the idea of an inversion, as defined in (2.33), of the unit sphere $B_1(0)$ in $\mathbb{R}^n$ with $|x_o| < 1$. From (2.33):

$$x^* - x_o = -\frac{1 - |x_o|^2}{|x - x_o|^2}(x - x_o),$$

we know that x^* lies on the ray starting from x through x_o. A good perspective to see how the inversion changes the position of a point $x \in B_1(0)$ is to start with the extreme situation. First, when $x \in \partial B_1(0)$, by (a) and similar triangles, one can see that $x^* \in \partial B_1(0)$ (see Figure 2.2). Then as x approaches toward the center x_o, from (a) one can see that x^* must be moving away from

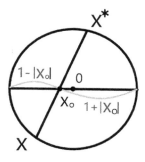

Fig. 2.2

x_o. In other words, when $|x| < 1$, we have $|x^*| > 1$. Similarly, when $|x| > 1$, we have $|x^*| < 1$. To conclude, the inversion maps $\partial B_1(0)$ to itself, $B_1(0)$ to $B_1^c(0)$, and $B_1^c(0)$ to $B_1(0)$.

Riesz noted that the above inversion can turn a measure $\lambda(x)$ with $supp\, \lambda \subset B_1(0)$, into another measure $\lambda^*(x)$ with $supp\, \lambda^* \subset B_1^c(0)$, and turn the potential $U_\alpha^\lambda(y)$ into another closely related potential

$$U_\alpha^{\lambda_{x_o}^*}(y^*) = |y^* - x_o|^{\alpha-n} U_\alpha^\lambda(y).$$

From the capacity theory, one knows that there is the so-called equilibrium measure $\lambda(x)$, for which $U_\alpha^\lambda(y)$ is identical to constant for $|y| < 1$, thus in this case $U_\alpha^{\lambda_{x_o}^*}(y^*)$ coincides with the fundamental solution in $B_1^c(0)$. This suggests that $U_\alpha^{\lambda_{x_o}^*}(y^*)$ may make a good candidate for $h(\cdot, \cdot)$, the α-harmonic part in the Green's function for the fractional Laplacian in $B_1(0)$. Moreover, from the construction, one can see that if both x_o and y^* are in $B_1(0)$, then

$$U_\alpha^{\lambda_{x_o}^*}(y^*) = \int_{|z|>1} \frac{\lambda_{x_o}^*(z)}{|y^* - z|^{n-\alpha}} dz$$

is really α-harmonic, since in the integrand, one always has $y^* \neq z$. Hence

$$U_\alpha^{\lambda_{x_o}^*}(y^*) = h(x_o, y^*)$$

is indeed the second part of the Green's function that we are looking for.

Let the Kelvin transform of $d\lambda(x)$ be

$$d\lambda_{x_o}^*(x^*) = |x - x_o|^{\alpha-n} d\lambda(x), \qquad (2.34)$$

under the inversion introduced in (2.33). It follows from (2.34) and (d) that

$$
\begin{aligned}
U_\alpha^{\lambda_{x_o}^*}(y^*) &= \int_{|x^*|>1} \frac{d\lambda_{x_o}^*(x^*)}{|y^* - x^*|^{n-\alpha}} \\
&= \frac{1}{|y^* - x_o|^{n-\alpha}} \int_{|x|<1} \frac{d\lambda(x)}{|x - y|^{n-\alpha}} \\
&= \frac{1}{|y^* - x_o|^{n-\alpha}} U_\alpha^\lambda(y).
\end{aligned}
\qquad (2.35)
$$

Now what's left is to seek an equilibrium measure of $B_1(0)$, and in fact it is well-known that such a measure takes the form

$$\lambda(x) = \begin{cases} (1 - |x|^2)^{-\alpha/2}, & |x| < 1, \\ 0, & |x| \geq 1. \end{cases} \tag{2.36}$$

In the following, we verify this by showing that for any $|y| < 1$, there exists a constant depending on n and α such that

$$U_\alpha^\lambda(y) = C_{n,\alpha}. \tag{2.37}$$

Let y be the center of the inversion defined in (2.33). Replacing x_o by y in (a)-(c) and it yields

$$(\bar{a})\ |x - y||x^* - y| = 1 - |y|^2,$$

$$(\bar{b})\ dx = \left(\frac{1 - |y|^2}{|x^* - y|^2}\right)^n dx^*,$$

$$(\bar{c})\ 1 - |x|^2 = \frac{(|x^*|^2 - 1)(1 - |y|^2)}{|x^* - y|^2}.$$

Therefore,

$$U_\alpha^\lambda(y)$$

$$= \int_{|x|<1} \frac{dx}{(1 - |x|^2)^{\alpha/2}|x - y|^{n-\alpha}}$$

$$= (1 - |y|^2)^{\alpha/2} \int_{|x^*|>1} \frac{dx^*}{(|x^*|^2 - 1)^{\alpha/2}|x^* - y|^n}$$

$$= (1 - |y|^2)^{\alpha/2} \int_1^\infty \int_0^\pi \frac{\omega_{n-2}(\rho \sin\theta)^{n-2}\rho \, d\rho \, d\theta}{(\rho^2 + |y|^2 - 2\rho|y|\cos\theta)^{n/2}(\rho^2 - 1)^{\alpha/2}}$$

$$= \omega_{n-2}(1 - |y|^2)^{\alpha/2} \int_1^\infty \int_0^\pi \frac{\sin^{n-2}\theta \, d\theta}{((\frac{\rho}{|y|})^2 - 2\frac{\rho}{|y|}\cos\theta + 1)^{n/2}|y|^n} \frac{\rho^{n-1}d\rho}{(\rho^2 - 1)^{\alpha/2}}$$

$$\tag{2.38}$$

$$= \omega_{n-2}(1 - |y|^2)^{\alpha/2} \int_1^\infty \frac{\int_0^\pi \sin^{n-2}\beta \, d\beta}{\rho^{n-2}(\rho^2 - |y|^2)} \frac{\rho^{n-1}d\rho}{(\rho^2 - 1)^{\alpha/2}}, \qquad t = \frac{\rho^2 - 1}{1 - |y|^2}$$

$$\tag{2.39}$$

$$= C_n \int_0^\infty \frac{dt}{t^{\alpha/2}(t + 1)}$$

$$= C_n B(1 - \frac{\alpha}{2}, \frac{\alpha}{2})$$

$$:= C_{n,\alpha}$$

where ω_{n-2} is the hyper-volume of $(n-2)$-dimensional unit sphere, and $B(x, y)$ is the Beta function:

$$B(x, y) = \int_0^\infty \frac{t^{x-1}}{(1 + t)^{x+y}} dt = \int_0^1 (1 - t)^{x-1}t^{y-1} dt.$$

In the calculation above, the inversion defined in (2.33) transforms $|x - y|^{n-\alpha}$ in $U_\alpha^\lambda(y)$ into $|x^* - y|^n$. It thus enables us to apply a known result from [L] which states that for $r > 1$,

$$\int_0^\pi \frac{\sin^{n-2}\theta \, d\theta}{(r^2 - 2r\cos\theta + 1)^{n/2}} = \frac{\int_0^\pi \sin^{n-2}\beta \, d\beta}{r^{n-2}(r^2 - 1)}. \tag{2.40}$$

This leads to (2.39).

It thus proves that for $|x_o| < 1$ and $|y^*| > 1$ (correspondingly $|y| < 1$),

$$h(x_o, y^*) = \frac{C_{n,\alpha}}{|x_o - y^*|^{n-\alpha}}. \tag{2.41}$$

While for $|x_o| < 1$ and $|y^*| < 1$, we still have

$$h(x_o, y^*) = \frac{1}{|x_o - y^*|^{n-\alpha}} U_\alpha^\lambda(y).$$

Note that $U_\alpha^\lambda(y)$ is no longer constant for $|y| > 1$. Since its expression in terms of n-dimensional integral may not be convenient in practical use, in the following we derive its equivalent forms via integrals of a single variable. For this purpose, we employ another inversion centered at y that lies outside the unit ball:

$$x^* = y + \frac{|y|^2 - 1}{|x - y|^2}(x - y). \tag{2.42}$$

Many properties of this inversion are quite similar to the inversion defined in (2.33). In particular, one can easily verify that (a)-(c) for (2.33) hold by replacing x_o by y:

$(\tilde{a})$ $|x - y||x^* - y| = |y|^2 - 1$,

$(\tilde{b})$ $dx = \left(\dfrac{|y|^2 - 1}{|x^* - y|^2}\right)^n dx^*$,

$(\tilde{c})$ $1 - |x|^2 = \dfrac{(1 - |x^*|^2)(|y|^2 - 1)}{|x^* - y|^2}$.

From (2.42):

$$x^* - y = \frac{|y|^2 - 1}{|x - y|^2}(x - y),$$

we know that x^* lies on the ray starting from y through x. When $x \in \partial B_1(0)$, by $(\tilde{a})$ and similar triangles, one can see that $x^* \in \partial B_1(0)$ (see Figure 2.3). Then as x moves closer to the center y, from $(\tilde{a})$ one can see that x^* must be moving away from y. In other words, when $|x| < 1$, we have $|x^*| < 1$. Similarly, $|x| > 1$, we have $|x^*| > 1$. To conclude, the inversion maps $\partial B_1(0)$ to itself, $B_1(0)$ to $B_1(0)$, and $B_1^c(0)$ to $B_1^c(0)$.

Similarly, we have

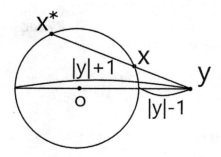

Fig. 2.3

$U_\alpha^\lambda(y)$

$$= \int_{|x|<1} \frac{dx}{(1-|x|^2)^{\alpha/2}|x-y|^{n-\alpha}}$$

$$= (|y|^2-1)^{\alpha/2} \int_{|x^*|<1} \frac{dx^*}{(1-|x^*|^2)^{\alpha/2}|x^*-y|^n}$$

$$= (|y|^2-1)^{\alpha/2} \int_0^1 \int_0^\pi \frac{\omega_{n-2}(\rho\sin\theta)^{n-2}\rho\,d\rho\,d\theta}{(\rho^2+|y|^2-2\rho|y|\cos\theta)^{n/2}(1-\rho^2)^{\alpha/2}}$$

$$= \omega_{n-2}(|y|^2-1)^{\alpha/2} \int_0^1 \int_0^\pi \frac{\sin^{n-2}\theta\,d\theta}{((\frac{|y|}{\rho})^2-2\frac{|y|}{\rho}\cos\theta+1)^{n/2}\rho^n} \frac{\rho^{n-1}d\rho}{(1-\rho^2)^{\alpha/2}}$$

$$= \omega_{n-2}(|y|^2-1)^{\alpha/2} \int_0^1 \frac{\int_0^\pi \sin^{n-2}\beta\,d\beta}{|y|^{n-2}(|y|^2-\rho^2)} \frac{\rho^{n-1}d\rho}{(1-\rho^2)^{\alpha/2}} \tag{2.43}$$

$$= C_n \frac{(|y|^2-1)^{\alpha/2}}{|y|^{n-2}} \int_0^1 \frac{\rho^{n-1}d\rho}{(|y|^2-\rho^2)(1-\rho^2)^{\alpha/2}}. \tag{2.44}$$

Recall that the α-harmonic part satisfies

$$h(x_o, y^*) = U_\alpha^{\lambda^*_{x_o}}(y^*)$$

$$= \frac{1}{|y^*-x_o|^{n-\alpha}} U_\alpha^\lambda(y).$$

It's better if we rewrite (2.44) as a function of x_o and y^*. To do this, we come back to the inversion defined in (2.33) centered at x_o. Then by (c), it holds that

$$|y|^2 = 1 + \frac{(1-|x_o|^2)(1-|y^*|^2)}{|x_o-y^*|^2} := 1 + \frac{t}{s}.$$

And (2.44) becomes

$$\frac{\left(\frac{t}{s}\right)^{\frac{\alpha}{2}}}{(1 + \frac{t}{s})^{\frac{n-2}{2}}} \int_0^1 \frac{\rho^{n-1}d\rho}{(1 - \rho^2 + \frac{t}{s})(1 - \rho^2)^{\alpha/2}}.$$

$$= \frac{\left(\frac{t}{s}\right)^{\frac{\alpha}{2}}}{2(1 + \frac{t}{s})^{\frac{n-2}{2}}} \int_0^1 \frac{(1 - \tau)^{\frac{n-2}{2}} d\tau}{\tau^{\frac{\alpha}{2}}(\tau + \frac{t}{s})} \qquad (\tau = 1 - \rho^2)$$

$$= \frac{1}{2(1 + \frac{t}{s})^{\frac{n-2}{2}}} \int_0^{s/t} \frac{(1 - \frac{t}{s}b)^{\frac{n-2}{2}} db}{b^{\frac{\alpha}{2}}(b + 1)} \qquad (bt/s = \tau)$$

$$= \frac{1}{2(s + t)^{\frac{n-2}{2}}} \int_0^{s/t} \frac{(s - tb)^{\frac{n-2}{2}} db}{b^{\frac{\alpha}{2}}(b + 1)}.$$

Therefore,

$$h(x_o, y^*) = \frac{1}{|y^* - x_o|^{n-\alpha}} \frac{1}{2(s + t)^{\frac{n-2}{2}}} \int_0^{s/t} \frac{(s - tb)^{\frac{n-2}{2}} db}{b^{\frac{\alpha}{2}}(b + 1)}. \qquad (2.45)$$

Expression (2.45) was first found in [Ku].
In [BGR], the authors obtained another more simplified form of h.

Lemma 2.2.2 *For $0 < \alpha < 2$ and $|y| \geq 1$,*

$$U_\alpha^\lambda(y) = \frac{C_{n,\alpha}}{B(\frac{\alpha}{2}, \frac{n-\alpha}{2})} \int_{|y|^2 - 1}^\infty \frac{b^{\alpha/2-1} db}{(b + 1)^{n/2}}. \qquad (2.46)$$

Proof. For $|y| > 1$, under the following inversion

$$I \ : \ x \to x^* = y - \frac{1 - |y|^2}{|x - y|^2}(x - y),$$

we have

$$\int_{|x| \leq 1} \frac{1}{(1 - |x|^2)^{\alpha/2}|x - y|^{n-\alpha}} dx$$

$$= (|y|^2 - 1)^{\alpha/2} \int_{|x^*| \leq 1} \frac{1}{(1 - |x^*|^2)^{\alpha/2}|x^* - y|^n} dx^*$$

$$= \frac{(|y|^2 - 1)^{\frac{\alpha}{2}}}{|y|^{n-2}} \int_0^1 \frac{r^{n-1}}{(1 - r^2)^{\alpha/2}} \frac{1}{|y|^2 - r^2} dr. \qquad (2.47)$$

Notice that for $\alpha \in (0, n)$,

$$B(\frac{n - \alpha}{2}, \frac{\alpha}{2}) = \int_0^1 (1 - t)^{\frac{n-\alpha}{2} - 1} t^{\frac{\alpha}{2} - 1} dt < \infty,$$

thus

$$B(\frac{n-\alpha}{2},\frac{\alpha}{2})\int_0^1 \frac{r^{n-1}}{(1-r^2)^{\alpha/2}}\frac{1}{|y|^2-r^2}dr$$

$$=\int_0^1 B(\frac{n-\alpha}{2},\frac{\alpha}{2})\frac{r^{n-1}}{(1-r^2)^{\alpha/2}}\frac{1}{|y|^2-r^2}dr$$

$$=\int_0^1\int_0^1 (1-t)^{\frac{n-\alpha}{2}-1}t^{\alpha/2-1}\,dt\frac{r^{n-1}}{(1-r^2)^{\alpha/2}}\frac{1}{|y|^2-r^2}dr$$

$$=\int_0^1\int_0^{r^2}\tau^{\frac{n-\alpha}{2}-1}(r^2-\tau)^{\alpha/2-1}d\tau\frac{r}{(1-r^2)^{\alpha/2}}\frac{1}{|y|^2-r^2}dr\quad(1-t=\frac{\tau}{r^2})$$

$$=\int_0^1\int_{\sqrt{\tau}}^1 \frac{r(r^2-\tau)^{\alpha/2-1}}{(1-r^2)^{\alpha/2}}\frac{1}{|y|^2-r^2}dr\,\tau^{\frac{n-\alpha}{2}-1}d\tau$$

$$=\frac{1}{2}\int_0^1\int_0^\infty \frac{ds}{s^{\alpha/2}[s(|y|^2-\tau)+|y|^2-1]}\tau^{\frac{n-\alpha}{2}-1}d\tau\quad(s=\frac{1-r^2}{r^2-\tau})$$

$$=\frac{1}{2}\int_0^1\int_0^\infty \frac{d\tilde{s}}{\tilde{s}^{\alpha/2}(\tilde{s}+1)}\frac{(|y|^2-\tau)^{\alpha/2-1}}{(|y|^2-1)^{\alpha/2}}\tau^{\frac{n-\alpha}{2}-1}d\tau\quad(\tilde{s}(|y|^2-1)=s(|y|^2-\tau))$$

$$=\frac{1}{2}\frac{B(1-\frac{\alpha}{2},\frac{\alpha}{2})}{(|y|^2-1)^{\alpha/2}}\int_0^1 (|y|^2-\tau)^{\alpha/2-1}\tau^{\frac{n-\alpha}{2}-1}d\tau$$

$$=\frac{1}{2}\frac{B(1-\frac{\alpha}{2},\frac{\alpha}{2})}{(|y|^2-1)^{\alpha/2}}\int_0^{\frac{1}{|y|^2}} |y|^{n-2}(1-\tilde{\tau})^{\alpha/2-1}\tilde{\tau}^{\frac{n-\alpha}{2}-1}d\tilde{\tau}\quad(\tau=\tilde{\tau}|y|^2)$$

$$=\frac{1}{2}B(1-\frac{\alpha}{2},\frac{\alpha}{2})\frac{|y|^{n-2}}{(|y|^2-1)^{\alpha/2}}\int_{|y|^2-1}^\infty \frac{b^{\alpha/2-1}\,db}{(b+1)^{n/2}}.\quad(b=\frac{1}{\tilde{\tau}}-1)\qquad(2.48)$$

Combining (2.47) and (2.48), it yields (2.46) and this proves the lemma.

Therefore, the Green's function on B_1 takes the form of

$$G_\alpha(x,y)=\frac{C_{n,\alpha}}{|x-y|^{n-\alpha}}-\frac{1}{|x-y|^{n-\alpha}}U_\alpha^\lambda(y^*)$$

$$=\frac{C_{n,\alpha}}{|x-y|^{n-\alpha}}\left(1-\frac{1}{B(\frac{\alpha}{2},\frac{n-\alpha}{2})}\int_{|y^*|^2-1}^\infty \frac{b^{\alpha/2-1}\,db}{(b+1)^{n/2}}\right)$$

$$=\frac{C_{n,\alpha}}{|x-y|^{n-\alpha}}\frac{1}{B(\frac{\alpha}{2},\frac{n-\alpha}{2})}\int_0^{|y^*|^2-1}\frac{b^{\alpha/2-1}\,db}{(b+1)^{n/2}}$$

$$=\frac{C_{n,\alpha}}{|x-y|^{n-\alpha}}\frac{1}{B(\frac{\alpha}{2},\frac{n-\alpha}{2})}\int_0^{\frac{(1-|y|^2)(1-|x|^2)}{|x-y|^2}}\frac{b^{\alpha/2-1}\,db}{(b+1)^{n/2}}.$$

Moreover, for $|y|<1$ and $|x_o|<1$,

$$U_\alpha^{\lambda_{x_o}^*}(y) = \int_{|z|>1} \frac{d\lambda_{x_o}^*(z)}{|y-z|^{n-\alpha}}$$

$$= \int_{|z|>1} \frac{U_\alpha^{\lambda_y^*}(z)}{C_{n,\alpha}} d\lambda_{x_o}^*(z)$$

$$= C_{n,\alpha}^{-1} \int_{|z|>1} \int_{|x|>1} \frac{d\lambda_y^*(x)}{|x-z|^{n-\alpha}} d\lambda_{x_o}^*(z)$$

$$= C_{n,\alpha}^{-1} \int_{|x|>1} \int_{|z|>1} \frac{d\lambda_{x_o}^*(z)}{|x-z|^{n-\alpha}} d\lambda_y^*(x)$$

$$= C_{n,\alpha}^{-1} \int_{|x|>1} U_\alpha^{\lambda_{x_o}^*}(x) d\lambda_y^*(x)$$

$$= \int_{|x|>1} \frac{d\lambda_y^*(x)}{|x_o-x|^{n-\alpha}}$$

$$= U_\alpha^{\lambda_y^*}(x_o). \tag{2.49}$$

This proves the symmetry of $h(x,y)$, and hence the symmetry of $G_\alpha(x,y)$.

3

Maximum Principles for the Fractional Laplacian

3.1 The Maximum Principle

In this section, we introduce a *maximum principle* for α-super harmonic functions, which plays important roles in the analysis of equations involving the fractional Laplacian, and it will be used again and again in the later chapters. We give two kinds of proofs for this *maximum principle*. One is Silvestre's [Si] proof based on a family of functions that approaches the Delta measure. The other is a direct approach through the integral definition of the fractional Laplacian operator. Though the former requires much more effort than the latter, it sheds light on some frequently used ideas in constructive proofs, and it also requires weaker regularity assumptions.

3.1.1 Indirect Proof Via the "Green-Like" Functions

Theorem 3.1.1 *Let Ω be a bounded domain in $\mathbb{R}^n$. Let $u \in L_\alpha$ be a lower semi-continuous function on $\bar{\Omega}$ satisfying*

$$\begin{cases} (-\triangle)^{\frac{\alpha}{2}} u \geq 0 & in\ \Omega, \\ u \geq 0 & on\ \mathbb{R}^n \backslash \Omega, \end{cases} \tag{3.1}$$

in the sense of distribution, then $u \geq 0$ in Ω.

We say that

$$(-\triangle)^{\frac{\alpha}{2}} u(x) \geq 0, \quad in\ \Omega,$$

holds in the distributional sense, if for all $\varphi \in C_0^\infty(\Omega)$ and $\varphi \geq 0$ in Ω,

$$\int_{\mathbb{R}^n} u(x)(-\triangle)^{\frac{\alpha}{2}} \varphi dx \geq 0. \tag{3.2}$$

Recall that for the usual Laplacian operator, we have the following well-known *maximum principle*.

Assume that $u \in C^2(\bar{\Omega})$ is a solution of

$$\begin{cases} -\triangle u \geq 0 & \text{in } \Omega, \\ u \geq 0 & \text{on } \partial\Omega. \end{cases} \tag{3.3}$$

Then

$$u \geq 0 \text{ in } \Omega. \tag{3.4}$$

There are many ways to prove the above principle. One of them is using the Green's function $G(x, y)$ for $-\triangle$, which is defined as

$$\begin{cases} -\triangle G(x, y) = \delta(x - y) & x, y \in \Omega, \\ G(x, y) = 0 & x \text{ or } y \in \partial\Omega. \end{cases} \tag{3.5}$$

Multiplying (3.3) by $G(x, y)$, integrating by parts, we have

$$0 \leq \int_\Omega -\triangle u(y)G(x, y)dy$$
$$= \int_\Omega u(y)(-\triangle G(x, y))dy + \int_{\partial\Omega} \left(u(y)\frac{\partial G}{\partial\nu} - G(x, y)\frac{\partial u}{\partial\nu} \right) d\sigma$$
$$= \int_\Omega u(y)\delta(x - y)dy + \int_{\partial\Omega} \left(u(y)\frac{\partial G}{\partial\nu} - G(x, y)\frac{\partial u}{\partial\nu} \right) d\sigma$$
$$= u(x) + \int_{\partial\Omega} u\frac{\partial G}{\partial\nu}d\sigma$$
$$\leq u(x).$$

Here we have used the fact that

$$\frac{\partial G}{\partial\nu} \leq 0 \text{ and } u \geq 0 \text{ on } \partial\Omega.$$

This proves (3.4).

For $(-\triangle)^{\frac{\alpha}{2}}$, it is known that such a Green's function $G_\alpha(x, y)$ exists and satisfies

$$\begin{cases} (-\triangle)^{\frac{\alpha}{2}} G_\alpha(x, y) = \delta(x - y), & x, y \in \Omega, \\ G_\alpha(x, y) = 0, & x \text{ or } y \in \mathbb{R}^n \backslash \Omega. \end{cases} \tag{3.6}$$

We may try to apply the same idea to derive the maximum principle for the fractional Laplacian. However, we do not have the corresponding integration by parts formula. From (3.2), it seems that if we replace φ by $G_\alpha(x, y)$, then we will have

$$\int_{\mathbb{R}^n} u(y)(-\triangle)^{\frac{\alpha}{2}} G_\alpha(x, y)dy \geq 0. \tag{3.7}$$

Unfortunately, $G_\alpha(x, y)$ does not belong to $C_0^\infty(\Omega)$ (this can be weakened to $C_0^{1,1}(\Omega)$). However, we can still utilize this idea and try to find some smooth function which plays the role of $G_\alpha(x, y)$ here.

Let

$$\psi(x) = \frac{C_n}{|x|^{n-\alpha}},$$

and

$$\Gamma(x) = \begin{cases} \psi(x), & |x| \geq 1, \\ \text{paraboloid}, & |x| < 1. \end{cases} \tag{3.8}$$

Let $\Gamma_\lambda(x) = \frac{1}{\lambda^{n-\alpha}} \Gamma(\frac{x}{\lambda})$, then

$$\Gamma_\lambda(x) = \begin{cases} \psi(x), & |x| \geq \lambda, \\ \text{paraboloid}, & |x| < \lambda; \end{cases} \qquad (3.9)$$

since one can verify that

$$\frac{1}{\lambda^{n-\alpha}} \psi(\frac{x}{\lambda}) = \psi(x).$$

Notice that for every $x \neq 0$,

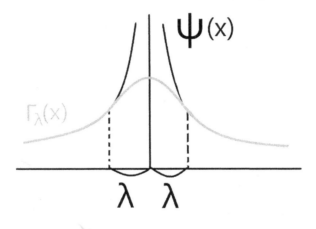

$$\psi(x)$$

$$\Gamma_\lambda(x)$$

$$\lambda \quad \lambda$$

Fig. 3.1

$$\Gamma_\lambda(x) \to \psi(x) \quad \text{as} \quad \lambda \to 0,$$

hence we can guess that

$$(-\triangle)^{\frac{\alpha}{2}} \Gamma_\lambda(x) \to (-\triangle)^{\frac{\alpha}{2}} \psi(x) = \delta(x), \quad \text{as} \quad \lambda \to 0,$$

in the sense of distributions. More precisely, we have the following propositions from [Si].

Proposition 3.1.1 *For $\alpha \in (0,2)$, $(-\triangle)^{\frac{\alpha}{2}} \Gamma$ is a positive continuous function in L^1, and*

$$(-\triangle)^{\frac{\alpha}{2}} \Gamma > 0, \quad x \in \mathbb{R}^n.$$

In addition, $\int_{\mathbb{R}^n} (-\triangle)^{\frac{\alpha}{2}} \Gamma(x) dx = 1$.

Proposition 3.1.2 *Let $\gamma_\lambda = (-\triangle)^{\alpha/2} \Gamma_\lambda$. The family γ_λ is an approximation of the delta function as $\lambda \to 0$ in the sense that*

$$u * \gamma_\lambda(x) = \int_{\mathbb{R}^n} u(y) \gamma_\lambda(x-y) dy \to u(x) \ a.e. \ as \ \lambda \to 0. \qquad (3.10)$$

We will prove the above two propositions right after the following proof of the *maximum principle*.

Proof. Now we have

$$\psi(x) - \Gamma_\lambda(x) = 0, \text{ for } |x| \geq \lambda.$$

Let $x^o \in \Omega$, one can choose λ small such that $B_\lambda(x^o) \subset \Omega$, thus we have

$$\psi(x - x^o) - \Gamma_\lambda(x - x^o) = 0, \text{ outside } B_\lambda(x^o).$$

However, ψ is not smooth. We will use another function

$$\Gamma_{\lambda_1}(x - x^o) - \Gamma_\lambda(x - x^o), 0 < \lambda_1 < \lambda,$$

which is 0 outside $B_\lambda(x^o)$ and is in $C^{1,1}(\mathbb{R}^n)$, as a test function in place of $G_\alpha(x, y)$ in (3.7).

Since $(-\triangle)^{\frac{\alpha}{2}} u(x) \geq 0$ in Ω, by definition

$$\int_{\mathbb{R}^n} u(x)(-\triangle)^{\frac{\alpha}{2}}[\Gamma_{\lambda_1}(x - x^o) - \Gamma_\lambda(x - x^o)]dx \geq 0,$$

i.e.

$$\int_{\mathbb{R}^n} u(x)(-\triangle)^{\frac{\alpha}{2}}\Gamma_{\lambda_1}(x - x^o)dx \geq \int_{\mathbb{R}^n} u(x)(-\triangle)^{\frac{\alpha}{2}}\Gamma_\lambda(x - x^o)dx. \quad (3.11)$$

Let $\lambda_1 \to 0$ in (3.11), it follows from Proposition 3.1.2 that

$$u(x^o) \geq \int_{\mathbb{R}^n} u(x)(-\triangle)^{\alpha/2}\Gamma_\lambda(x - x^o)dx. \quad (3.12)$$

Before proceeding, we would like to explain the meaning of (3.12). Consider the case where $\alpha = 2$ in (3.12) and we see that

$$u(x^o) \geq \int_{\mathbb{R}^n} u(x)(-\triangle)\Gamma_\lambda(x - x^o)dx. \quad (3.13)$$

When $\alpha = 2$,

$$\Gamma_\lambda(x - x^o) = \begin{cases} \frac{C_n}{|x - x^o|^{n-2}}, & |x - x^o| \geq \lambda, \\ \text{paraboloid}, & |x - x^o| < \lambda, \end{cases}$$

and

$$-\triangle\Gamma_\lambda(x - x^o) = \begin{cases} 0, & |x - x^o| > \lambda, \\ C_\lambda, & |x - x^o| < \lambda. \end{cases}$$

Therefore, Eq. (3.13) becomes

$$u(x^o) \geq \int_{B_\lambda(x^o)} C_\lambda u(x)dx.$$

Since

$$\int_{\mathbb{R}^n} -\triangle \Gamma_\lambda(x - x^o)dx = 1,$$

it implies that $C_\lambda = \frac{1}{|B_\lambda(x^o)|}$ and

$$u(x^o) \geq \frac{1}{|B_\lambda(x^o)|} \int_{B_\lambda(x^o)} u(x)dx.$$

From here, one can see that the right hand side of (3.12) is some kind of average value of u. We will use (3.12) to show that $u \geq 0$ in Ω, by deriving a contradiction at a negative minimum of u. The condition that u is lower semi-continuous guarantees the existence of such a minimum.

To see this, we first show that

$$\inf u(x) > -\infty, \quad x \in \bar{\Omega}. \tag{3.14}$$

If (3.14) does not hold, then there exists a sequence of points $\{x^k\} \subseteq \bar{\Omega}$ such that

$$\lim_{k \to \infty} u(x^k) \to -\infty. \tag{3.15}$$

Since $\bar{\Omega}$ is bounded, $\{x^k\}$ has a subsequence, still denoted by $\{x^k\}$, that converges to $x^o \in \bar{\Omega}$. Due to the lower semi-continuity of u, it holds that

$$u(x^o) \leq \lim_{k \to \infty} u(x^k). \tag{3.16}$$

Together with (3.15), it gives

$$u(x^o) \leq -\infty,$$

which is impossible. This proves (3.14).

From (3.14), we deduce that there is a minimizing sequence $\{x^k\} \in \bar{\Omega}$ of u such that

$$\lim_{k \to \infty} u(x^k) = \inf_{\bar{\Omega}} u. \tag{3.17}$$

Since $\bar{\Omega}$ is bounded, $\{x^k\}$ has a subsequence that converges to $x^o \in \bar{\Omega}$ and

$$u(x^o) \geq \lim_{k \to \infty} u(x^k). \tag{3.18}$$

Next we show that x^o is the minimum point of u. From the lower semi-continuity of u we know that

$$u(x^o) \leq \lim_{k \to \infty} u(x^k). \tag{3.19}$$

Together with (3.17) and (3.18), we have

$$u(x^o) = \inf_{\bar{\Omega}} u. \tag{3.20}$$

Hence

$$u(x^o) = \min_{\bar{\Omega}} u(x). \tag{3.21}$$

Case i) If $x^o \in \partial\Omega$, then $u(x^o) \geq 0$ on $\bar{\Omega}$. We are done.

Case ii) If $x^o \in \Omega$, then we choose λ sufficiently small such that $B_\lambda(x^o) \subset \Omega$, hence by (3.12), we have

$$0 \leq \int_{\mathbb{R}^n} [u(x^o) - u(x)]\gamma_\lambda(x - x^o)dx \tag{3.22}$$

$$= \int_{\mathbb{R}^n \backslash \Omega} [u(x^o) - u(x)]\gamma_\lambda(x - x^o)dx$$

$$+ \int_\Omega [u(x^o) - u(x)]\gamma_\lambda(x - x^o)dx. \tag{3.23}$$

By Proposition 3.1.1 and the assumption that $u \geq 0$ in $\mathbb{R}^n \backslash \Omega$, we can see that the first integral in (3.23) is negative and the second one is non-positive. A contradiction with (3.22). Therefore, we must have

$$u(x) \geq 0 \text{ in } \Omega.$$

Now we give the proof of Proposition 3.1.1.

Proof. Since Γ is $C^{1,1}(\mathbb{R}^n) \cap L_\alpha$, $(-\triangle)^{\alpha/2}\Gamma$ is well-defined.

Case i) Recall that for $|x| \geq 1$, $\Gamma(x) = \psi(x)$; and for $|x| < 1$, $\Gamma(x) < \psi(x)$. If $x^o \notin B_1$, since ψ is the fundamental solution, we have

$$(-\triangle)^{\frac{\alpha}{2}}\Gamma(x^o) = C_{n,\alpha}PV \int_{\mathbb{R}^n} \frac{\Gamma(x^o) - \Gamma(y)}{|x^o - y|^{n+\alpha}}dy$$

$$> C_{n,\alpha}PV \int_{\mathbb{R}^n} \frac{\psi(x^o) - \psi(y)}{|x^o - y|^{n+\alpha}}dy = 0.$$

Case ii) If $x^o \in B_1 \backslash \{0\}$, there exist an x^1 and a positive δ such that $\psi(x - x^1) + \delta$ touches Γ from above at the point x^o. Using the singular integral representation, it gives

$$(-\triangle)^{\frac{\alpha}{2}}\Gamma(x^o) = C_{n,\alpha}PV \int_{\mathbb{R}^n} \frac{\Gamma(x^o) - \Gamma(y)}{|x^o - y|^{n+\alpha}}dy$$

$$> C_{n,\alpha}PV \int_{\mathbb{R}^n} \frac{\psi(x^o - x^1) + \delta - \psi(y - x^1) - \delta}{|x^o - y|^{n+\alpha}}dy = 0$$

since $(-\triangle)^{\frac{\alpha}{2}}(\psi(x^o - x^1) + \delta) = 0$.

If $x^o = 0$, then Γ attains its maximum at x^o,

$$(-\triangle)^{\frac{\alpha}{2}}\Gamma(x^o) = C_{n,\alpha}PV \int_{\mathbb{R}^n} \frac{\Gamma(x^o) - \Gamma(y)}{|x^o - y|^{n+\alpha}}dy > 0, \tag{3.24}$$

because we are integrating a positive function.

To show that $\int_{\mathbb{R}^n} (-\triangle)^{\frac{\alpha}{2}}\Gamma(x)dx = 1$, we consider a smooth cutoff function η such that

$$0 \leq \eta(x) \leq 1 \text{ for every } x \in \mathbb{R}^n,$$

$\eta(x) = 1$ for every $x \in B_1(0)$, and $supp\, \eta \subset B_2(0)$.

Let $\eta_R(x) = \eta(\frac{x}{R})$, then we have

$$\int_{\mathbb{R}^n} (-\triangle)^{\frac{\alpha}{2}} \Gamma(x) dx - 1 = \lim_{R \to \infty} \langle (-\triangle)^{\frac{\alpha}{2}} \Gamma - (-\triangle)^{\frac{\alpha}{2}} \psi, \eta_R \rangle$$

$$= \lim_{R \to \infty} \langle \Gamma - \psi, (-\triangle)^{\frac{\alpha}{2}} \eta_R \rangle = 0$$

since clearly $(-\triangle)^{\frac{\alpha}{2}} \eta_R$ goes to 0 uniformly on each compact set, and $\Gamma - \psi$ is an L^1-function with compact support. To see this, for each fixed x, by choosing R sufficiently large (larger than $|x|$), we have

$$\left| (-\triangle)^{\alpha/2} \eta_R(x) \right|$$

$$= \left| C_{n,\alpha} PV \int_{\mathbb{R}^n} \frac{\eta(\frac{x}{R}) - \eta(\frac{y}{R})}{|x-y|^{n+\alpha}} dy \right|$$

$$\leq C_{n,\alpha} \left(PV \int_{B_R(0)} \frac{1-1}{|x-y|^{n+\alpha}} dy + \int_{\mathbb{R}^n \setminus B_R(0)} \frac{1}{|x-y|^{n+\alpha}} dy \right)$$

$$= C_{n,\alpha} \int_{\mathbb{R}^n \setminus B_R(0)} \frac{1}{|x-y|^{n+\alpha}} dy \to 0, \text{ as } R \to \infty.$$

Next, we prove Proposition 3.1.2.

Proof. First of all, notice that $u(y)\gamma_\lambda(x-y)$ is integrable for every given x since $u \in L_\alpha$ and $\gamma_\lambda(x)$ decays as $1/(1+|x|^{n+\alpha})$. Indeed, for $|x|$ large, since $\Gamma_\lambda(x) = \psi(x)$ and $\psi(y) - \Gamma_\lambda(y)$ is a compactly supported function in L^1, then

$$\gamma_\lambda(x) = C_{n,\alpha} PV \int_{\mathbb{R}^n} \frac{\Gamma_\lambda(x) - \Gamma_\lambda(y)}{|x-y|^{n+\alpha}} dy$$

$$= C_{n,\alpha} PV \int_{\mathbb{R}^n} \frac{\psi(x) - \psi(y)}{|x-y|^{n+\alpha}} dy + C_{n,\alpha} PV \int_{\mathbb{R}^n} \frac{\psi(y) - \Gamma_\lambda(y)}{|x-y|^{n+\alpha}} dy$$

$$= C_{n,\alpha} PV \int_{\mathbb{R}^n} \frac{\psi(y) - \Gamma_\lambda(y)}{|x-y|^{n+\alpha}} dy$$

$$\cong \frac{1}{|x|^{n+\alpha}}. \tag{3.25}$$

Now we check the re-scaling property of $\gamma_\lambda = (-\triangle)^{\alpha/2} \Gamma_\lambda$.

$$\gamma_\lambda(x) = (-\triangle)^{\alpha/2} \Gamma_\lambda(x) = (-\triangle)^{\alpha/2} \left(\frac{\Gamma(x/\lambda)}{\lambda^{n-\alpha}} \right)$$

$$= \frac{1}{\lambda^n} [(-\triangle)^{\alpha/2} \Gamma](x/\lambda)$$

$$= \frac{1}{\lambda^n} \gamma_1(x/\lambda).$$

Using this re-scaling property, we first verify (3.10) for $\phi \in C_0^\infty(\mathbb{R}^n)$, that is

$$\lim_{\lambda \to 0} \int_{\mathbb{R}^n} \phi(y) \gamma_\lambda(x-y) \, dy = \phi(x). \qquad (3.26)$$

In fact,

$$\lim_{\lambda \to 0} \int_{\mathbb{R}^n} \phi(y) \gamma_\lambda(x-y) \, dy = \lim_{\lambda \to 0} \int_{\mathbb{R}^n} \phi(y) \frac{1}{\lambda^n} \gamma_1(\frac{x-y}{\lambda}) \, dy$$

$$= \lim_{\lambda \to 0} \int_{\mathbb{R}^n} \phi(-\lambda z + x) \gamma_1(z) \, dz$$

$$= \phi(x) \int_{\mathbb{R}^n} \gamma_1(z) dz = \phi(x).$$

To prove (3.10) for $u \in L_\alpha$, it suffices to show that $\forall \phi \in C_0^\infty(\mathbb{R}^n)$,

$$\int_{\mathbb{R}^n} \phi(x) \lim_{\lambda \to 0} \int_{\mathbb{R}^n} u(y) \gamma_\lambda(x-y) \, dy \, dx = \int_{\mathbb{R}^n} \phi(x) u(x) dx. \qquad (3.27)$$

By (3.25), for each fixed x and R sufficiently large,

$$\gamma_\lambda(x-y) \cong \frac{1}{|x-y|^{n+\alpha}}, \quad y \in B_R^c.$$

Thus

$$\int_{B_R^c} u(y) \gamma_\lambda(x-y) \, dy \leq g_2(x).$$

Since $\gamma_\lambda, u \in L_{loc}^1$,

$$\int_{B_R} u(y) \gamma_\lambda(x-y) \, dy \leq g_1(x).$$

With $\phi \in C_0^\infty$, we can apply the Lebesgue's dominated convergence theorem and the Fubini's theorem to obtain

$$\text{LHS of } (3.27) = \lim_{\lambda \to 0} \int_{\mathbb{R}^n} \phi(x) \int_{\mathbb{R}_n} u(y) \gamma_\lambda(x-y) \, dy \, dx$$

$$= \lim_{\lambda \to 0} \int_{\mathbb{R}^n} u(y) \int_{\mathbb{R}^n} \phi(x) \gamma_\lambda(x-y) \, dx \, dy$$

$$= \int_{\mathbb{R}^n} u(y) \lim_{\lambda \to 0} \int_{\mathbb{R}^n} \phi(x) \gamma_\lambda(x-y) \, dx \, dy$$

$$= \int_{\mathbb{R}^n} \phi(y) u(y) dy.$$

This proves (3.27) and hence completes the proof of the proposition.

3.1.2 A Much Simpler Proof by Using the Defining Integral

Notice that, with an additional regularity assumption that $u \in L_\alpha \cap C_{loc}^{1,1}(\Omega)$, one can greatly simplify the proof of the *maximum principle* by using the integral defining the fractional Laplacian.

Theorem 3.1.2 (Maximum Principle) *Let Ω be a bounded domain in $\mathbb{R}^n$. Assume that $u \in L_\alpha \cap C^{1,1}_{loc}(\Omega)$ and is lower semi-continuous on $\bar{\Omega}$. If*

$$\begin{cases} (-\triangle)^{\alpha/2}u(x) \geq 0 & in \ \Omega, \\ u(x) \geq 0 & in \ \mathbb{R}^n \backslash \Omega, \end{cases} \tag{3.28}$$

then

$$u(x) \geq 0 \ in \ \Omega. \tag{3.29}$$

If $u = 0$ at some point in Ω, then

$$u(x) = 0 \ in \ \mathbb{R}^n.$$

These conclusions hold for unbounded region Ω if we further assume that

$$\varliminf_{|x|\to\infty} u(x) \geq 0.$$

Proof. If (3.29) does not hold, then the lower semi-continuity of u on $\bar{\Omega}$ indicates that there exists a $x^0 \in \bar{\Omega}$ such that

$$u(x^0) = \min_{\bar{\Omega}} u < 0.$$

And one can further deduce from condition (3.28) that x^0 is in the interior of Ω.

Then it follows that

$$(-\triangle)^{\alpha/2}u(x^0) = C_{n,\alpha}PV \int_{\mathbb{R}^n} \frac{u(x^0) - u(y)}{|x^0 - y|^{n+\alpha}}dy$$
$$\leq C_{n,\alpha} \int_{\mathbb{R}^n\backslash\Omega} \frac{u(x^0) - u(y)}{|x^0 - y|^{n+\alpha}}dy$$
$$< 0,$$

which contradicts inequality (3.28). This verifies (3.29).

If at some point $x^o \in \Omega$, $u(x^o) = 0$, then from

$$0 \leq (-\triangle)^{\alpha/2}u(x^o) = C_{n,\alpha}PV \int_{\mathbb{R}^n} \frac{-u(y)}{|x^o - y|^{n+\alpha}}dy$$

and $u \geq 0$, we must have

$$u(x) = 0 \ \text{almost everywhere in } \mathbb{R}^n.$$

For unbounded region Ω, the condition

$$\varliminf_{|x|\to\infty} u(x) \geq 0$$

guarantees that the negative minimum of u must be attained at some point x^0, then we can derive the same contradiction as the above.

This completes the proof.

4

Poisson Representations

4.1 Introducing the Poisson Representation

As one can see in Chapter 2, with the help of Green's function $G(x, y)$, we are able to express the solution of the homogeneous Dirichlet problem

$$\begin{cases} (-\triangle)^{\alpha/2} u(x) = f(x), \ x \in B_r(0), \\ u(x) = 0, \qquad\qquad x \in B_r^c(0), \end{cases} \tag{4.1}$$

as

$$u(x) = \int_{B_r(0)} G(x, y) f(y) dy.$$

Now consider the non-homogeneous Dirichlet problem

$$\begin{cases} (-\triangle)^{\alpha/2} u(x) = f(x), \ x \in B_r(0), \\ u(x) = g(x), \qquad\qquad x \in B_r^c(0). \end{cases} \tag{4.2}$$

We split the solution into two parts $u = u_1 + u_2$, where u_1 is the solution of (4.1), and u_2 satisfies

$$\begin{cases} (-\triangle)^{\alpha/2} u_2(x) = 0, \ x \in B_r(0), \\ u_2(x) = g(x), \qquad x \in B_r^c(0). \end{cases} \tag{4.3}$$

In this chapter, we will show that

$$u_2(x) = \int_{|y|>r} P_r(y, x) g(y) dy, \quad \forall\, x \in B_r(0), \tag{4.4}$$

where

$$P_r(y, x) = \begin{cases} \frac{\Gamma(n/2)}{\pi^{\frac{n}{2}+1}} \sin\frac{\pi\alpha}{2} \left(\frac{r^2-|x|^2}{|y|^2-r^2}\right)^{\frac{\alpha}{2}} \frac{1}{|x-y|^n}, & |y| > r, \\ 0, & |y| < r, \end{cases} \tag{4.5}$$

is the so-called Poisson kernel and (4.4) the Poisson representation of $u_2(x)$.

Now, the solution of (4.2) can be represented by

$$u(x) = \int_{|y|<r} G(x,y)f(y)dy + \int_{|y|>r} P_r(y,x)g(y)dy, \quad \forall\, x \in B_r(0).$$

From representation (4.4), we can see that, due to the nonlocal nature of the fractional Laplacian, in order to guarantee the uniqueness of solution, one must prescribe the Dirichlet condition on the whole complement of $B_r(0)$ instead of just on the boundary. If we only require

$$u_2(x) = g(x), \quad x \in \partial B_r(0),$$

then by varying the values of $g(x)$ beyond the domain, we can get infinitely many solutions from (4.4).

From (4.4), one can also see that if u is α-harmonic in the whole space, then

$$u(x) = \int_{|y|>r} P_r(y,x)u(y)dy, \quad \forall\, x \in B_r(0).$$

This Poisson representation of u by itself will enable us to prove the smoothness of α-harmonic functions, and to obtain the uniqueness–Liouville type theorems for α-harmonic functions in the next chapter.

Recall that we say a function v is α-harmonic, if $v \in C_{loc}^{1,1} \cap L_\alpha$ and

$$(-\triangle)^{\alpha/2}v(x) \equiv C_{n,\alpha}PV \int_{\mathbb{R}^n} \frac{v(x)-v(y)}{|x-y|^{n+\alpha}}dy = 0.$$

For convenience, in the following, we denote

$$\hat{u}(x) = \begin{cases} \int_{|y|>r} P_r(y,x)u(y)dy, & |x| < r, \\ u(x), & |x| \geq r. \end{cases} \qquad (4.6)$$

We prove

Theorem 4.1.1 *Let $u \in L_\alpha$ and $\hat{u}(x)$ be defined by (4.6). Then $\hat{u}$ is α-harmonic in $B_r(0)$.*

Remark 4.1.1 *The above theorem actually implies that $\hat{u}(x)$ is a solution of the Dirichlet problem:*

$$\begin{cases} (-\triangle)^{\alpha/2}\hat{u}(x) = 0, \; x \in B_r(0), \\ \hat{u}(x) = u(x), \qquad x \in B_r^c(0). \end{cases}$$

The proof of Theorem 4.1.1 consists of two parts. First we show that $\hat{u}$ is harmonic in the average sense (Lemma 4.1.1), then we show that it is α-harmonic by the Pizzetti type formula (Lemma 4.1.2).

Let

$$\varepsilon_\alpha^{(r)}(x) = \begin{cases} 0, & |x| < r. \\ \frac{\Gamma(n/2)}{\pi^{\frac{n}{2}+1}} \sin \frac{\pi\alpha}{2} \frac{r^\alpha}{(|x|^2-r^2)^{\frac{\alpha}{2}}|x|^n}, & |x| > r. \end{cases} \qquad (4.7)$$

We say that u is α-harmonic in the average sense (see [L]) if for all suffi-ciently small r,

$$\varepsilon_\alpha^{(r)} * u(x) = u(x),$$

where $*$ denotes the convolution.

One can verify via direct calculation that

$$\int_{|x|>r} \varepsilon_\alpha^{(r)}(x)dx = 1.$$

Hence $\varepsilon_\alpha^{(r)} * u(x)$ is some kind of weighted average of u in the complement of $B_r(x)$. And its counterpart for the regular Laplacian $-\triangle$ is the average of u on the boundary $\partial B_r(x)$.

Lemma 4.1.1 *Assume that*

$$u \text{ is locally bounded and } u \in L_\alpha. \tag{4.8}$$

Then $\hat{u}(x)$ is α-harmonic in the average sense in $B_r(0)$.

To relate the average of u to $-\triangle u$, we recall the well-known Pizzetti's formula:

$$\lim_{r \to 0} \frac{1}{r^2}\left[u(x) - \frac{1}{|B_r(x)|}\int_{B_r(x)} u(y)dy\right] = c\,(-\triangle)u(x).$$

Similarly, for the fractional Laplacian, we will prove

Lemma 4.1.2 (Pizzetti's formula) *Let $u \in L_\alpha$ and $C^{1,1}$ in a neighborhood of x. Then we have*

$$\lim_{r \to 0} \frac{1}{r^\alpha}\left[u(x) - \varepsilon_\alpha^{(r)} * u(x)\right] = c\,(-\triangle)^{\frac{\alpha}{2}}u(x) \tag{4.9}$$

where $c = \frac{\Gamma(n/2)}{\pi^{\frac{n}{2}+1}}\sin\frac{\pi\alpha}{2}$.

This formula implies immediately that if u is α-harmonic in the average sense, then it is also α-harmonic. It was obtained in [L]. Here we present a different and more direct proof.

Combining the above two lemmas and the smoothness of $\hat{u}$, we derive

Theorem 4.1.2 *If $u \in L_\alpha$ is α-harmonic in the sense of distributions, then it is smooth, and it is also α-harmonic in the sense of average, and hence α-harmonic. The converse is also true. Now being α-harmonic in all the three senses are equivalent.*

4.2 $\hat{u}_k$ is α-Harmonic in the Average Sense

Proof of Lemma 4.1.1. The outline is as follows.

i) Approximate u by a sequence of smooth, compactly supported functions $\{u_k\}$, such that $u_k(x) \to u(x)$ in the sense of

$$\int_{|z|>r} \frac{|u_k(z) - u(z)|}{|z|^n (|z|^2 - r^2)^{\frac{\alpha}{2}}} dz \to 0.$$

This is possible under our assumption (4.8).

ii) For each u_k, find a signed measure ν_k such that $supp\, \nu_k \subset B_r^c$ and

$$u_k(x) = U_\alpha^{\nu_k}(x), \quad |x| > r.$$

Then

$$\hat{u}_k(x) = U_\alpha^{\nu_k}(x), \quad x \in \mathbb{R}^n.$$

iii) Verify that $\hat{u}_k(x)$ is α-harmonic in the average sense for $|x| < r$. That is, for each fixed small $\delta > 0$,

$$(\varepsilon_\alpha^{(\delta)} * \hat{u}_k)(x) = \hat{u}_k(x).$$

iv) Show that as $k \to \infty$

$$\varepsilon_\alpha^{(\delta)} * \hat{u}_k \to \varepsilon_\alpha^{(\delta)} * \hat{u},$$

and

$$\hat{u}_k \to \hat{u}.$$

Then we arrive at

$$(\varepsilon_\alpha^{(\delta)} * \hat{u})(x) = \hat{u}(x), \quad |x| < r.$$

Now we fill in the details.

i) There are several ways to construct such a sequence $\{u_k\}$. One is to use the mollifier. Let

$$j(x) = \begin{cases} C\, exp(-\frac{1}{1-|x|^2}), & |x| < 1, \\ 0, & |x| \geq 1. \end{cases}$$

Then $j(x)$ satisfies the following properties:

(a) It's compactly supported;

(b) $\lim_{\varepsilon \to 0} j_\varepsilon(x) := \lim_{\varepsilon \to 0} \varepsilon^{-n} j(x/\varepsilon) = \delta(x)$, for a properly chosen C.

Let

$$u|_{B_k}(x) = \begin{cases} u(x), & |x| < k, \\ 0, & |x| \geq k, \end{cases} \tag{4.10}$$

and

$$J_\varepsilon(u|_{B_k})(x) = \int_{\mathbb{R}^n} j_\varepsilon(x - y) u|_{B_k}(y) dy.$$

Then for any $\varepsilon > 0$,

$$J_\epsilon(u|_{B_k})(x) = 0, \quad |x| > k + \varepsilon. \tag{4.11}$$

By (b), for each such k, we can choose ϵ_k small such that

$$\int_{B_{k+1} \setminus B_r} \frac{|u_k(z) - u|_{B_k}(z)|}{|z|^n (|z|^2 - r^2)^{\frac{\alpha}{2}}} dz$$

$$= \int_{B_{k+1} \setminus B_r} \frac{o(1)}{|z|^n (|z|^2 - r^2)^{\frac{\alpha}{2}}} dz < \frac{\delta}{2}, \tag{4.12}$$

where

$$u_k := J_{\epsilon_k}(u|_{B_k}).$$

By (4.11), we know that

$$I := \int_{B_r^c} \frac{|u_k(z) - u(z)|}{|z|^n (|z|^2 - r^2)^{\frac{\alpha}{2}}} dz$$

$$= \int_{B_{k+1}^c} \frac{|u_k(z) - u(z)|}{|z|^n (|z|^2 - r^2)^{\frac{\alpha}{2}}} dz + \int_{B_{k+1} \setminus B_r} \frac{|u_k(z) - u(z)|}{|z|^n (|z|^2 - r^2)^{\frac{\alpha}{2}}} dz$$

$$= \int_{B_{k+1}^c} \frac{|u(z)|}{|z|^n (|z|^2 - r^2)^{\frac{\alpha}{2}}} dz + \int_{B_{k+1} \setminus B_r} \frac{|u_k(z) - u(z)|}{|z|^n (|z|^2 - r^2)^{\frac{\alpha}{2}}} dz$$

$$:= I_1 + I_2. \tag{4.13}$$

With (4.10), we have

$$I_2 := \int_{B_{k+1} \setminus B_r} \frac{|u_k(z) - u(z)|}{|z|^n (|z|^2 - r^2)^{\frac{\alpha}{2}}} dz$$

$$< \int_{B_{k+1} \setminus B_r} \frac{|u_k(z) - u|_{B_k}(z)| + |u|_{B_k}(z) - u(z)|}{|z|^n (|z|^2 - r^2)^{\frac{\alpha}{2}}} dz$$

$$= \int_{B_{k+1} \setminus B_k} \frac{|u_k(z) - u|_{B_k}(z)| + |0 - u(z)|}{|z|^n (|z|^2 - r^2)^{\frac{\alpha}{2}}} dz$$

$$+ \int_{B_k \setminus B_r} \frac{|u_k(z) - u|_{B_k}(z)| + |u(z) - u(z)|}{|z|^n (|z|^2 - r^2)^{\frac{\alpha}{2}}} dz$$

$$= \int_{B_{k+1} \setminus B_r} \frac{|u_k(z) - u|_{B_k}(z)|}{|z|^n (|z|^2 - r^2)^{\frac{\alpha}{2}}} dz$$

$$+ \int_{B_{k+1} \setminus B_k} \frac{|u(z)|}{|z|^n (|z|^2 - r^2)^{\frac{\alpha}{2}}} dz$$

$$= I_{21} + I_{22}.$$

From the assumption that $u \in L_\alpha$, we know that for any $\delta > 0$ and k sufficiently large (larger than r), it holds that

$$I_1 + I_{22} = \int_{|z| \geq k} \frac{|u(z)|}{|z|^n (|z|^2 - r^2)^{\frac{\alpha}{2}}} dz < \frac{\delta}{2}. \tag{4.14}$$

Combining (4.12), (4.13) and (4.14), it yields that

$$I < I_1 + I_{21} + I_{22} < \frac{\delta}{2} + \frac{\delta}{2} = \delta.$$

Due to the arbitrariness of δ, it proves that as $k \to \infty$,

$$\int_{|z|>r} \frac{|u_k(z) - u(z)|}{|z|^n (|z|^2 - r^2)^{\frac{\alpha}{2}}} dz \to 0. \qquad (4.15)$$

ii) In this step, we want to find a signed measure ν_k with support in B_r^c, such that

$$u_k(x) = U_\alpha^{\nu_k}(x), \quad |x| > r,$$

and hence it would follow that

$$\hat{u}_k(x) = U_\alpha^{\nu_k}(x), \quad x \in \mathbb{R}^n.$$

Since $u_k \in C_0^\infty(\mathbb{R}^n)$, there exists a signed measure ψ_k such that

$$u_k(x) = U_\alpha^{\psi_k}(x). \qquad (4.16)$$

Indeed, let $\psi_k(x) = C(-\triangle)^{\frac{\alpha}{2}} u_k(x)$, then

$$U_\alpha^{\psi_k}(x) = \int_{\mathbb{R}^n} \frac{C}{|x-y|^{n-\alpha}} (-\triangle)^{\frac{\alpha}{2}} u_k(y) dy$$

$$= \int_{\mathbb{R}^n} \delta(x-y) u_k(y) dy = u_k(x).$$

Here we have used the fact that $\frac{C}{|x-y|^{n-\alpha}}$ is the fundamental solution of $(-\triangle)^{\alpha/2}$.

Because the support of ψ_k is unknown, we cannot take it as ν_k. To find ν_k, let $\psi_k|_{B_r}$ be the restriction of ψ_k on B_r. If we set

$$\nu_k = \psi_k - \psi_k|_{B_r},$$

then *supp* $\nu_k \subseteq B_r^c$. This is what we want. But what we are not yet satisfied with is

$$U_\alpha^{\nu_k} = U_\alpha^{\psi_k} - U_\alpha^{\psi_k|_{B_r}}, \qquad (4.17)$$

which is not $U_\alpha^{\psi_k}(= u_k)$ as we desired. Hence we need another measure $\tilde{\psi}_k$, whose support is also in B_r^c, to counteract $U_\alpha^{\psi_k|_{B_r}}$ in (4.17). This can be done through the Poisson kernel, which "transports" a point outside B_r to its inside, or inside out. To be more precise, using the fact (see (2.32) in Chapter 2 or (1.6.12′) in [L]) that

$$\frac{1}{|y-x|^{n-\alpha}} = \int_{|z|>r} \frac{P_r(z,y)}{|z-x|^{n-\alpha}} dz, \quad |y| < r < |x|, \qquad (4.18)$$

for $|x| > r$ we have

$$U_\alpha^{\psi_k|_{B_r}}(x)$$

$$= \int_{|y|<r} \frac{\psi_k|_{B_r}(y)}{|x-y|^{n-\alpha}}\,dy$$

$$= \int_{|y|<r} \int_{|z|>r} \frac{P_r(z,y)}{|x-z|^{n-\alpha}}\,dz\,\psi_k|_{B_r}(y)\,dy$$

$$= \int_{|z|>r} \int_{|y|<r} P_r(z,y)\psi_k|_{B_r}(y)\,dy\,\frac{1}{|x-z|^{n-\alpha}}\,dz. \qquad (4.19)$$

Let

$$\tilde{\psi}_k(z) := \int_{|y|<r} P_r(z,y)\psi_k|_{B_r}(y)dy. \qquad (4.20)$$

Then (4.19) becomes

$$U_\alpha^{\psi_k|_{B_r}}(x) = \int_{|z|>r} \frac{\tilde{\psi}_k(z)}{|x-z|^{n-\alpha}}dz = U_\alpha^{\tilde{\psi}_k}(x).$$

On the other hand, (4.5) guarantees that $supp\,\tilde{\psi}_k \in B_r^c$. It turns out that this $\tilde{\psi}_k$ is the "balancing" measure we want.

Let $\nu_k = \psi_k - \psi_k|_{B_r} + \tilde{\psi}_k$. Then $supp\,\nu_k \subset B_r^c$ and

$$U_\alpha^{\nu_k}(x) = U_\alpha^{\psi_k}(x) + U_\alpha^{\tilde{\psi}_k}(x) - U_\alpha^{\psi_k|_{B_r}}(x) = U_\alpha^{\psi_k}(x), \quad |x| > r.$$

Together with (4.16), it shows that

$$u_k(x) = U_\alpha^{\nu_k}(x), \quad |x| > r.$$

Hence for $|x| < r$,

$$\hat{u}_k(x) = \int_{|y|>r} P_r(y,x)u_k(y)dy$$

$$= \int_{|y|>r} P_r(y,x)U_\alpha^{\nu_k}(y)dy$$

$$= \int_{|y|>r} P_r(y,x) \int_{|z|>r} \frac{\nu_k(z)}{|y-z|^{n-\alpha}}dz\,dy$$

$$= \int_{|z|>r} \int_{|y|>r} \frac{P_r(y,x)}{|y-z|^{n-\alpha}}dy\,\nu_k(z)\,dz$$

$$= \int_{|z|>r} \frac{\nu_k(z)}{|x-z|^{n-\alpha}}dz$$

$$= U_\alpha^{\nu_k}(x).$$

Together with (4.6), we have

$$\hat{u}_k(x) = U_\alpha^{\nu_k}(x), \quad x \in \mathbb{R}^n.$$

iii)   Assume that $supp\,\nu \subset B_r^c(0)$. We show that, for small $\delta > 0$,

$$(\varepsilon_\alpha^{(\delta)} * U_\alpha^\nu)(x) = U_\alpha^\nu(x), \quad |x| < r.$$

In other words, $U_\alpha^\nu(x)$ is α-harmonic in the average sense in $B_r(0)$. Actually, for any given x and $r > 0$, choose $\delta > 0$ small such that $\delta \leq r - |x|$, then

$$
\begin{aligned}
&(\varepsilon_\alpha^{(\delta)} * U_\alpha^\nu)(x) \\
&= \int_{|y-x|>\delta} \varepsilon_\alpha^{(\delta)}(|y-x|) U_\alpha^\nu(y) dy \\
&= \int_{|y-x|>\delta} \varepsilon_\alpha^{(\delta)}(y-x) \int_{|z|>r} \frac{\nu(z)}{|z-y|^{n-\alpha}} dz\, dy \\
&= \int_{|z|>r} \nu(z) \int_{|y-x|>\delta} \varepsilon_\alpha^{(\delta)}(|y-x|) \frac{1}{|z-y|^{n-\alpha}} dy\, dz. \qquad (4.21)
\end{aligned}
$$

To continue, we first show that

$$\int_{|y-x|>\delta} \varepsilon_\alpha^{(\delta)}(|y-x|) \frac{1}{|z-y|^{n-\alpha}} dy = \frac{1}{|z-x|^{n-\alpha}}, \quad |z-x| > \delta, \qquad (4.22)$$

i.e., the well-known fundamental solution $\Gamma(|z-x|) = \frac{C_{n,\alpha}}{|z-x|^{n-\alpha}}$ is α-harmonic in the sense of average for $z \neq x$.

To see (4.22), we start from

$$\int_{|y|>\delta} P_\delta(y,x) \frac{1}{|y-z|^{n-\alpha}} dy = \frac{1}{|x-z|^{n-\alpha}}, \quad |x| < \delta < |z|,$$

or

$$C \int_{|y|>\delta} \frac{(\delta^2 - |x|^2)^{\alpha/2}}{(|y|^2 - \delta^2)^{\alpha/2} |y-x|^n} \frac{1}{|y-z|^{n-\alpha}} dy = \frac{1}{|x-z|^{n-\alpha}}.$$

First let $x = 0$,

$$C \int_{|y|>\delta} \frac{\delta^\alpha}{(|y|^2 - \delta^2)^{\alpha/2} |y|^n} \frac{1}{|y-z|^{n-\alpha}} dy = \frac{1}{|z|^{n-\alpha}}, \quad |z| > \delta.$$

Then for $|x| < r$, let $y = \bar{y} - x$,

$$C \int_{|\bar{y}-x|>\delta} \frac{\delta^\alpha}{(|\bar{y}-x|^2 - \delta^2)^{\alpha/2} |\bar{y}-x|^n} \frac{1}{|\bar{y}-x-z|^{n-\alpha}} d\bar{y} = \frac{1}{|z|^{n-\alpha}}.$$

Finally, let $z = \bar{z} - x$,

$$C \int_{|\bar{y}-x|>\delta} \frac{\delta^\alpha}{(|\bar{y}-x|^2 - \delta^2)^{\alpha/2} |\bar{y}-x|^n} \frac{1}{|\bar{y}-\bar{z}|^{n-\alpha}} d\bar{y} = \frac{1}{|\bar{z}-x|^{n-\alpha}}, \quad |\bar{z}-x| > \delta.$$

This verifies (4.22).

Given $r > 0$, for $|x| > r$ and $|z| > r$, choose $0 < \delta \leqslant r - |x|$, then for all such δ, (4.22) holds. In particular, (4.22) is valid for all $|z| > r$.

Combining (4.21) and (4.22), we have

$$(\varepsilon_\alpha^{(\delta)} * U_\alpha^\nu)(x) = \int_{|z|>r} \nu(z) \frac{1}{|z-x|^{n-\alpha}} dz$$
$$= U_\alpha^\nu(x).$$

iv) In the previous step, we have shown that each $\hat{u}_k$ is α-harmonic in the average sense:

$$(\varepsilon_\alpha^{(\delta)} * \hat{u}_k)(x) = \hat{u}_k(x), \quad |x| < r, \text{ for sufficiently small } \delta.$$

To see this is also true for $\hat{u}(x)$, we take limit of both sides of the above identity.

For each fixed x, we first show

$$\hat{u}_k(x) \to \hat{u}(x), \quad |x| < r.$$

In fact, by (4.15),

$$\hat{u}_k(x) - \hat{u}(x) = \int_{|y|>r} P_r(y,x)[u_k(y) - u(y)]dy$$
$$= C \int_{|y|>r} \frac{(r^2 - |x|^2)^{\frac{\alpha}{2}}[u_k(y) - u(y)]}{(|y|^2 - r^2)^{\frac{\alpha}{2}}|x-y|^n} dy \to 0.$$

Next, we show that, for each fixed $\delta > 0$ and fixed x,

$$(\varepsilon_\alpha^{(\delta)} * \hat{u}_k)(x) \to (\varepsilon_\alpha^{(\delta)} * \hat{u})(x). \tag{4.23}$$

To estimate (4.23), let

$$(\varepsilon_\alpha^{(\delta)} * \hat{u}_k)(x) - (\varepsilon_\alpha^{(\delta)} * \hat{u})(x)$$
$$= C \int_{|y-x|>\delta} \frac{\delta^\alpha[\hat{u}_k(y) - \hat{u}(y)]}{(|x-y|^2 - \delta^2)^{\frac{\alpha}{2}}|x-y|^n} dy$$
$$= C\{ \int_{\substack{|y-x|>\delta \\ |y|<r-\eta}} \frac{\delta^\alpha[\hat{u}_k(y) - \hat{u}(y)]}{(|x-y|^2 - \delta^2)^{\frac{\alpha}{2}}|x-y|^n} dy$$
$$+ \int_{\substack{|y-x|>\delta \\ r-\eta<|y|<r}} \frac{\delta^\alpha[\hat{u}_k(y) - \hat{u}(y)]}{(|x-y|^2 - \delta^2)^{\frac{\alpha}{2}}|x-y|^n} dy$$
$$+ \int_{\substack{|y-x|>\delta \\ |y|>r}} \frac{\delta^\alpha[\hat{u}_k(y) - \hat{u}(y)]}{(|x-y|^2 - \delta^2)^{\frac{\alpha}{2}}|x-y|^n} dy \}$$
$$:= C(I_1 + I_2 + I_3).$$

For each fixed x with $|x| < r$, choose δ and η such that

$$B_\delta(x) \cap B_{r-2\eta}^c(0) = \emptyset.$$

In the following calculations, one can see that such division of integrating region separates the two types of possible singularities, $|y| = r$ and $|x-y| = \delta$, from each other. This makes it easier for us to estimate the integral.

It follows from (4.15) that as $k \to \infty$

$$I_3 = \int_{\substack{|y-x|>\delta \\ |y|>r}} \frac{\delta^\alpha [u_k(y) - u(y)]}{(|x-y|^2 - \delta^2)^{\frac{\alpha}{2}} |x-y|^n} dy \to 0.$$

$$I_2 = \int_{\substack{|y-x|>\delta \\ r-\eta<|y|<r}} \frac{\delta^\alpha \int_{|z|>r} P_r(z,y)[u_k(z) - u(z)]dz}{(|x-y|^2 - \delta^2)^{\frac{\alpha}{2}} |x-y|^n} dy$$

$$= C\delta^\alpha \int_{|z|>r} \frac{u_k(z) - u(z)}{(|z|^2 - r^2)^{\frac{\alpha}{2}}} \int_{\substack{|y-x|>\delta \\ r-\eta<|y|<r}} \frac{(r^2 - |y|^2)^{\frac{\alpha}{2}} dy}{(|x-y|^2 - \delta^2)^{\frac{\alpha}{2}} |x-y|^n |z-y|^n} dz$$

$$= C\delta^\alpha \int_{|z|>r} \frac{u_k(z) - u(z)}{(|z|^2 - r^2)^{\frac{\alpha}{2}}} \cdot I_{21}(x,z)dz.$$

Noting that in the ring $r - \eta < |y| < r$, we have

$$|x - y| > \eta + \delta.$$

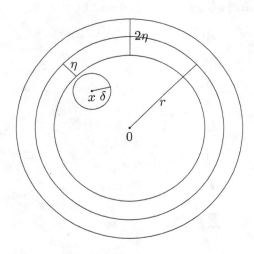

It then follows that

$$I_{21}(x,z)$$

$$\leq \frac{1}{(2\eta\delta + \eta^2)^{\frac{\alpha}{2}} (\eta + \delta)^n} \int_{r-\eta<|y|<r} \frac{(r^2 - |y|^2)^{\frac{\alpha}{2}} dy}{|z-y|^n}$$

$$= C \int_{r-\eta}^r (r^2 - \tau^2)^{\frac{\alpha}{2}} \left\{ \int_{S_\tau} \frac{1}{|z-y|^n} d\sigma_y \right\} d\tau$$

$$= C \int_{r-\eta}^{r} (r^2 - \tau^2)^{\frac{\alpha}{2}} \left\{ \int_0^\pi \frac{\omega_{n-2}(\tau \sin \theta)^{n-2} \tau d\theta}{(\tau^2 + |z|^2 - 2\tau |z| \cos \theta)^{\frac{n}{2}}} \right\} d\tau$$

$$= C \int_{r-\eta}^{r} (r^2 - \tau^2)^{\frac{\alpha}{2}} \frac{1}{\tau^n} \int_0^\pi \frac{\tau^{n-1} \sin^{n-2} \theta d\theta}{((\frac{|z|}{\tau})^2 - 2\frac{|z|}{\tau} \cos \theta + 1)^{\frac{n}{2}}} d\tau \qquad (4.24)$$

$$= C \int_{r-\eta}^{r} \frac{(r^2 - \tau^2)^{\frac{\alpha}{2}}}{\tau} \frac{d\tau}{(\frac{|z|}{\tau})^{n-2}((\frac{|z|}{\tau})^2 - 1)} \int_0^\pi \sin^{n-2} \beta d\beta \qquad (4.25)$$

$$< \frac{Cr^{n-1}}{|z|^{n-2}} \int_{r-\eta}^{r} \frac{(r^2 - \tau^2)^{\frac{\alpha}{2}}}{|z|^2 - \tau^2} d\tau$$

$$= \frac{Cr^{n-1}}{|z|^{n-2}} \cdot J.$$

In deriving (4.25) from (4.24), we made the following substitution (See Appendix in [L]):

$$\frac{\sin \theta}{\sqrt{(\frac{|z|}{\tau})^2 - 2\frac{|z|}{\tau} \cos \theta + 1}} = \frac{\sin \beta}{\frac{|z|}{\tau}}.$$

To estimate the last integral J, we consider

(a)   For $r < |z| < r + 1$,

$$J \leq \int_{r-\eta}^{r} \frac{(r+\tau)^{\frac{\alpha}{2}-1}}{(r-\tau)^{1-\frac{\alpha}{2}}} d\tau \leq C_{\alpha,r}.$$

(b)   For $|z| \geq r + 1$, obviously,

$$J \sim \frac{1}{|z|^2}, \quad \text{for } |z| \text{ large}.$$

In summary,

$$I_{21}(x, z) \sim \begin{cases} 1, & \text{for } |z| \text{ near } r, \\ |z|^{-n}, & \text{for } |z| \text{ large}. \end{cases}$$

Therefore, by (4.15), as $k \to \infty$,

$$I_2 = \delta^\alpha \int_{|z|>r} \frac{u_k(z) - u(z)}{(|z|^2 - r^2)^{\alpha/2}} I_{21}(x, z) dz \to 0.$$

Now what remains is to estimate

$$I_1 = \delta^\alpha \int_{|z|>r} \frac{u_k(z) - u(z)}{(|z|^2 - r^2)^{\frac{\alpha}{2}}} I_{11}(x, z) dz,$$

where

$$I_{11}(x,z) = \int_{\substack{|y-x|>\delta \\ |y|<r-\eta}} \frac{(r^2 - |y|^2)^{\frac{\alpha}{2}} dy}{(|x-y|^2 - \delta^2)^{\frac{\alpha}{2}} |x-y|^n |z-y|^n}.$$

$$I_{11}(x,z) \leq \frac{r^\alpha}{\delta^n} \int_{\substack{|y-x|>\delta \\ |y|<r-\eta}} \frac{dy}{(|x-y|^2 - \delta^2)^{\frac{\alpha}{2}} |z-y|^n}$$

$$\leq \frac{r^\alpha}{\delta^n (|z| - r + \eta)^n} \int_{\delta<|y-x|<2r} \frac{dy}{(|x-y|^2 - \delta^2)^{\frac{\alpha}{2}}}$$

$$= \frac{r^\alpha}{\delta^n (|z| - r + \eta)^n} \int_\delta^{2r} \frac{\omega_{n-1} \tau^{n-1} d\tau}{(\tau^2 - \delta^2)^{\frac{\alpha}{2}}}$$

$$\leq \frac{C}{|z|^n}.$$

By (4.15), as $k \to \infty$, we have $I_1 \to 0$. This verifies (4.23) and hence completes the proof.

4.3 The Fractional Version of Pizzetti's Formula

Proof of Lemma 4.1.2. By using the property

$$\int_{|y-x|>r} \varepsilon_\alpha^{(r)}(x-y) dy = 1,$$

we have

$$\frac{1}{r^\alpha} \left[u(x) - \varepsilon_\alpha^{(r)} * u(x) \right]$$

$$= \frac{1}{r^\alpha} u(x) - c \int_{|y-x|>r} \frac{u(y)}{(|x-y|^2 - r^2)^{\frac{\alpha}{2}} |x-y|^n} dy$$

$$= c \int_{|y-x|>r} \frac{u(x) - u(y)}{(|x-y|^2 - r^2)^{\frac{\alpha}{2}} |x-y|^n} dy. \tag{4.26}$$

Comparing (4.26) with

$$(-\triangle)^{\frac{\alpha}{2}} u(x) = \lim_{r \to 0} \int_{|y-x|>r} \frac{u(x) - u(y)}{|x-y|^{\alpha+n}} dy,$$

one may expect that

$$\lim_{r \to 0} \int_{|y-x|>r} \frac{u(x) - u(y)}{|x-y|^{\alpha+n}} dy = \lim_{r \to 0} \int_{|y-x|>r} \frac{u(x) - u(y)}{(|x-y|^2 - r^2)^{\frac{\alpha}{2}} |x-y|^n} dy.$$

Indeed, consider

$$\int_{|y-x|>r} \frac{u(x) - u(y)}{|x - y|^n} \left(\frac{1}{(|x - y|^2 - r^2)^{\frac{\alpha}{2}}} - \frac{1}{|x - y|^\alpha} \right) dy$$

$$= \int_{r<|y-x|<1} \frac{u(x) - u(y)}{|x - y|^n} \left(\frac{1}{(|x - y|^2 - r^2)^{\frac{\alpha}{2}}} - \frac{1}{|x - y|^\alpha} \right) dy$$

$$+ \int_{|y-x|\geq 1} \frac{u(x) - u(y)}{|x - y|^n} \left(\frac{1}{(|x - y|^2 - r^2)^{\frac{\alpha}{2}}} - \frac{1}{|x - y|^\alpha} \right) dy$$

$$= I_1 + I_2. \tag{4.27}$$

It is easy to see that as $r \to 0$, I_2 tends to zero. Actually, same conclusion is true for I_1.

$$|I_1| \leq \left| \int_{r<|y-x|<1} \frac{-\nabla u(x) \cdot (y - x) + O(|y - x|^2)}{|x - y|^n} \left(\frac{1}{(|x - y|^2 - r^2)^{\frac{\alpha}{2}}} - \frac{1}{|x - y|^\alpha} \right) dy \right| \tag{4.28}$$

$$\leq C \int_{r<|y-x|<1} \frac{|x - y|^2}{|x - y|^n} \left(\frac{1}{(|x - y|^2 - r^2)^{\frac{\alpha}{2}}} - \frac{1}{|x - y|^\alpha} \right) dy \tag{4.29}$$

$$= C \int_r^1 \frac{\tau^2}{\tau^n} \left(\frac{1}{(\tau^2 - r^2)^{\frac{\alpha}{2}}} - \frac{1}{\tau^\alpha} \right) \tau^{n-1} d\tau \qquad \tau = |y - x|$$

$$\leq C \int_1^\infty \left(\frac{1}{r^\alpha (s^2 - 1)^{\frac{\alpha}{2}}} - \frac{1}{r^\alpha s^\alpha} \right) sr^2 ds \qquad s = \frac{\tau}{r}$$

$$= C r^{2-\alpha} \int_1^\infty \left(\frac{s^\alpha - (s^2 - 1)^{\frac{\alpha}{2}}}{(s^2 - 1)^{\frac{\alpha}{2}} s^\alpha} \right) s \, ds. \tag{4.30}$$

Equation (4.28) follows from the Taylor expansion. Due to symmetry, we have

$$\int_{r<|y-x|<1} \frac{\nabla u(x) \cdot (y - x)}{|x - y|^n} \left(\frac{1}{(|x - y|^2 - r^2)^{\frac{\alpha}{2}}} - \frac{1}{|x - y|^\alpha} \right) dy = 0$$

and arrive at (4.29). It is easy to see that the integral in (4.30) converges near 1. To see that it also converges near infinity, we estimate

$$s^\alpha - (s^2 - 1)^{\frac{\alpha}{2}}.$$

Let $f(t) = t^{\alpha/2}$. By the *mean value theorem*,

$$f(s^2) - f(s^2 - 1) = f'(\xi)(s^2 - (s^2 - 1))$$
$$= \frac{\alpha}{2} \xi^{\frac{\alpha}{2}-1} \sim s^{\alpha-2}, \text{ for } s \text{ sufficiently large.}$$

This implies that

$$\frac{s^\alpha - (s^2 - 1)^{\frac{\alpha}{2}}}{(s^2 - 1)^{\frac{\alpha}{2}} s^\alpha} s \sim \frac{s^{\alpha-2} s}{(s^2 - 1)^{\frac{\alpha}{2}} s^\alpha} \sim \frac{1}{s^{1+\alpha}}.$$

Now it is obvious that (4.30) converges near infinity. Thus we have

$$\int_1^\infty \left(\frac{s^\alpha - (s^2-1)^{\frac{\alpha}{2}}}{(s^2-1)^{\frac{\alpha}{2}} s^\alpha} \right) s\,ds < \infty.$$

Since $0 < \alpha < 2$, as $r \to 0$, (4.30) goes to zero and I_1 converges to zero. Together with (4.26) and (4.27), we obtain (4.9). This concludes the proof.

4.4 Equivalence in Being α-Harmonic among Three Senses

Proof of Theorem 4.1.2.

Assume that $u \in L_\alpha$ is α-harmonic in the distributional sense, that is

$$< (-\triangle)^{\alpha/2} u, \phi > \equiv \int_{\mathbb{R}^n} u(x)(-\triangle)^{\alpha/2}\phi(x)dx = 0, \quad \forall \phi \in C_0^\infty(\mathbb{R}^n).$$

In the Poisson's expression of $\hat{u}$, by the *Lebesgue's dominated convergence theorem*, one can differentiate the integrand to show that $\hat{u}$ is smooth in $B_r(0)$ for any $r > 0$. Lemma 4.1.1 indicates that $\hat{u}$ is α-harmonic in the average sense, hence by Lemma 4.1.2, it is also α-harmonic. That is we have

$$\begin{cases} (-\triangle)^{\alpha/2}(\hat{u} - u)(x) = 0, & x \in B_r(0), \\ \hat{u} - u = 0, & x \notin B_r(0), \end{cases}$$

in the distributional sense. Now by the *maximum principle* (see Chapter 3 or [Si]), $u(x) = \hat{u}(x)$, hence u is also α-harmonic in the average sense, hence α-harmonic.

The converse can be derived from Lemma 4.1.2 immediately.

5

Liouville Type Theorems for α-Harmonic Functions

5.1 A Liouville Type Theorem in $\mathbb{R}^n$

The well-known classical Liouville's theorem states that

Any harmonic function bounded below in all of $\mathbb{R}^n$ is constant.

One of its important applications is the proof of the Fundamental Theorem of Algebra. It is also a key ingredient in deriving a priori estimates for solutions in PDE analysis.

This Liouville theorem has been generalized to the fractional Laplacian.

Proposition 5.1.1 ([BKN], [ZCCY]) *Let $0 < \alpha < 2$ and $n \geq 2$. Assume that u is a solution of*

$$\begin{cases} (-\triangle)^{\alpha/2} u(x) = 0, \, x \in \mathbb{R}^n, \\ u(x) \geq 0, \qquad\quad x \in \mathbb{R}^n. \end{cases}$$

Then u must be constant.

One interesting application of this Liouville theorem is to obtain integral representations of solutions for nonlinear equations and systems involving the fractional Laplacian.

Same result preserves when the condition of being lower-bounded in the proposition above is weakened.

Theorem 5.1.1 *Let $0 < \alpha < 2$. Assume that $u \in L_\alpha$ and is lower semi-continuous. If u is a solution of*

$$(-\triangle)^{\frac{\alpha}{2}} u(x) = 0, \quad x \in \mathbb{R}^n,$$

and for some $0 \leq \gamma \leq 1$ and $\gamma < \alpha$,

$$\lim_{|x| \to \infty} \frac{u(x)}{|x|^\gamma} \geq 0, \tag{5.1}$$

then

$$u(x) \equiv C, \ x \in \mathbb{R}^n.$$

The proof will be conducted by using the Poisson representation of α-harmonic functions introduced in the previous chapter. This theorem is actually a consequence of the following more general result.

Theorem 5.1.2 *Let $0 < \alpha < 2$, and $u \in L_\alpha$ be α-harmonic. Then u is affine. In particular, if $\alpha \leq 1$, then u is constant.*

The proof of Theorem 5.1.2 is based on Fourier analysis.

By using Theorem 5.1.1, we will be able to establish an integral representation for solutions of semi-linear equations involving the fractional Laplacian.

Theorem 5.1.3 *Assume $u \in L_\alpha$ is lower semi-continuous. Suppose u is a locally bounded positive solution of*

$$\begin{cases} (-\triangle)^{\frac{\alpha}{2}} u(x) = u^p(x), & x \in \mathbb{R}^n, \\ u(x) > 0, & x \in \mathbb{R}^n, \end{cases}$$

in the sense of distribution.

Then it is also a solution of

$$u(x) = \int_{\mathbb{R}^n} \frac{C_{n,\alpha}}{|x - y|^{n-\alpha}} u^p(y) dy. \tag{5.2}$$

The converse is also true.

The significance of the integral expression is that it will enable one to apply the *method of moving planes in integral forms* to investigate the properties of the solutions, such as symmetry, monotonicity, and regularity, as we will see in the next chapter.

5.1.1 The Classical Proof of the Liouville Theorem

We prove Theorem 5.1.1.

Step 1.

We first show that, for $|x| < r$,

$$u(x) = \int_{|y| > r} P_r(y, x) u(y) dy, \tag{5.3}$$

where $P_r(y, x)$ is some kind of Poisson kernel for $|x| < r$:

$$P_r(y, x) = \begin{cases} \frac{\Gamma(n/2)}{\pi^{\frac{n}{2}+1}} \sin \frac{\pi \alpha}{2} \left[\frac{r^2 - |x|^2}{|y|^2 - r^2} \right]^{\frac{\alpha}{2}} \frac{1}{|x-y|^n}, & |y| > r, \\ 0, & |y| < r. \end{cases}$$

Let

$$\hat{u}(x) = \begin{cases} \int_{|y|>r} P_r(y,x)u(y)dy, & |x| < r, \\ u(x), & |x| \geq r. \end{cases}$$

In previous chapter, we have proved that $\hat{u}$ is α-harmonic in the ball $B_r(0)$.
Let $w(x) = \hat{u}(x) - u(x)$, then

$$\begin{cases} (-\triangle)^{\frac{\alpha}{2}} w(x) = 0, & |x| < r, \\ w(x) = 0, & |x| \geq r. \end{cases}$$

To show that $w \equiv 0$, we employ the following *maximum principle* (also see Chapter 3).

Lemma 5.1.1 ([Si])   *Let Ω be a bounded domain in $\mathbb{R}^n$, and assume that v is a lower semi-continuous function on $\bar{\Omega}$ satisfying*

$$\begin{cases} (-\triangle)^{\frac{\alpha}{2}} v \geq 0 \text{ in } \Omega, \\ v \geq 0 \qquad \text{ on } \mathbb{R}^n \backslash \Omega. \end{cases}$$

Then $v \geq 0$ in Ω.

Applying this lemma to both $v = w$ and $v = -w$, we conclude that

$$w(x) \equiv 0.$$

Hence

$$\hat{u}(x) \equiv u(x).$$

This verifies (5.3).
    Step 2.
    We will show that, for each fixed $x \in \mathbb{R}^n$, and for any unit vector ν, we have

$$\frac{\partial u}{\partial \nu}(x) \geq 0. \tag{5.4}$$

It follows from the arbitrariness of ν that

$$\nabla u(x) = 0,$$

and therefore

$$u \equiv C \text{ in } \mathbb{R}^n.$$

Now what's left is to prove (5.4). Through an elementary calculation, one can derive that

$$\frac{\partial u}{\partial x_i}(x) = -\int_{|y|>r} P_r(y,x) \left[\frac{\alpha x_i}{r^2 - |x|^2} + \frac{n(x_i - y_i)}{|y - x|^2} \right] u(y)dy,$$

and consequently,

$$\frac{\partial u}{\partial \nu}(x) = -\int_{|y|>r} P_r(y,x) \left[\frac{\alpha x \cdot \nu}{r^2 - |x|^2} + \frac{n(x - y) \cdot \nu}{|y - x|^2} \right] u(y)dy.$$

By (5.1), we see that for any fixed $\epsilon > 0$, when $|x|$ is sufficiently large, it holds

$$u(x) \geq -\epsilon |x|^{\gamma}. \tag{5.5}$$

Otherwise, there exists an ϵ_o and a sequence $\{x^k\}$ with $|x^k| \to \infty$ as $k \to \infty$ such that

$$u(x^k) < -\epsilon_o |x^k|^{\gamma},$$

and hence

$$\lim_{k \to \infty} \frac{u(x^k)}{|x^k|^{\gamma}} < -\epsilon_o.$$

This is a contradiction with our assumption (5.1).

Now write

$$\frac{\partial u}{\partial \nu}(x) = -\int_{|y|>r} P_r(y, x) \left[\frac{\alpha x \cdot \nu}{r^2 - |x|^2} + \frac{n(x-y) \cdot \nu}{|y-x|^2} \right] [u(y) + \epsilon |y|^{\gamma}] dy$$

$$+ \int_{|y|>r} P_r(y, x) \left[\frac{\alpha x \cdot \nu}{r^2 - |x|^2} + \frac{n(x-y) \cdot \nu}{|y-x|^2} \right] \epsilon |y|^{\gamma} dy$$

$$= H_1 + H_2. \tag{5.6}$$

For each fixed x, let r be sufficiently large to ensure both (5.5) and (5.3) and the following

$$\left| \frac{\alpha x \cdot \nu}{r^2 - |x|^2} \right| \leq \frac{1}{r}; \quad \text{and for } |y| > r, \quad \left| \frac{n(x-y) \cdot \nu}{|y-x|^2} \right| \leq \frac{n}{|y-x|} \leq \frac{2n}{r}.$$

Then we have

$$H_1 \geq -\frac{2n+1}{r} \int_{|y|>r} P_r(y, x) [u(y) + \epsilon |y|^{\gamma}] dy$$

$$= -\frac{2n+1}{r} u(x) - \frac{2n+1}{r} \int_{|y|>r} P_r(y, x) \epsilon |y|^{\gamma} dy$$

$$= -\frac{2n+1}{r} u(x) - H_{11}. \tag{5.7}$$

Obviously, the first term on the right hand side of the above inequality approaches 0 as $r \to \infty$. Next we show that one also has

$$H_{11} < C\epsilon, \quad \text{for all sufficiently large } r. \tag{5.8}$$

Here and below, the letter C stands for various constants. In fact,

$$H_{11} = \frac{2n+1}{r} \int_{|y|>r} P_r(y,x)\epsilon|y|^\gamma dy$$

$$= \frac{2n+1}{r} \frac{\Gamma(n/2)}{\pi^{\frac{n}{2}+1}} \sin\frac{\pi\alpha}{2} (r^2 - |x|^2)^{\frac{\alpha}{2}} \epsilon \int_{|y|>r} \frac{|y|^\gamma}{(|y|^2 - r^2)^{\frac{\alpha}{2}}|x-y|^n} dy$$

$$= C\epsilon \frac{(r^2 - |x|^2)^{\frac{\alpha}{2}}}{r} \int_{|y|>r} \frac{|y|^\gamma}{(|y|^2 - r^2)^{\frac{\alpha}{2}}|x-y|^n} dy$$

$$\leq C\epsilon \frac{(r^2 - |x|^2)^{\frac{\alpha}{2}}}{r} \int_{|y|>r} \frac{|y|^\gamma}{(|y|^2 - r^2)^{\frac{\alpha}{2}}(|y| - |x|)^n} dy$$

$$= C\epsilon \frac{(r^2 - |x|^2)^{\frac{\alpha}{2}}}{r} \int_r^\infty \frac{\tau^{\gamma+n-1}}{(\tau^2 - r^2)^{\frac{\alpha}{2}}(\tau - |x|)^n} d\tau \tag{5.9}$$

$$= C\epsilon \frac{(r^2 - |x|^2)^{\frac{\alpha}{2}}}{r^{\alpha-\gamma+1}} \int_1^\infty \frac{s^{\gamma+n-1}}{(s^2 - 1)^{\frac{\alpha}{2}}(s - \frac{|x|}{r})^n} ds \tag{5.10}$$

$$\leq \frac{C\epsilon}{r^{1-\gamma}} \int_1^\infty \frac{s^{\gamma+n-1}}{(s^2 - 1)^{\frac{\alpha}{2}}(s - \frac{|x|}{r})^n} ds. \tag{5.11}$$

We derive (5.9) and (5.10) by letting $|y| = \tau$ and $\tau = rs$ respectively. From the assumption that $\gamma < \alpha$ and $0 < \alpha < 2$, it is not difficult to see that

$$\int_1^\infty \frac{s^{\gamma+n-1}}{(s^2 - 1)^{\frac{\alpha}{2}}(s - \frac{|x|}{r})^n} ds < \infty.$$

Noticing that $\gamma \leq 1$, and by (5.11), we arrive at (5.8).

Through an identical argument, we can deduce

$$|H_2| \leq \frac{2n+1}{r} \int_{|y|>r} P_r(y,x)\epsilon|y|^\gamma dy,$$

and

$$|H_2| < C\epsilon, \quad \text{for sufficiently large } r. \tag{5.12}$$

From (5.6), (5.7), (5.8), and (5.12), letting $r \to \infty$ we conclude that

$$\frac{\partial u}{\partial \nu} \geq -C\epsilon.$$

The fact that ϵ is arbitrary establishes (5.4), hence proves the theorem.

5.1.2 Proof via Fourier Analysis

We prove Theorem 5.1.2.

A function $u \in L_\alpha$ is a tempered distribution, and hence u admits a Fourier transform $\mathcal{F}u$. If we prove that $\mathcal{F}u$ has support at one point (the origin) then $\mathcal{F}u$ is a finite combination of the Dirac's delta measure and its derivatives. Therefore u is a polynomial. The only polynomials belonging to L_α are affine functions and eventually constants (depending on α).

From

$$\langle(-\triangle)^{\alpha/2}u, \psi\rangle = \int_{\mathbb{R}^n} u(x)(-\triangle)^{\alpha/2}\psi(x)dx \qquad \forall \psi \in C_0^{\infty}(\mathbb{R}^n),$$

and the fact that

$$\mathcal{F}((-\triangle)^{\alpha/2}\psi)(\xi) = |\xi|^{\alpha}\mathcal{F}(\psi)(\xi) \qquad \text{for } \psi \in \mathcal{S},$$

we observe that $(-\triangle)^{\alpha/2}u = 0$ means that for any $\psi \in \mathcal{S}$

$$0 = \langle(-\triangle)^{\alpha/2}u, \psi\rangle = \int_{\mathbb{R}^n} u(x)(-\triangle)^{\alpha/2}\psi(x)dx$$

$$= \int_{\mathbb{R}^n} u(x)\mathcal{F}^{-1}(|\xi|^{\alpha}\mathcal{F}\psi)(x)dx. \tag{5.13}$$

We claim that

$$\langle\mathcal{F}u, \phi\rangle = 0 \qquad \text{for any } \phi \in C_0^{\infty}(\mathbb{R}^n\backslash\{0\}). \tag{5.14}$$

Indeed let $\phi \in C_0^{\infty}(\mathbb{R}^n\backslash\{0\})$. The function $\phi(\xi)/|\xi|^{\alpha}$ belongs to $C_0^{\infty}(\mathbb{R}^n\backslash\{0\}) \subset \mathcal{S}$. Therefore there exists $\psi \in \mathcal{S}$ such that $\mathcal{F}(\psi)(\xi) = \phi(\xi)/|\xi|^{\alpha}$.

Now, since u is a tempered distribution and from (5.13), we have

$$\langle\mathcal{F}u, \phi\rangle = \langle\mathcal{F}u, |\xi|^{\alpha}\mathcal{F}\psi\rangle =$$

$$= \langle u, \mathcal{F}^{-1}(|\xi|^{\alpha}\mathcal{F}\psi)\rangle = \int_{\mathbb{R}^n} u(x)\mathcal{F}^{-1}(|\xi|^{\alpha}\mathcal{F}\psi)(x)dx = 0.$$

That is the claim.

From (5.14) we infer that $\mathcal{F}u$ has support in $\{0\}$. Then $\mathcal{F}u$ is a finite combination of the Dirac's delta measure at the origin and its derivatives. Therefore, we conclude that u is a polynomial (see for instance [G]). Since $u \in L_{\alpha}$, we obtain that u has at most a linear growth. This concludes the proof.

The same idea can be applied to prove that the polyharmonic tempered distributions are polynomials.

Theorem 5.1.4 *Let $u \in \mathcal{S}'(\mathbb{R}^n)$, m an integer and $(-\triangle)^m u = 0$. Then u is a polynomial.*

The proof is similar. Actually the proof is simpler since the symbol of $(-\triangle)^m$, $|\xi|^{2m}$, being a polynomial, is a good multiplier for tempered distribution. We omit the details.

Remark 5.1.1 *With the same technique one can prove that if $u \in L_{\alpha} \cap L_{\beta}$, with $0 < \alpha, \beta < 2$ and u solves*

$$(-\triangle)^{\alpha/2}u(x) + (-\triangle)^{\beta/2}u(x) = 0,$$

then u is affine (and constant if $\min\{\alpha, \beta\} \leq 1$).

A more general result can be obtained with some hypothesis on the symbol of the operator involved. We leave the details to the interested readers.

5.1.3 Equivalence Between Two Equations in $\mathbb{R}^n$

Theorem 5.1.5 *Assume $u \in L_\alpha$ is lower semi-continuous. Suppose u is a locally bounded solution of*

$$\begin{cases} (-\triangle)^{\frac{\alpha}{2}} u(x) = u^p(x), & x \in \mathbb{R}^n, \\ u(x) > 0, & x \in \mathbb{R}^n, \end{cases} \tag{5.15}$$

in the sense of distribution.

Then it is also a solution of

$$u(x) = \int_{\mathbb{R}^n} \frac{C_{n,\alpha}}{|x-y|^{n-\alpha}} u^p(y) dy. \tag{5.16}$$

The converse is also true.

Proof. Assume that u is a positive solution of (5.15). We first show that

$$\int_{\mathbb{R}^n} \frac{u^p(y)}{|x-y|^{n-\alpha}} dy < \infty.$$

In Chapter 2 we know the Green's function in $B_1(0)$ takes the form of

$$G_1(x,y) = \frac{C_{n,\alpha}}{|x-y|^{n-\alpha}} \int_0^{\frac{(1-|x|^2)(1-|y|^2)}{|x-y|^2}} \frac{b^{\alpha/2-1}}{(1+b)^{\frac{n}{2}}} db.$$

Let $s_R = \frac{|x-y|^2}{R^2}$ and $t_R = \left(1-|\frac{x}{R}|^2\right)\left(1-|\frac{y}{R}|^2\right)$. Through re-scaling we obtain the Green's function on $B_R(0)$ as

$$G_R(x,y) = \frac{G_1(\frac{x}{R}, \frac{y}{R})}{R^{n-\alpha}} = \frac{C_{n,\alpha}}{|x-y|^{n-\alpha}} \int_0^{\frac{t_R}{s_R}} \frac{b^{\alpha/2-1}}{(1+b)^{\frac{n}{2}}} db.$$

Let

$$v_R(x) = \int_{B_R(0)} G_R(x,y) u^p(y) dy.$$

Our assumption $u \in L_{loc}^\infty(\mathbb{R}^n)$ guarantees that for each given $R > 0$, $v_R(x)$ is well-defined and continuous. It's easy to see that

$$\begin{cases} (-\triangle)^{\frac{\alpha}{2}} v_R(x) = u^p(x), & x \in B_R(0), \\ v_R(x) = 0, & x \notin B_R(0). \end{cases}$$

Let

$$w_R(x) = u(x) - v_R(x).$$

Then

$$\begin{cases} (-\triangle)^{\frac{\alpha}{2}} w_R(x) = 0, & x \in B_R(0), \\ w_R(x) \geq 0, & x \notin B_R(0). \end{cases}$$

By the *maximum principle*, we have

$$w_R(x) \geq 0, \quad x \in B_R(0).$$

As $R \to \infty$,
$$s_R \to 0, \quad t_R \to 1,$$
$$\int_0^{\frac{t_R}{s_R}} \frac{b^{\alpha/2-1}}{(1+b)^{\frac{n}{2}}} \, db \to \int_0^\infty \frac{b^{\alpha/2-1}}{(1+b)^{\frac{n}{2}}} \, db = C.$$

It thus implies that
$$G_R(x,y) \to G(x,y) = \frac{C_{n,\alpha}}{|x-y|^{n-\alpha}}, \quad R \to \infty,$$
$$v_R(x) \to v(x) := \int_{\mathbb{R}^n} \frac{C_{n,\alpha}}{|x-y|^{n-\alpha}} u^p(y) \, dy,$$

and
$$w_R(x) \to w(x) := u(x) - v(x).$$

It's easy to see that
$$\begin{cases} (-\triangle)^{\frac{\alpha}{2}} w(x) = 0, & x \in \mathbb{R}^n, \\ w \geq 0, & x \in \mathbb{R}^n. \end{cases}$$

By the *Liouville's theorem*, we deduce that
$$w(x) \equiv c, \quad x \in \mathbb{R}^n,$$

and
$$u(x) = w(x) + v(x) \geq c.$$

If $c > 0$, then
$$u(x) \geq v(x) = \int_{\mathbb{R}^n} \frac{C_{n,\alpha}}{|x-y|^{n-\alpha}} u^p(y) dy$$
$$\geq \int_{\mathbb{R}^n} \frac{C_{n,\alpha}}{|x-y|^{n-\alpha}} c^p dy$$
$$= \infty.$$

This is a contradiction, hence $c = 0$ and
$$u(x) = v(x) = \int_{\mathbb{R}^n} \frac{C_{n,\alpha}}{|x-y|^{n-\alpha}} u^p(y) dy.$$

5.2 A Liouville Type Theorem on $\mathbb{R}_+^n$

We consider the Dirichlet problem for α-harmonic functions
$$\begin{cases} (-\triangle)^{\alpha/2} u(x) = 0, \ u(x) \geq 0, & x \in \mathbb{R}_+^n, \\ u(x) \equiv 0, & x \notin \mathbb{R}_+^n. \end{cases} \tag{5.17}$$

It is well-known that
$$u(x) = \begin{cases} C x_n^{\alpha/2}, & x \in \mathbb{R}_+^n, \\ 0, & x \notin \mathbb{R}_+^n, \end{cases}$$

is a family of solutions for problem (5.17) with any positive constant C.

Then one may naturally ask: *Are these the only solutions?*

Our main objective here is to answer this question affirmatively and prove

Theorem 5.2.1 *Let* $0 < \alpha < 2$, $u \in L_\alpha$. *Assume u is a solution of (5.17) in the sense of distribution, then either $u \equiv 0$ or*

$$u(x) = \begin{cases} C x_n^{\alpha/2}, & x \in \mathbb{R}^n_+, \\ 0, & x \notin \mathbb{R}^n_+, \end{cases} \tag{5.18}$$

for some positive constant C.

Remark 5.2.1 In Theorem 1.5 in [RS1], Ros-Oton and Serra obtained a similar result. Their operator is more general, however, they required stronger growth condition on the solution u:

$$\|u\|_{L^\infty(B_R)} \le C R^\beta \quad \text{for all } R \ge 1; \text{ for some } \beta < \alpha.$$

Obviously, our condition $u \in L_\alpha$ is much weaker.

Our proof is mainly based on the Poisson representation of the solutions.

Theorem 5.2.2 *Assume that $u \in L_\alpha$ and is lower semi-continuous on $\overline{\mathbb{R}^n_+}$. If u is a solution of (5.17) in the sense of distribution, then for $|x - x_r| < r$,*

$$u(x) = \int_{|y-x_r|>r} P_r(x - x_r, y - x_r) u(y) dy,$$

where $x_r = (0, \cdots, 0, r)$, *and* $P_r(x - x_r, y - x_r)$ *is the Poisson kernel for* $|x - x_r| < r$:

$$P_r(x - x_r, y - x_r)$$
$$= \begin{cases} \frac{\Gamma(n/2)}{\pi^{\frac{n}{2}+1}} \sin \frac{\pi\alpha}{2} \left[\frac{r^2 - |x - x_r|^2}{|y - x_r|^2 - r^2} \right]^{\frac{\alpha}{2}} \frac{1}{|x-y|^n}, & |y - x_r| > r, \\ 0, & elsewhere. \end{cases} \tag{5.19}$$

The proof is quite similar to that in $\mathbb{R}^n$ in Chapter 4. Hence we leave it out.

As a consequence, we obtain an integral representation of solutions for the Dirichlet problem on the half space (see [ZCLC]).

Theorem 5.2.3 *Assume $u \in L_\alpha$ is lower semi-continuous. If u is a locally bounded positive solution of*

$$\begin{cases} (-\triangle)^{\frac{\alpha}{2}} u(x) = u^p(x), & x \in \mathbb{R}^n_+, \\ u(x) = 0, & x \notin \mathbb{R}^n_+, \end{cases}$$

in the sense of distribution, then it is also a solution of

$$u(x) = \int_{\mathbb{R}^n_+} G_+(x, y) u^p(y) dy, \tag{5.20}$$

where $G_+(x, y)$ is the Green's function associated with $(-\triangle)^{\frac{\alpha}{2}}$ in $\mathbb{R}^n_+$.

This integral representation will be employed to study the properties of solutions in the next chapter.

5.2.1 The Proof of the Liouville Theorem

Proof. To begin with, we show that for each fixed $x \in \mathbb{R}^n_+$,

$$\frac{\partial u}{\partial x_i}(x) = 0, \qquad i = 1, 2, \cdots, n - 1. \tag{5.21}$$

This implies that $u(x) = u(x_n)$. Then we calculate

$$\frac{\partial u}{\partial x_n}(x) = \frac{\alpha}{2x_n} u(x). \tag{5.22}$$

Together with (5.21), we derive that

$$u(x) = C x_n^{\alpha/2},$$

therefore

$$u(x) = \begin{cases} C x_n^{\alpha/2}, & x \in \mathbb{R}^n_+, \\ 0, & x \notin \mathbb{R}^n_+. \end{cases}$$

Now we prove (5.21) and (5.22). Through an elementary calculation, one can derive that, for $i = 1, 2, \cdots, n - 1$,

$$\begin{aligned} \frac{\partial u}{\partial x_i}(x) &= \int_{|y-x_r|>r} \left(\frac{-\alpha x_i}{r^2 - |x - x_r|^2} + \frac{n(y_i - x_i)}{|y - x|^2} \right) P_r(x - x_r, y - x_r) u(y) dy \\ &= \int_{|y-x_r|>r} \frac{-\alpha x_i}{r^2 - |x - x_r|^2} P_r(x - x_r, y - x_r) u(y) dy \\ &\quad + \int_{|y-x_r|>r} \frac{n(y_i - x_i)}{|y - x|^2} P_r(x - x_r, y - x_r) u(y) dy \\ &:= I_1 + I_2. \end{aligned} \tag{5.23}$$

For each fixed $x \in B_r(x_r) \subset \mathbb{R}^n_+$ and for any given $\epsilon > 0$, for r sufficiently large, we have

$$\begin{aligned} |I_1| &= \left| \int_{|y-x_r|>r} \frac{-\alpha x_i}{r^2 - |x - x_r|^2} P_r(x - x_r, y - x_r) u(y) dy \right| \\ &\leq \int_{|y-x_r|>r} \left| \frac{-\alpha x_i}{r^2 - |x - x_r|^2} \right| P_r(x - x_r, y - x_r) u(y) dy \\ &= \left| \frac{\alpha x_i}{r^2 - |x - x_r|^2} \right| \int_{|y-x_r|>r} P_r(x - x_r, y - x_r) u(y) dy \\ &= \left| \frac{\alpha x_i}{2x_n r - |x|^2} \right| u(x) \\ &\leq \frac{C}{r} \\ &< \epsilon. \end{aligned} \tag{5.24}$$

To estimate I_2, we divide the region $|y - x_r| > r$ into two parts: one inside the ball $|y| < R$ and one outside the ball.

$$\begin{aligned}
|I_2| &= \left| \int_{|y-x_r|>r} \frac{n(y_i - x_i)}{|y - x|^2} P_r(x - x_r, y - x_r) u(y) dy \right| \\
&\leq \int_{|y-x_r|>r} \left| \frac{n(y_i - x_i)}{|y - x|^2} \right| P_r(x - x_r, y - x_r) u(y) dy \\
&= \int_{\substack{|y-x_r|>r \\ |y|>R}} \left| \frac{n(y_i - x_i)}{|y - x|^2} \right| P_r(x - x_r, y - x_r) u(y) dy \\
&\quad + \int_{\substack{|y-x_r|>r \\ |y|\leq R}} \left| \frac{n(y_i - x_i)}{|y - x|^2} \right| P_r(x - x_r, y - x_r) u(y) dy \\
&:= I_{21} + I_{22}.
\end{aligned} \tag{5.25}$$

For any $\epsilon > 0$, we can choose $R = R(\epsilon)$ sufficiently large such that when $|y| > R$,

$$\left| \frac{n(y_i - x_i)}{|y - x|^2} \right| \leq \frac{n}{|y - x|} \leq \frac{C}{R} < \epsilon,$$

thus

$$\begin{aligned}
I_{21} &< \int_{\substack{|y-x_r|>r \\ |y|>R}} \epsilon P_r(x - x_r, y - x_r) u(y) dy \\
&\leq \epsilon \int_{|y-x_r|>r} P_r(x - x_r, y - x_r) u(y) dy \\
&= \epsilon u(x) \\
&\leq C\epsilon.
\end{aligned} \tag{5.26}$$

To estimate I_{22}, we employ the expression of the Poisson kernel.

$$\begin{aligned}
I_{22} &= C \int_{\substack{|y-x_r|>r \\ |y|\leq R}} \left[\frac{r^2 - |x - x_r|^2}{|y - x_r|^2 - r^2} \right]^{\alpha/2} \frac{u(y)}{|x - y|^n} \left| \frac{n(y_i - x_i)}{|y - x|^2} \right| dy \\
&\leq C \int_{\substack{|y-x_r|>r \\ |y|\leq R}} \left[\frac{2x_n r - |x|^2}{|y|^2 - 2y_n r} \right]^{\alpha/2} \frac{u(y)}{|x - y|^n} \frac{1}{|y - x|} dy \\
&\leq C \int_{\substack{|y-x_r|>r \\ |y|\leq R}} \left[\frac{2x_n r - |x|^2}{|y|^2 - 2y_n r} \right]^{\alpha/2} \frac{u(y)}{|x - y|^{n+1}} dy \\
&= C \int_{\substack{|y-x_r|>r \\ |y|\leq R, \, y_n>0}} \left[\frac{2x_n r - |x|^2}{|y|^2 - 2y_n r} \right]^{\alpha/2} \frac{u(y)}{|x - y|^{n+1}} dy \\
&\leq C_R \int_{\substack{|y-x_r|>r \\ |y|\leq R, \, y_n>0}} \left[\frac{2x_n r - |x|^2}{|y|^2 - 2y_n r} \right]^{\alpha/2} \frac{1}{|x - y|^{n+1}} dy.
\end{aligned} \tag{5.27}$$

In (5.27), we have used the fact that the α-harmonic function u is bounded in the region

$$D_{R,r} = \{y = (y', y_n) \mid |y - x_r| > r, \; |y| < R, \; y_n > 0\}.$$

The bound depends on R, however is independent of r, since $D_{R,r_1} \subset D_{R,r_2}$ for $r_1 > r_2$. For each such fixed open domain $D_{R,r}$, the bound of the α-harmonic function u can be derived from the interior smoothness (see, for instance [BKN] and [FW]) and the estimate up to the boundary (see [RS]).

Set $y = (y', y_n), \sigma = |y'|$, for fixed x and sufficiently large r, we have

$$\left[\frac{2x_n r - |x|^2}{|y|^2 - 2y_n r}\right]^{\alpha/2} \leq \frac{Cr^{\alpha/2}}{||y|^2 - 2y_n r|^{\alpha/2}}$$

$$= \frac{Cr^{\alpha/2}}{|\sigma^2 + y_n^2 - 2y_n r|^{\alpha/2}}$$

$$= \frac{Cr^{\alpha/2}}{|(y_n - r)^2 + \sigma^2 - r^2|^{\alpha/2}}. \tag{5.28}$$

and

$$\frac{1}{|x - y|^{n+1}} \leq \frac{C}{(1 + |y|)^{n+1}}$$

$$\leq \frac{C}{(1 + |y'|)^{n+1}}$$

$$= \frac{C}{(1 + \sigma)^{n+1}}. \tag{5.29}$$

For convenience of estimation, we amplify the domain $D_{R,r}$ a little bit. Define

$$\hat{D}_{R,r} = \left\{y = (y', y_n) \in \mathbb{R}^n_+ \mid |y - x_r| > r, \; |y'| \leq R, 0 < y_n < \bar{y}_n\right\},$$

where $\bar{y} = (y', \bar{y}_n) \in \partial B_r(x_r)$:

$$(\bar{y}_n - r)^2 + \sigma^2 = r^2. \tag{5.30}$$

Then it is easy to see that

$$D_{R,r} \subset \hat{D}_{R,r}. \tag{5.31}$$

From (5.30), for sufficiently large r (much larger than R), we have

$$\bar{y}_n = r - \sqrt{r^2 - \sigma^2}.$$

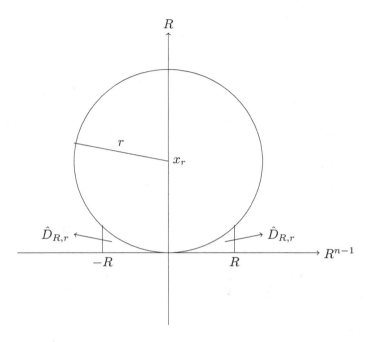

Set

$$y_n = r - s\sqrt{r^2 - \sigma^2}. \tag{5.32}$$

Then for $0 < y_n < \bar{y}_n$,

$$1 < s < \frac{r}{\sqrt{r^2 - \sigma^2}}, \tag{5.33}$$

and

$$dy_n = -\sqrt{r^2 - \sigma^2}ds. \tag{5.34}$$

Continuing from the right side of (5.27), we integrate in the direction of y_n first, and then with respect to y'. By (5.27) through (5.29) and (5.31) through (5.34), we derive

$$I_{22} \le C \int_{\hat{D}_{R,r}} \left[\frac{2x_n r - |x|^2}{|y|^2 - 2y_n r}\right]^{\alpha/2} \frac{1}{|x-y|^{n+1}} dy$$

$$\le C \int_{|y'|<R} \int_0^{\bar{y}_n} \frac{r^{\alpha/2}}{|(y_n-r)^2 + \sigma^2 - r^2|^{\alpha/2}} dy_n \frac{1}{(1+\sigma)^{n+1}} dy'$$

$$\le C \int_0^R \int_0^{\bar{y}_n} \frac{r^{\alpha/2}}{|(y_n-r)^2 + \sigma^2 - r^2|^{\alpha/2}} dy_n \frac{\sigma^{n-2}}{(1+\sigma)^{n+1}} d\sigma$$

$$\leq C \int_0^R \int_1^{\frac{r}{\sqrt{r^2 - \sigma^2}}} \frac{r^{\alpha/2}\sqrt{r^2 - \sigma^2}}{[(s\sqrt{r^2 - \sigma^2})^2 - (r^2 - \sigma^2)]^{\alpha/2}} ds \frac{\sigma^{n-2}}{(1+\sigma)^{n+1}} d\sigma$$

$$= C \int_0^R r^{\alpha/2}(r^2 - \sigma^2)^{\frac{1-\alpha}{2}} \int_1^{\frac{r}{\sqrt{r^2 - \sigma^2}}} \frac{1}{(s^2 - 1)^{\alpha/2}} ds \frac{\sigma^{n-2}}{(1+\sigma)^{n+1}} d\sigma$$

$$\leq C \int_0^R r^{\alpha/2}(r^2 - \sigma^2)^{\frac{1-\alpha}{2}} \int_1^{\frac{r}{\sqrt{r^2 - \sigma^2}}} \frac{1}{(s-1)^{\alpha/2}} ds \frac{\sigma^{n-2}}{(1+\sigma)^{n+1}} d\sigma \qquad (5.35)$$

$$\leq C \int_0^R r^{\alpha/2} r^{1-\alpha} \int_1^{\frac{r}{\sqrt{r^2 - \sigma^2}}} \frac{1}{(s-1)^{\alpha/2}} ds \frac{\sigma^{n-2}}{(1+\sigma)^{n+1}} d\sigma \qquad (5.36)$$

$$= C \int_0^R r^{1-\alpha/2} \left(\frac{r}{\sqrt{r^2 - \sigma^2}} - 1 \right)^{1-\alpha/2} \frac{\sigma^{n-2}}{(1+\sigma)^{n+1}} d\sigma$$

$$= C \int_0^R r^{1-\alpha/2} \left(\frac{\sigma^2}{(r + \sqrt{r^2 - \sigma^2})\sqrt{r^2 - \sigma^2}} \right)^{1-\alpha/2} \frac{\sigma^{n-2}}{(1+\sigma)^{n+1}} d\sigma$$

$$\leq C \int_0^R r^{1-\alpha/2} \left(\frac{1}{r^2} \right)^{1-\alpha/2} \frac{\sigma^{n-2+2-\alpha}}{(1+\sigma)^{n+1}} d\sigma$$

$$= C r^{\alpha/2 - 1} \int_0^R \frac{\sigma^{n-\alpha}}{(1+\sigma)^{n+1}} d\sigma$$

$$\leq C r^{\alpha/2 - 1}$$

$$= \frac{C}{r^{1-\alpha/2}}. \qquad (5.37)$$

Equation (5.35) is valid because

$$\frac{1}{(s^2 - 1)^{\alpha/2}} = \frac{1}{(s+1)^{\alpha/2}} \frac{1}{(s-1)^{\alpha/2}}$$

$$\leq \frac{1}{(1+1)^{\alpha/2}} \frac{1}{(s-1)^{\alpha/2}}$$

$$= \frac{1}{2^{\alpha/2}} \frac{1}{(s-1)^{\alpha/2}}$$

$$\leq \frac{1}{(s-1)^{\alpha/2}}.$$

Since R is fixed and $\sigma^2 \leq R^2$, when r is sufficiently large (much larger than R), we have $r^2 - \sigma^2 > 0$, and the value of $(r^2 - \sigma^2)^{\frac{1-\alpha}{2}}$ can be dominated by $(r^2)^{\frac{1-\alpha}{2}}$ (i.e. $r^{1-\alpha}$), this verifies (5.36).

For any $\epsilon > 0$ and $R = R(\epsilon)$, since $0 < \alpha < 2$, for sufficiently large r (5.37) implies that

$$I_{22} \leq C \frac{1}{r^{1-\alpha/2}} < \epsilon. \tag{5.38}$$

Combining (5.23) through (5.26), and (5.38), we arrive at

$$\left| \frac{\partial u}{\partial x_i}(x) \right| < C\epsilon,$$

for sufficiently large $R = R(\epsilon)$ and much larger r.

The fact that ϵ is arbitrary implies

$$\frac{\partial u}{\partial x_i}(x) = 0, \qquad i = 1, 2, \cdots, n-1. \tag{5.39}$$

This proves (5.21).

Now, let's prove (5.22). Similarly, for fixed $x \in B_r(x_r) \subset \mathbb{R}^n_+$, one can derive that

$$
\begin{aligned}
\frac{\partial u}{\partial x_n}(x) &= \int_{|y-x_r|>r} \left(\frac{\alpha(r-x_n)}{r^2 - |x-x_r|^2} + \frac{n(y_n - x_n)}{|y-x|^2} \right) P_r(x - x_r, y - x_r) u(y) dy \\
&= \int_{|y-x_r|>r} \frac{\alpha(r-x_n)}{r^2 - |x-x_r|^2} P_r(x - x_r, y - x_r) u(y) dy \\
&\quad + \int_{|y-x_r|>r} \frac{n(y_n - x_n)}{|y-x|^2} P_r(x - x_r, y - x_r) u(y) dy \\
&:= J_1 + J_2.
\end{aligned}
\tag{5.40}
$$

Similar to I_2, for sufficiently large r, we can derive

$$|J_2| \leq C\epsilon,$$

for any $\epsilon > 0$. That is

$$J_2 \to 0, \quad \text{as } r \to \infty. \tag{5.41}$$

Now we estimate J_1.

$$
\begin{aligned}
J_1 &= \frac{\alpha(r-x_n)}{r^2 - |x-x_r|^2} \int_{|y-x_r|>r} P_r(x - x_r, y - x_r) u(y) dy \\
&= \frac{\alpha(r-x_n)}{2x_n r - |x|^2} u(x).
\end{aligned}
$$

It follows that

$$J_1 \to \frac{\alpha}{2x_n} u(x), \quad \text{as } r \to \infty. \tag{5.42}$$

By (5.40) through (5.42), for each fixed $x \in B_r(x_r) \subset \mathbb{R}^n_+$, letting $r \to \infty$, we arrive at

$$\frac{\partial u}{\partial x_n}(x) = \frac{\alpha}{2x_n} u(x).$$

This verifies (5.22), and hence completes the proof of Theorem 5.2.1. $\square$

5.2.2 Equivalence Between Two Equations in $\mathbb{R}^n_+$

Theorem 5.2.4 *Assume $u \in L_\alpha$ is lower semi-continuous. If u is a locally bounded positive solution of*

$$\begin{cases} (-\triangle)^{\frac{\alpha}{2}} u(x) = u^p(x), & x \in \mathbb{R}^n_+, \\ u(x) = 0, & x \notin \mathbb{R}^n_+, \end{cases} \tag{5.43}$$

in the sense of distribution, then it is also a solution of

$$u(x) = \int_{\mathbb{R}^n_+} G_+(x,y) u^p(y) dy, \tag{5.44}$$

where $G_+(x,y)$ is the Green's function associated with $(-\triangle)^{\frac{\alpha}{2}}$ in $\mathbb{R}^n_+$.

Proof. Recall that the Green's function on $B_R(P_R)$ assumes the form of

$$\begin{aligned} G_R(x,y) &= \frac{G_1(\frac{x-P_R}{R}, \frac{y-P_R}{R})}{R^{n-\alpha}} \\ &= \frac{C_{n,\alpha}}{|x-y|^{n-\alpha}} \int_0^{\frac{(R-\frac{|x-P_R|^2}{R})(R-\frac{|y-P_R|^2}{R})}{|x-y|^2}} \frac{b^{\alpha/2-1}}{(1+b)^{\frac{n}{2}}} db, \end{aligned} \tag{5.45}$$

where $P_R = (0, \cdot, 0, R)$. Let $R \to \infty$, we derive the Green's function in $\mathbb{R}^n_+$:

$$G_+(x,y) = \frac{C_{n,\alpha}}{|x-y|^{n-\alpha}} \int_0^{\frac{4x_n y_n}{|x-y|^2}} \frac{b^{\alpha/2-1}}{(1+b)^{\frac{n}{2}}} db.$$

Fix x, for $|y| > R$ with R sufficiently large,

$$\frac{4x_n y_n}{|x-y|^2} \sim \frac{y_n}{|y|^2},$$

and

$$\frac{b^{\alpha/2-1}}{(1+b)^{\frac{n}{2}}} \sim b^{\alpha/2-1},$$

therefore,

$$\int_0^{\frac{4x_n y_n}{|x-y|^2}} \frac{b^{\alpha/2-1}}{(1+b)^{\frac{n}{2}}} db \sim \int_0^{\frac{y_n}{|y|^2}} b^{\alpha/2-1} db = \frac{2}{\alpha} \left(\frac{y_n}{|y|^2} \right)^{\alpha/2}. \tag{5.46}$$

Now, we are ready to prove the equivalence. Assume that u is a positive solution of (5.43), we first show that

$$\int_{\mathbb{R}^n_+} G_+(x,y) u^p(y) dy < \infty.$$

Set

$$v_R(x) = \int_{B_R(P_R)} G_R(x,y)u^p(y)dy.$$

From the local boundedness assumption on u, one can see that, for each $R > 0$, $v_R(x)$ is well-defined and continuous. Moreover

$$\begin{cases} (-\triangle)^{\frac{\alpha}{2}} v_R(x) = u^p(x), & x \in B_R(P_R), \\ v_R(x) = 0, & x \notin B_R(P_R). \end{cases}$$

Let

$$w_R(x) = u(x) - v_R(x);$$

then

$$\begin{cases} (-\triangle)^{\frac{\alpha}{2}} w_R(x) = 0, & x \in B_R(P_R), \\ w_R(x) \geq 0, & x \notin B_R(P_R). \end{cases}$$

From the *maximum principle*, we have

$$w_R(x) \geq 0, \quad x \in B_R(P_R).$$

Letting $R \to \infty$, we arrive at

$$u(x) \geq \int_{\mathbb{R}^n_+} G_+(x,y)u^p(y)dy. \tag{5.47}$$

Let

$$v(x) = \int_{\mathbb{R}^n_+} G_+(x,y)u^p(y)dy.$$

Then

$$\begin{cases} (-\triangle)^{\frac{\alpha}{2}} v(x) = u^p(x), & x \in \mathbb{R}^n_+, \\ v(x) = 0, & x \notin \mathbb{R}^n_+. \end{cases}$$

Set $w = u - v$, we have

$$\begin{cases} (-\triangle)^{\frac{\alpha}{2}} w(x) = 0, w \geq 0, & x \in \mathbb{R}^n_+, \\ w(x) = 0, & x \notin \mathbb{R}^n_+. \end{cases}$$

Then by Theorem 5.2.1, we deduce that either

$$w(x) \equiv 0, \quad x \in \mathbb{R}^n,$$

or there is a constant $c_0 > 0$, such that

$$w(x) = c_0 x_n^{\frac{\alpha}{2}}, \quad x_n > 0.$$

We derive a contradiction in the latter case.

Notice that

$$u(x) = w(x) + v(x) \geq w(x) = c_0(x_n)^{\frac{\alpha}{2}}.$$

Let $x = (x', x_n)$, $y = (y', y_n)$, $r^2 = |x' - y'|^2$ and $a^2 = |x_n - y_n|^2$. It follows from (5.46) that for each fixed x and for sufficiently large R,

$$u(x) \geq v(x) = \int_{\mathbb{R}^n_+} G_+(x,y)u^p(y)dy$$

$$\geq c \int_{\mathbb{R}^n_+} G_+(x,y)(y_n^{\frac{\alpha}{2}})^p dy$$

$$\geq c \int_{\mathbb{R}^n_+ \setminus B_R^+(0)} \frac{y_n^{\alpha/2}}{|y|^\alpha}(y_n^{\alpha/2})^p dy$$

$$\geq c \int_R^\infty y_n^{\frac{\alpha(p-1)}{2}} dy_n$$

$$= \infty.$$

This contradicts the local boundedness of u. Therefore, we must have $w \equiv 0$, that is

$$u(x) = v(x) = \int_{\mathbb{R}^n_+} G_+(x,y)u^p(y)dy.$$

On the other hand, assume that $u(x)$ is a solution of the integral equation (5.17) in the sense of distribution. Then for any $\phi \in C_0^\infty(\mathbb{R}^n_+)$, we have

$$< (-\triangle)^{\frac{\alpha}{2}}u, \phi > = < \int_{\mathbb{R}^n_+} G_+(x,y)u^p(y)dy, (-\triangle)^{\frac{\alpha}{2}}\phi(x) >$$

$$= \int_{\mathbb{R}^n_+} \left\{ \int_{\mathbb{R}^n_+} G_+(x,y)u^p(y)dy \right\}(-\triangle)^{\frac{\alpha}{2}}\phi(x)dx$$

$$= \int_{\mathbb{R}^n_+} \left\{ \int_{\mathbb{R}^n_+} G_+(x,y)(-\triangle)^{\frac{\alpha}{2}}\phi(x)dx \right\}u^p(y)dy$$

$$= \int_{\mathbb{R}^n_+} \left\{ \int_{\mathbb{R}^n_+} (-\triangle)^{\frac{\alpha}{2}}G_+(x,y)\phi(x)dx \right\}u^p(y)dy$$

$$= \int_{\mathbb{R}^n_+} \left\{ \int_{\mathbb{R}^n_+} \delta(x,y)\phi(x)dx \right\}u^p(y)dy$$

$$= \int_{\mathbb{R}^n_+} \phi(y)u^p(y)dy$$

$$= < u^p, \phi >$$

This shows that u also satisfies (5.43).

6

A Direct Method of Moving Planes for the Fractional Order Equations

6.1 Overview of the Method of Moving Planes

The *method of the moving planes* was invented by the Soviet mathematician Alexanderoff in the early 1950's. Decades later, it was further developed by Serrin [Se], Gidas, Ni, and Nirenberg [GNN], Caffarelli, Gidas, and Spruck [CGS], Lin [Lin], Li [Li], Chen and Li [CL4], [CL8], Chang and Yang [CY], and many others. The method has been applied to free boundary problems, semi-linear differential equations, and other problems. Particularly for semi-linear differential equations, there were many significant contributions. We refer to [F] for more descriptions.

Then what is the *method of moving planes*?

How can one use it to catch the underlying symmetry of a solution?

Here we use two simple examples and the corresponding graphs to illustrate the essence of this method.

Assume that u is a positive solution of some equation defined in a symmetric domain Ω and it equals zero on the boundary. Meanwhile, the equation is symmetric with respect to Ω. Then one would naturally guess that u must also possess the same symmetry as the domain Ω. One classical approach to deal with such symmetry problem is the *method of moving planes*. To get a general idea of how the method works, we first consider one simple example on a bounded domain in one dimensional Euclidean space $\mathbb{R}^1$.

Let $\Omega = (-1, 1)$. Suppose that $u(x)$ is a positive solution of a symmetric equation in Ω and $u(-1) = 0 = u(1)$. In one dimension, the moving plane reduces to a point:

$$T_\lambda = \{x \mid x = \lambda\}.$$

Let

$$\Sigma_\lambda = \{x \mid -1 < x < \lambda\}$$

be the region to the left of T_λ in Ω, and

$$x^\lambda = 2\lambda - x$$

be the reflection of x about T_λ.

We compare $u(x^\lambda)$ and $u(x)$. For simplicity, set $w_\lambda(x) = u(x^\lambda) - u(x)$.

To roughly illustrate the idea, let's first look at Figure 6.1, where $T = T_\lambda$ and $y = x^\lambda$. Since $u = 0$ on the boundary and $u > 0$ inside, we may expect that when T_λ is sufficiently close to -1 (the T_1 position), we have

$$w_\lambda(x) \geq 0, \ \forall x \in \Sigma_\lambda. \tag{6.1}$$

This provides a starting point for us.

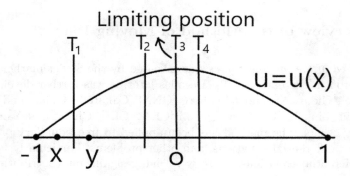

Fig. 6.1: Moving Planes 1-Dimension

Then we move the plane T continuously to the right as long as inequality (6.1) holds until its limiting position and prove that u must be symmetric about the limiting plane. From Fig. 6.1, when the plane T is moved to T_2 position, inequality (6.1) is still valid, hence we can keep moving it. The T_3 position is the limiting one, because after passing it, say at T_4 position, inequality (6.1) is violated. One can see that u is symmetric about the limiting position, the origin.

To be more rigorous and precise, in general we carry out the *method of moving planes* in the following two steps.

Step 1. We try to show that when λ is sufficiently close to -1,

$$w_\lambda(x) \geq 0, \ x \in \Sigma_\lambda. \tag{6.2}$$

This provides a starting point to move the plane T_λ.

Step 2. We move the plane T_λ to the right continuously as long as (6.2) is true until the limiting position T_{λ_0}, where

$$\lambda_0 = \sup\{\lambda \mid w_\rho(x) \geq 0, \forall x \in \Sigma_\rho, \rho \leq \lambda\}.$$

We prove that u must be symmetric about T_{λ_0}.

From the reasoning above, one can see that the key is to obtain (6.2). This is usually proved by a *maximum principle*.

Now let's take a look at a higher-dimensional case where Ω is an unbounded region, for instance, $\Omega = \mathbb{R}^n$. Assume that $u(x)$ is a positive solution to some radially symmetric equation, with $\lim\limits_{|x|\to\infty} u(x) = 0$. In order to show that u is also radially symmetric, we first choose any direction to be the x_1-direction and let

$$T_\lambda = \{x \in \mathbb{R}^n \mid x_1 = \lambda\}, \ \ \Sigma_\lambda = \{x \in \mathbb{R}^n \mid x_1 < \lambda\},$$

and

$$x^\lambda = \{(2\lambda - x_1, x') \mid x = (x_1, x') \in \mathbb{R}^n\}$$

be the reflection of x about the plane T_λ (see Fig. 6.2).

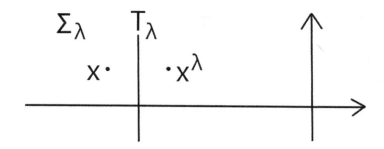

Fig. 6.2: Moving Planes Higher-Dimensions

Set

$$w_\lambda(x) = u(x^\lambda) - u(x).$$

We carry out the *method of moving planes* in two steps identical to those in $\mathbb{R}^1$.

Step 1. Show that for λ sufficiently negative,

$$w_\lambda(x) \geq 0, \ x \in \Sigma_\lambda. \tag{6.3}$$

Step 2. Move the plane T_λ continuously to the right as long as (6.3) holds until its limiting position T_{λ_0}, and prove that

$$w_{\lambda_0}(x) \equiv 0, \ \ \text{in } \Sigma_{\lambda_0}.$$

It follows that u is symmetric about the plane T_{λ_0}. Since x_1 direction can be chosen arbitrarily, we conclude that u is radially symmetric about some point.

Now we have a rough understanding of how the method works. In the following sections, we give more specific examples and more rigorous proofs to further explain the method.

6.2 Problems Involving the Laplacian

6.2.1 A Simple Example on a Bounded Domain

We start with an equation having some good monotonicity on the right-hand side.

$$\begin{cases} -\triangle u(x) = \frac{1}{1+u(x)}, & x \in B_1(0) \subset \mathbb{R}^n, \\ u = 0, & x \in \partial B_1(0). \end{cases} \tag{6.4}$$

Assume that $u(x)$ is a classical positive solution of (6.4). We will show that $u(x)$ is radially symmetric about the origin, i.e. $u(x) = u(|x|)$.

Proof. To start, we choose a direction to be the x_1-direction and let

$$T_\lambda = \{x \in \mathbb{R}^n \mid x_1 = \lambda\}, \quad \Sigma_\lambda = \{x \in B_1(0) \mid x_1 < \lambda\},$$

and the reflection of $x = (x_1, x') \in B_1(0)$ about T_λ be

$$x^\lambda = (2\lambda - x_1, x').$$

(See Fig. 6.3.)

Let $u_\lambda(x) = u(x^\lambda)$ and $w_\lambda(x) = u_\lambda(x) - u(x)$. It follows from (6.4) that

$$-\triangle w_\lambda(x) = \frac{1}{1 + u_\lambda(x)} - \frac{1}{1 + u(x)}. \tag{6.5}$$

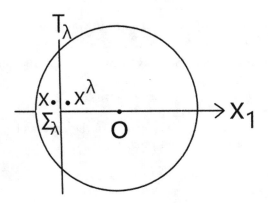

Fig. 6.3: Moving Planes on the Unit Ball

What we want to derive is that

$$w_\lambda(x) \geq 0, \ x \in \Sigma_\lambda. \tag{6.6}$$

Usually, a contradiction argument is employed to derive (6.6). Suppose (6.6) is violated, then there exists some $x^0 \in \Sigma_\lambda$ such that

$$w_\lambda(x^0) = \min_{\Sigma_\lambda} w_\lambda < 0.$$

On one hand,

$$-\triangle w_\lambda(x^0) = \frac{1}{1 + u_\lambda(x^0)} - \frac{1}{1 + u(x^0)} > 0.$$

While on the other hand, because x^0 is a negative minimum of $w_\lambda(x)$, it yields

$$-\triangle w_\lambda(x^0) \leq 0.$$

This contradiction implies that our original supposition is false. Therefore, (6.6) holds for any $\lambda \leq 0$. In particular, we have

$$w_0(x) \geq 0, \ x \in \Sigma_0.$$

Similarly, we can derive that

$$w_0(x) \leq 0, \ x \in \Sigma_0.$$

Hence

$$w_0(x) \equiv 0, \ x \in \Sigma_0.$$

This proves that $u(x)$ is symmetric about T_0. Because the direction of x_1 can be chosen arbitrarily, we have actually shown that $u(x)$ is radially symmetric about the origin.

The above serves as a simple case when the right-hand side of the equation, $f(u) = \frac{1}{1+u(x)}$, has the needed monotonicity, i.e., $f(\cdot)$ is decreasing in u. As one can see, the *method of moving planes* has not come into play in deriving the symmetry due to the decreasing nature of $f(\cdot)$. However, we have applied the *maximum principle*, a constant player in deriving inequality (6.6) when using the *method of moving planes*. The essence of the above approach is the following.

Let $\Sigma_\lambda^- = \{x \in \Sigma_\lambda | w_\lambda < 0\}$.

Suppose that

$$\Sigma_\lambda^- \neq \emptyset, \tag{6.7}$$

then, due to the decreasing nature of $f(\cdot)$, we have

$$\begin{cases} -\triangle w_\lambda(x) > 0, \ x \in \Sigma_\lambda^-, \\ w_\lambda(x) = 0, \quad\ \ x \in \partial\Sigma_\lambda^-. \end{cases} \tag{6.8}$$

By the *maximum principle*, we have

$$w_\lambda(x) \geq 0, \ x \in \Sigma_\lambda^-. \tag{6.9}$$

This contradicts with our assumption (6.7). From here, one may see $-\triangle$ as a "positive" operator in the sense that (6.8) implies (6.9).

6.2.2 General Equations on Bounded Domains

Now we consider a more general case where the right-hand side may not be monotone.

$$\begin{cases} -\triangle u(x) = f(u(x)), \ x \in B_1(0) \subset \mathbb{R}^n, \\ u = 0, \qquad\qquad\quad x \in \partial B_1(0). \end{cases} \tag{6.10}$$

Using the notation introduced in the previous example, (6.10) gives

$$-\triangle w_\lambda(x) = f(u_\lambda) - f(u).$$

Recall that in the beginning of this chapter we always focus on the equation about $w_\lambda(x)$. So here we deliberately rewrite the equation above as

$$-\triangle w_\lambda(x) + c(x)w_\lambda(x) = 0, \tag{6.11}$$

where $c(x) = -\frac{f(u_\lambda)-f(u)}{u_\lambda-u}$.

At the end of the subsection above, we have seen the "positiveness" of the $-\triangle$ operator in the sense that if

$$\begin{cases} -\triangle w_\lambda(x) \geq 0, \ x \in \Omega, \\ w_\lambda(x) \geq 0, \qquad x \in \partial\Omega, \end{cases}$$

then

$$w_\lambda(x) \geq 0, \ x \in \Omega.$$

Apparently, if $c(x) \geq 0$, then $-\triangle + c(x)$ is also "positive". Actually, one may expect more, that is, if $c(x)$ is not "too" negative, $-\triangle + c(x)$ is still "positive". More precisely, one can show that, if $c(x) > -\lambda_1$, where λ_1 is the first nonzero eigenvalue of the $-\triangle$ operator, then

$$-\triangle + c(x) \text{ is still "positive".}$$

A well-known fact is that the narrower the region, the larger the first eigenvalue λ_1. And hence the more negative $c(x)$ is allowed to be, as illustrated in the following.

Theorem 6.2.1 (Narrow Region Principle) *Assume that Ω is a narrow region. Without loss of generality, we may assume that Ω is contained in the slab $\{x \in \mathbb{R}^n \mid 0 < x_1 < \delta\}$ as shown in the following figure.*

Consider

$$\begin{cases} -\triangle w(x) + c(x)w(x) \geq 0, \ x \in \Omega, \\ w(x) \geq 0, \qquad\qquad\qquad x \in \partial\Omega. \end{cases} \tag{6.12}$$

If $c(x)$ is bounded from below, then for sufficiently small δ, we have

$$w(x) \geq 0, \ x \in \Omega. \tag{6.13}$$

Proof. Let $\bar{w}(x) = \frac{w(x)}{\phi(x)}$. Then through an elementary computation we arrive at

$$-\triangle\bar{w} - \frac{2\nabla\bar{w}\nabla\phi}{\phi} + \left(c(x) - \frac{\triangle\phi}{\phi}\right)\bar{w} \geq 0. \tag{6.14}$$

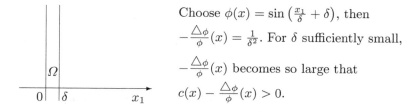

Choose $\phi(x) = \sin\left(\frac{x_1}{\delta} + \delta\right)$, then $-\frac{\triangle\phi}{\phi}(x) = \frac{1}{\delta^2}$. For δ sufficiently small, $-\frac{\triangle\phi}{\phi}(x)$ becomes so large that $c(x) - \frac{\triangle\phi}{\phi}(x) > 0.$

If (6.13) does not hold, then w is negative somewhere, so is $\bar{w}$, hence there exists an $x^0 \in \Omega$ such that

$$\bar{w}(x^0) = \min_\Omega \bar{w} < 0,$$

then at this point, $\nabla\bar{w} = 0$, and consequently,

$$-\triangle\bar{w} - \frac{2\nabla\bar{w}\nabla\phi}{\phi} + \left(c(x) - \frac{\triangle\phi}{\phi}\right)\bar{w}$$
$$= -\triangle\bar{w}(x^0) + \left(c(x^0) - \frac{\triangle\phi(x^0)}{\phi(x^0)}\right)\bar{w}(x^0)$$
$$< 0.$$

This is a contradiction with (6.14). Therefore, (6.13) must be valid, and this proves the theorem.

Now we are ready to show that the positive solutions of (6.10) are radially symmetric about the origin. Actually, this is an elegant result obtained by Gidas, Ni, and Nirenberg in [GNN].

Theorem 6.2.2 *(Gidas-Ni-Nirenberg) Assume that $f(\cdot)$ is a Lipschitz continuous function such that*

$$|f(p) - f(q)| \leq C_o|p - q| \tag{6.15}$$

for some constant C_o. Then every positive classical solution u of (6.10) is radially symmetric and monotone decreasing about the origin.

Proof. Condition (6.15) implies that $c(x)$ in (6.11) satisfies

$$|c(x)| \equiv |c(x, \lambda)| = |\frac{f(u(x^\lambda)) - f(u(x))}{u(x^\lambda) - u(x)}| \leq C_o. \qquad (6.16)$$

Here T_λ, Σ_λ and x^λ are the same as those defined for (6.4).

Step 1. Begin moving the plane from near the left end of $B_1(0)$ along the x_1-axis.

Apparently, Σ_λ is a narrow region in the x_1-direction for λ very close to -1. Also,

$$w_\lambda(x) \geq 0, \ x \in \partial\Sigma_\lambda.$$

Then an application of the *narrow region principle* gives

$$w_\lambda(x) \geq 0, \ x \in \Sigma_\lambda, \qquad (6.17)$$

for λ near -1.

Step 2. Move the plane continuously to the right until its limiting position as long as (6.17) holds.

In this part, we show that the moving plane will not come to a halt until it hits the origin. To be more precise, define

$$\bar\lambda = \sup\{\lambda \leq 0 \mid w_\mu(x) \geq 0, \ \forall x \in \Sigma_\mu, \mu \leq \lambda\}.$$

We claim that

$$\bar\lambda = 0. \qquad (6.18)$$

If (6.18) is violated, then we would be able to move T_λ further to the right a little bit, and this contradicts the definition of $\bar\lambda$.

Suppose that $\bar\lambda < 0$, then the reflection of the curved part of $\partial\Sigma_{\bar\lambda}$ falls inside $B_1(0)$ and

$$w_{\bar\lambda}(x) \geq 0, \ x \in \partial\Sigma_{\bar\lambda}.$$

It follows from the *strong maximum principle* that

$$w_{\bar\lambda}(x) > 0, \ x \in \Sigma_{\bar\lambda}. \qquad (6.19)$$

Set d_o to be the maximum width of the narrow regions so that we are able to employ the *narrow region principle*. Let $\delta = \min\{\frac{d_o}{2}, -\bar\lambda\}$. We study function $w_{\bar\lambda+\delta}(x)$ on the narrow region

$$\Omega_\delta = \Sigma_{\bar\lambda+\delta} \cap \{x \in B_1(0) \mid x_1 > \bar\lambda - \frac{d_o}{2}\}.$$

(See Figure 6.4.)

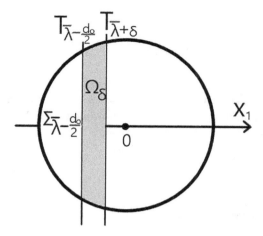

Fig. 6.4: Narrow Region Ω_δ

It's easy to obtain

$$\begin{cases} -\triangle w_{\bar{\lambda}+\delta}(x) + c(x, \bar{\lambda} + \delta)w_{\bar{\lambda}+\delta}(x) = 0, \, x \in \Omega_\delta, \\ w_{\bar{\lambda}+\delta}(x) \geq 0, \qquad\qquad\qquad\qquad x \in \partial\Omega_\delta. \end{cases} \qquad (6.20)$$

Regarding the boundary condition, we already know that $w_{\bar{\lambda}+\delta}(x) \geq 0$ is true on the two curved parts and one flat part of $\partial\Omega_\delta$ where $x_1 = \bar{\lambda}+\delta$. What's left is to verify that the same inequality holds on the rest of $\partial\Omega_\delta$ where $x_1 = \bar{\lambda}-\frac{d_\varrho}{2}$.

Indeed, equation (6.19) indicates that $w_{\bar{\lambda}}$ is positive and bounded away from 0 on $\Sigma_{\bar{\lambda}-\frac{d_\varrho}{2}}$. More precisely, there exists a constant $c > 0$ such that

$$w_{\bar{\lambda}}(x) \geq c, \, x \in \Sigma_{\bar{\lambda}-\frac{d_\varrho}{2}}.$$

Due to the continuity of w_λ with respect to λ, we have, for sufficiently small δ,

$$w_{\bar{\lambda}+\delta}(x) \geq 0, \, x \in \Sigma_{\bar{\lambda}-\frac{d_\varrho}{2}}.$$

Now we have the needed condition on the boundary of Ω_δ and hence are able to apply the *narrow region principle* to derive that

$$w_{\bar{\lambda}+\delta}(x) \geq 0, \, x \in \Omega_\delta.$$

And further,

$$w_{\bar{\lambda}+\delta}(x) \geq 0, \, x \in \Sigma_{\bar{\lambda}+\delta}.$$

This is a contradiction with the definition of $\bar{\lambda}$. Hence (6.18) is valid, or

$$u(-x_1, x') \leq u(x_1, x'), \, \forall\, x_1 \geq 0. \qquad (6.21)$$

If we move the plane from near $x_1 = 1$ to the left, by a similar argument, we can derive

$$u(-x_1, x') \geq u(x_1, x'), \ \forall \, x_1 \geq 0. \tag{6.22}$$

Combining (6.21) and (6.22), we drive that $u(x)$ is symmetric about the plane T_0. Moreover, the arbitrariness of the x_1-direction leads to the radial symmetry of $u(x)$ about the origin. The monotonicity comes directly from the argument. This completes the proof.

6.2.3 On Unbounded Domains

Let's consider classical positive solutions for the equation defined in the whole space

$$-\triangle u(x) = u^p(x), \ x \in \mathbb{R}^n. \tag{6.23}$$

Assume that $u(x)$ decays at certain rate at infinity, which we will determine later, this time we want to show that $u(x)$ is radially symmetric about some point in $\mathbb{R}^n$.

Let

$$\Sigma_\lambda = \{x = (x_1, \cdots, x_n) \in \mathbb{R}^n \mid x_1 < \lambda\}, \quad T_\lambda = \partial \Sigma_\lambda$$

and let x^λ be the reflection point of x about the plane T_λ, i.e.

$$x^\lambda = (2\lambda - x_1, x_2, \cdots, x_n)$$

(See Fig. 6.2.) Let

$$u_\lambda(x) = u(x^\lambda), \quad \text{and} \quad w_\lambda(x) = u_\lambda(x) - u(x).$$

Using the *method of moving planes*, we carry out the proof in three steps.

We begin by moving the plane T_λ from near $-\infty$ along the x_1-axis. We want to show that for λ sufficiently negative,

$$w_\lambda(x) \geq 0, \ x \in \Sigma_\lambda. \tag{6.24}$$

Recall that, previously, in deriving the counterpart of (6.24), we based our argument on the contradiction obtained at the negative minimum point of $w_\lambda(x)$ or of $\bar{w}_\lambda$. Because here the domain is unbounded, in order to employ such a contradiction argument, we need to ensure that all negative minima of w_λ are contained in a fixed ball, in other words, w_λ are nonnegative outside a fixed ball for all λ.

Theorem 6.2.3 *(Decay at Infinity) Assume that Ω is an unbounded region in $\mathbb{R}^n$ and $w(x)$ is a classical solution of*

$$\begin{cases} -\triangle w(x) + c(x)w(x) \geq 0, & x \in \Omega, \\ w(x) \geq 0, & x \in \partial\Omega. \end{cases} \tag{6.25}$$

Suppose there exist $0 < s < n - 2$ and $R_0 > 0$, such that

$$c(x) > -\frac{s(n - 2 - s)}{|x|^2}, \quad \forall\, |x| \geq R_0, \tag{6.26}$$

and

$$\lim_{|x| \to \infty} w(x)|x|^s = 0. \tag{6.27}$$

If Ω is contained in $B_{R_0}^c(0)$, the complement of $B_{R_0}(0)$. Then

$$w(x) \geq 0, \quad x \in \Omega. \tag{6.28}$$

Proof. Let $\bar{w}(x) = \frac{w(x)}{\phi(x)}$, then we derive easily

$$-\triangle\bar{w} - \frac{2\nabla\bar{w}\nabla\phi}{\phi} + \left(c(x) - \frac{\triangle\phi}{\phi}\right)\bar{w} \geq 0. \tag{6.29}$$

Set

$$\phi(x) = \frac{1}{|x|^s}.$$

Since

$$-\frac{\triangle\phi}{\phi} = \frac{s(n - 2 - s)}{|x|^2},$$

with condition (6.26) one can see that

$$c(x) - \frac{\triangle\phi}{\phi} > 0, \quad \forall\, |x| \geq R_0.$$

In the contrary of (6.28), w is negative somewhere in Ω, so is $\bar{w}$. Condition (6.27) guarantees that, there exists an $x^0 \in \Omega$ such that

$$\bar{w}(x^0) = \min_{\Omega} \bar{w} < 0.$$

Then at this point,

$$-\triangle\bar{w} - \frac{2\nabla\bar{w}\nabla\phi}{\phi} + \left(c(x) - \frac{\triangle\phi}{\phi}\right)\bar{w}\bigg|_{x=x^0}$$

$$= -\triangle\bar{w}(x^0) + \left(c(x^0) - \frac{\triangle\phi(x^0)}{\phi(x^0)}\right)\bar{w}(x^0)$$

$$< 0.$$

This is a contradiction with (6.29). Therefore, (6.28) must be valid and this proves the theorem.

Theorem 6.2.4 *Assume that*

$$u(x) = o(\frac{1}{|x|^{\frac{2}{p-1}}}), \ as \ |x| \to \infty. \tag{6.30}$$

Then for $p > 1$, the positive solutions of (6.23) must be radially symmetric about some point in $\mathbb{R}^n$.

Proof. We carry out the proof in two steps. To begin with, we show that for λ sufficiently negative,

$$w_\lambda(x) \geq 0, \ \forall x \in \Sigma_\lambda \tag{6.31}$$

with an application of the *Decay at Infinity*.

Next, we move the plane T_λ along the x_1-axis to the right as long as inequality (6.31) holds. The plane will eventually stop at some limiting position λ_o. Then we are able to claim that

$$w_{\lambda_o}(x) \equiv 0, \ \forall x \in \Sigma_{\lambda_o}.$$

The symmetry and monotone decreasing properties of solution u about T_{λ_o} follows naturally from the proof. Also, because of the arbitrariness of the x_1-axis, we conclude that u must be radially symmetric and monotone about some point.

Step 1. Start moving the plane T_λ along the x_1-axis from near $-\infty$ to the right.

By the Mean Value Theorem, it is easy to have

$$-\triangle w_\lambda = u_\lambda^p(x) - u^p(x) = p\psi_\lambda^{p-1}(x)w_\lambda(x),$$

where $\psi_\lambda(x)$ is in between $u_\lambda(x)$ and $u(x)$. By the *Decay at Infinity* argument (Theorem 6.2.3), it suffices to check the decay rate of $\psi_\lambda^{p-1}(x)$, and to be more precise, only at the points $\tilde{x}$ where w_λ is negative. Since

$$u_\lambda(\tilde{x}) < u(\tilde{x}),$$

we have

$$0 \leq u_\lambda(\tilde{x}) \leq \psi_\lambda(\tilde{x}) \leq u(\tilde{x}).$$

The decay assumption (6.30) instantly yields that

$$\psi_\lambda^{p-1}(\tilde{x}) = o\left(\frac{1}{|\tilde{x}|^2}\right).$$

Consequently, there exists $R_0 > 0$, such that

$$c(\tilde{x}, \lambda) \equiv p\psi_\lambda^{p-1}(\tilde{x}) > -\frac{s(n-2-s)}{|\tilde{x}|^2}, \ \forall |\tilde{x}| \geq R_0.$$

Now take $\Omega = \Sigma_\lambda$, then by Theorem 6.2.3, it's easy to conclude that for $\lambda \leq -R_0$, we must have (6.31). This completes the preparation for the moving of planes.

Step 2. Keep moving the plane to the limiting position T_{λ_0} as long as (6.31) holds.

Let
$$\lambda_0 = \sup\{\lambda \mid w_\mu(x) \geq 0, \ \forall x \in \Sigma_\mu, \mu \leq \lambda\}.$$

In this part, we will see that

$$w_{\lambda_0}(x) \equiv 0, \quad x \in \Sigma_{\lambda_0}. \tag{6.32}$$

Otherwise, the plane T_{λ_0} can still be moved further to the right. More rigorously, there exists a $\delta_0 > 0$ such that for all $0 < \delta < \delta_0$, we have

$$w_{\lambda_0+\delta}(x) \geq 0 , \ \ \forall x \in \Sigma_{\lambda_0+\delta}. \tag{6.33}$$

This would contradict the definition of λ_0, and hence (6.32) must be true. Below we prove (6.33).

Suppose (6.32) is false, then w_{λ_0} is positive somewhere in Σ_{λ_0}, and by the *strong maximum principle*,

$$w_{\lambda_0}(x) > 0, \quad x \in \Sigma_{\lambda_0}.$$

It follows that for any positive number σ,

$$w_{\lambda_0}(x) \geq C > 0, \quad x \in \overline{\Sigma_{\lambda_0-\sigma} \cap B_{R_0}(0)},$$

where R_0 is defined in *Step 1*. Since w_λ depends on λ continuously, for δ positive small, we have

$$w_{\lambda_0+\delta}(x) \geq 0, \quad x \in \overline{\Sigma_{\lambda_0-\sigma} \cap B_{R_0}(0)}. \tag{6.34}$$

Meanwhile, it follows from the argument of the *Narrow Region Principle* that for sufficiently small σ and δ,

$$w_{\lambda_0+\delta}(x) \geq 0, \ x \in (\Sigma_{\lambda_0+\delta}\backslash\Sigma_{\lambda_0-\sigma}) \cap B_{R_0}(0). \tag{6.35}$$

Finally, take $\Omega = \Sigma_{\lambda_0+\delta}\backslash B_{R_0}(0)$, and by Theorem 6.2.3 (*Decay at Infinity*), we conclude that

$$w_{\lambda_0+\delta}(x) \geq 0, \ x \in \Sigma_{\lambda_0+\delta}\backslash B_{R_0}(0). \tag{6.36}$$

A combination of (6.34) through (6.36) gives (6.33).
This completes the proof. □

6.3 On the Fractional Laplacian − Introduction

As one can see from the previous section, the key ingredients in carrying out the *method of moving planes* are the *Narrow Region Principle* and the

Decay at Infinity. In those proofs, we constructed an auxiliary function $\bar{w} = \frac{w}{\phi}$ satisfying

$$-\triangle\bar{w} - \frac{2\nabla\bar{w}\nabla\phi}{\phi} + \big(c(x) - \frac{\triangle\phi}{\phi}\big)\bar{w} \geq 0.$$

At a negative minimum x^0 of $\bar{w}$, we have $\nabla\bar{w}(x^0) = 0$, hence the middle term $\frac{2\nabla\bar{w}\nabla\phi}{\phi}$ does not contribute anything. However, in the case of the fractional Laplacian with

$$(-\triangle)^{\alpha/2}w(x) + c(x)w(x) \geq 0,$$

if we employ the same technique, we arrive at

$$(-\triangle)^{\alpha/2}\bar{w}(x)$$
$$-\frac{1}{\phi(x)}\int_{\mathbb{R}^n} \frac{(\bar{w}(x) - \bar{w}(y))(\phi(x) - \phi(y))}{|x - y|^{n+\alpha}}dy$$
$$+ \big(c(x) + \frac{(-\triangle)^{\alpha/2}\phi}{\phi}(x)\big)\bar{w}(x) \geq 0.$$

At a negative minimum point x^0 of $\bar{w}$, we have no control over the sign of the middle term. Hence a new approach is required.

6.3.1 The Extension and the Integral Equation Approaches

The nonlocality of the fractional Laplacian makes it difficult to study. To circumvent this difficulty, Caffarelli and Silvestre [CaS] introduced the *extension method* that reduced this nonlocal problem into a local one in higher dimensions as we explained in Chapter 1. Recall that for a function $u : \mathbb{R}^n \to \mathbb{R}$, consider the extension $U : \mathbb{R}^n \times [0, \infty) \to \mathbb{R}$ that satisfies

$$\begin{cases} div(y^{1-\alpha}\nabla U) = 0, \ (x, y) \in \mathbb{R}^n \times [0, \infty), \\ U(x, 0) = u(x). \end{cases}$$

Then

$$(-\triangle)^{\alpha/2}u = -C_{n,\alpha}\lim_{y\to 0^+} y^{1-\alpha}\frac{\partial U}{\partial y}.$$

This *extension method* has been applied successfully to study equations involving the fractional Laplacian, and a series of fruitful results have been obtained (see [BCPS] [CZ] and the references therein).

In [BCPS], among many interesting results, when the authors considered the properties of the positive solutions for

$$(-\triangle)^{\alpha/2}u = u^p(x), \ x \in \mathbb{R}^n, \tag{6.37}$$

they first used the above *extension method* to reduce the nonlocal problem into a local one for $U(x, y)$ in one higher dimensional half space $\mathbb{R}^n \times [0, \infty)$, then applied the *method of moving planes* to show the symmetry of $U(x, y)$ in x, and hence derived the non-existence in the subcritical case:

Proposition 6.3.1 *(Brandle-Colorado-Pablo-Sanchez) Let $1 \leq \alpha < 2$. Then the problem*

$$\begin{cases} div(y^{1-\alpha} \nabla U) = 0, & (x,y) \in \mathbb{R}^n \times [0,\infty), \\ -\lim_{y \to 0^+} y^{1-\alpha} \dfrac{\partial U}{\partial y} = U^p(x,0), \ x \in \mathbb{R}^n \end{cases} \tag{6.38}$$

has no positive bounded solution provided $p < (n+\alpha)/(n-\alpha)$.

They then took trace to obtain

Corollary 6.3.1 *Assume that $1 \leq \alpha < 2$ and $1 < p < \frac{n+\alpha}{n-\alpha}$. Then equation (6.37) possesses no bounded positive solution.*

A similar *extension method* was adopted in [CZ] to obtain the non-existence of positive solutions for an indefinite fractional problem:

Proposition 6.3.2 *(Chen-Zhu) Let $1 \leq \alpha < 2$ and $1 < p < \infty$. Then the equation*

$$(-\triangle)^{\alpha/2} u = x_1 u^p, \ \ x \in \mathbb{R}^n \tag{6.39}$$

possesses no positive bounded solutions.

The common restriction $\alpha \geq 1$ is due to the approach where they need to apply the *method of moving planes* on the solutions U of the extended problem

$$div(y^{1-\alpha} \nabla U) = 0, \ \ (x,y) \in \mathbb{R}^n \times [0,\infty). \tag{6.40}$$

Because of the monotonicity requirement on the coefficient $y^{1-\alpha}$, they have to assume $\alpha \geq 1$. It seems that this condition cannot be weakened if one wants to carry the *method of moving planes* on extended equation (6.40).

Then what happens in the case $0 < \alpha < 1$?

Actually, this case can be treated by considering the corresponding integral equations as we will see in the next chapter. In [CLO], the authors showed that if $u \in H^{\alpha/2}(\mathbb{R}^n)$ is a positive weak solution of (6.37), then it also satisfies the integral equation

$$u(x) = C \int_{\mathbb{R}^n} \frac{1}{|x-y|^{n-\alpha}} u^p(y) dy. \tag{6.41}$$

Applying the *method of moving planes in integral forms*, they obtained the radial symmetry in the critical case and non-existence in the subcritical case for positive solutions of (6.41).

Under the weaker condition that $u \in L_\alpha(\mathbb{R}^n)$, the equivalence between pseudo-differential equation (6.37) and integral equation was also established in [ZCCY] by employing a Liouville theorem for α-harmonic functions.

In the case of more general nonlinearity, for instance, when considering

$$(-\triangle)^{\alpha/2} u = f(x,u), \ \ x \in \mathbb{R}^n, \tag{6.42}$$

in order to show that a positive solution of (6.42) also solves

$$u(x) = C \int_{\mathbb{R}^n} \frac{1}{|x-y|^{n-\alpha}} f(y, u(y)) dy, \qquad (6.43)$$

so far one needs to assume that $f(x, u)$ is nonnegative, which is not satisfied by the right hand side of equation (6.39). Hence in this situation, the integral equation approach renders powerless.

Another technical restriction in carrying out the method of moving planes on the integral equation is that both $f(x, u)$ and $\frac{\partial f}{\partial u}$ must be monotone increasing in u, which may not be necessary if one directly works on pseudo-differential equation (6.42).

In summary, either by *extension* or by integral equations, one needs to impose extra conditions on the solutions, which would not be necessary if we consider the pseudo-differential equation directly.

Moreover, for equations involving the uniformly elliptic nonlocal operators

$$C_{n,\alpha} \lim_{\epsilon \to 0} \int_{\mathbb{R}^n \setminus B_\epsilon(x)} \frac{a(x-z)(u(x)-u(z))}{|x-z|^{n+\alpha}} dz = f(x, u), \qquad (6.44)$$

where

$$0 < c_0 \le a(y) \le C_1;$$

and for equations containing fully nonlinear nonlocal operators, such as

$$F_\alpha(u(x)) \equiv C_{n,\alpha} \lim_{\epsilon \to 0} \int_{\mathbb{R}^n \setminus B_\epsilon(x)} \frac{G(u(x)-u(z))}{|x-z|^{n+\alpha}} dz = f(x, u)$$

(see [CaS1] for the introductions of these operators), so far as we know, there has neither been any *extension method* nor *integral equation method* that work for these kinds of operators. This motivates us to come up with direct approaches on general nonlocal operators.

6.3.2 Key Ingredients in the Direct Application of the Method of Moving Planes

Recently, some success on the direct approach was achieved in [JW] by Jarohs and Weth. There they introduced *antisymmetric maximum principles* and applied them to carry out the *method of moving planes* directly on nonlocal problems to show the symmetry of solutions. The operators they considered are quite general, however, their maximum principles only work for bounded regions Ω, and they only studied weak solutions defined by $H^{\alpha/2}(\Omega)$ inner product.

Here we will develop a systematical approach to carry out the *method of moving planes* for nonlocal problems, either on bounded or unbounded domains. Recall that in the previous section when we focus on local elliptic operators, the argument is mainly based on the *Narrow Region Principle* and the *Decay at Infinity*. These two approaches were introduced decades ago in

[CL] and then summarized in the book [CL1]. Over the years, they have been used extensively by many researchers to solve various problems. For the fractional Laplacian and more general nonlocal operators, a parallel system is established in [CLL] by very elementary methods, so that it can be conveniently applied to various nonlocal problems. The main theorems and how they fit in the framework of the method of moving planes are illustrated in the following.

As usual, let

$$T_\lambda = \{x \in \mathbb{R}^n \mid x_1 = \lambda, \text{ for some } \lambda \in \mathbb{R}\}$$

be the moving planes,

$$\Sigma_\lambda = \{x \in \mathbb{R}^n \mid x_1 < \lambda\}$$

be the region to the left of the plane, and

$$x^\lambda = (2\lambda - x_1, x_2, \ldots, x_n)$$

be the reflection of x about the plane T_λ.

Assume that u is a solution of pseudo-differential equation (6.37) or (6.42). To compare the values of $u(x)$ with $u(x^\lambda)$, we denote

$$w_\lambda(x) = u(x^\lambda) - u(x).$$

The first step is to show that for λ sufficiently negative, we have

$$w_\lambda(x) \geq 0, \quad x \in \Sigma_\lambda. \tag{6.45}$$

This provides a starting point to move the plane. Then in the second step, we move the plane to the right as long as inequality (6.45) holds to its limiting position to show that u is symmetric about the limiting plane. A maximum principle is usually used to prove (6.45). Since w_λ is an anti-symmetric function:

$$w_\lambda(x) = -w_\lambda(x^\lambda),$$

we first prove (for simplicity of notation, in the following, we denote w_λ by w and Σ_λ by Σ).

Theorem 6.3.1 *(The Maximum Principle for the Anti-symmetric Functions) Let Ω be a bounded domain in Σ. Assume that $w \in L_\alpha \cap C^{1,1}_{loc}(\Omega)$ and is lower semi-continuous on $\bar{\Omega}$. If*

$$\begin{cases} (-\triangle)^{\alpha/2} w(x) \geq 0 & in \ \Omega, \\ w(x) \geq 0 & in \ \Sigma \backslash \Omega, \\ w(x^\lambda) = -w(x) & in \ \Sigma, \end{cases}$$

then

$$w(x) \geq 0 \ in \ \Omega.$$

If $w = 0$ at some point in Ω, then

$$w(x) = 0 \ \text{almost everywhere in } \mathbb{R}^n.$$

These conclusions hold for unbounded region Ω if we further assume that

$$\lim_{|x| \to \infty} w(x) \geq 0.$$

In many cases, w may not satisfy the equation

$$(-\triangle)^{\alpha/2} w \geq 0$$

as required in the previous theorem. However one can derive that

$$(-\triangle)^{\alpha/2} w + c(x) w(x) \geq 0$$

for some function $c(x)$ depending on u. Fortunately, in the process of moving planes, each time we only need to move T_λ a little bit to the right, hence the increment of Σ_λ is a narrow region, and a maximum principle is easier to hold in a narrow region as you will see below.

Theorem 6.3.2 *(Narrow Region Principle) Let Ω be a bounded narrow region in Σ, such that it is contained in*

$$\{x| \ \lambda - \delta < x_1 < \lambda\}$$

with small δ. Suppose that $w \in L_\alpha \cap C_{loc}^{1,1}(\Omega)$ and is lower semi-continuous on $\bar{\Omega}$. If $c(x)$ is bounded from below in Ω and

$$\begin{cases} (-\triangle)^{\alpha/2} w(x) + c(x) w(x) \geq 0 & \text{in } \Omega, \\ w(x) \geq 0 & \text{in } \Sigma \backslash \Omega, \\ w(x^\lambda) = -w(x) & \text{in } \Sigma, \end{cases}$$

then for sufficiently small δ, we have

$$w(x) \geq 0 \ \text{in } \Omega.$$

This conclusion holds for unbounded region Ω if we further assume that

$$\lim_{|x| \to \infty} w(x) \geq 0.$$

As one will see from the proof of this theorem, the contradiction arguments are conducted at a negative minimum of w. Hence when working on an unbounded domain, one needs to ensure that such minima would be contained in a fixed ball for all solutions w. This can be done when $c(x)$ decays faster than $1/|x|^\alpha$ near infinity.

Theorem 6.3.3 *(Decay at Infinity) Let Ω be an unbounded region in Σ. Assume*

$$\begin{cases} (-\triangle)^{\alpha/2} w(x) + c(x)w(x) \geq 0 & \text{in } \Omega, \\ w(x) \geq 0 & \text{in } \Sigma \backslash \Omega, \\ w(x^\lambda) = -w(x) & \text{in } \Sigma, \end{cases}$$

with

$$\lim_{|x| \to \infty} |x|^\alpha c(x) \geq 0,$$

then there exists a constant $R_0 > 0$ for all solutions w, such that if

$$w(x^0) = \min_\Omega w(x) < 0,$$

then

$$|x^0| \leq R_0.$$

In Sections 6.4, 6.5, and 6.7, we prove various maximum principles needed in carrying out the method of moving planes.

In Section 6.8, 6.9, and 6.10, we use a series of examples to illustrate how these maximum principles can be used to obtain symmetry, monotonicity, and non-existence of positive solutions.

6.4 Various Maximum Principles for the Fractional Laplacian in $\mathbb{R}^n$

6.4.1 Maximum Principle for the Anti-Symmetric Functions

Theorem 6.4.1 *Let T be a hyperplane in $\mathbb{R}^n$. Without loss of generality, we may assume that*

$$T = \{x \in \mathbb{R}^n \mid x_1 = \lambda, \text{ for some } \lambda \in \mathbb{R}\}.$$

Let

$$\tilde{x} = (2\lambda - x_1, x_2, \dots, x_n)$$

be the reflection of x about the plane T. Denote

$$H = \{x \in \mathbb{R}^n \mid x_1 < \lambda\} \quad \text{and} \quad \tilde{H} = \{x \mid \tilde{x} \in H\}.$$

Let Ω be a bounded domain in H. Assume that $w \in L_\alpha \cap C^{1,1}_{loc}(\Omega)$ and is lower semi-continuous on $\bar\Omega$. If

$$\begin{cases} (-\triangle)^{\alpha/2} w(x) \geq 0 & \text{in } \Omega, \\ w(x) \geq 0 & \text{in } H \backslash \Omega, \\ w(\tilde{x}) = -w(x) & \text{in } H, \end{cases} \tag{6.46}$$

then

$$w(x) \geq 0 \text{ in } \Omega. \tag{6.47}$$

If $w = 0$ at some point in Ω, then

$$w(x) = 0 \quad \text{almost everywhere in } \mathbb{R}^n.$$

These conclusions hold for unbounded region Ω if we further assume that

$$\lim_{|x| \to \infty} w(x) \geq 0.$$

Proof. If (6.47) does not hold, then the lower semi-continuity of w on $\bar{\Omega}$ indicates that there exists an $x^0 \in \bar{\Omega}$ such that

$$w(x^0) = \min_{\bar{\Omega}} w < 0.$$

And one can further deduce from condition (6.46) that x^0 is in the interior of Ω.

It follows that

$$
\begin{aligned}
(-\triangle)^{\alpha/2} w(x^0) &= C_{n,\alpha} PV \int_{\mathbb{R}^n} \frac{w(x^0) - w(y)}{|x^0 - y|^{n+\alpha}} dy \\
&= C_{n,\alpha} PV \left\{ \int_H \frac{w(x^0) - w(y)}{|x^0 - y|^{n+\alpha}} dy + \int_{\tilde{H}} \frac{w(x^0) - w(y)}{|x^0 - y|^{n+\alpha}} dy \right\} \\
&= C_{n,\alpha} PV \left\{ \int_H \frac{w(x^0) - w(y)}{|x^0 - y|^{n+\alpha}} dy + \int_H \frac{w(x^0) - w(\tilde{y})}{|x^0 - \tilde{y}|^{n+\alpha}} dy \right\} \\
&= C_{n,\alpha} PV \left\{ \int_H \frac{w(x^0) - w(y)}{|x^0 - y|^{n+\alpha}} dy + \int_H \frac{w(x^0) + w(y)}{|x^0 - \tilde{y}|^{n+\alpha}} dy \right\} \\
&\leq C_{n,\alpha} \int_H \left\{ \frac{w(x^0) - w(y)}{|x^0 - \tilde{y}|^{n+\alpha}} + \frac{w(x^0) + w(y)}{|x^0 - \tilde{y}|^{n+\alpha}} \right\} dy \\
&= C_{n,\alpha} \int_H \frac{2w(x^0)}{|x^0 - \tilde{y}|^{n+\alpha}} dy \\
&< 0,
\end{aligned}
$$

which contradicts with inequality (6.46). This verifies (6.47).

Now we have shown that $w \geq 0$ in H. If there is some point $x^o \in \Omega$, such that $w(x^o) = 0$, then

$$
\begin{aligned}
0 \leq (-\triangle)^{\alpha/2} w(x^o) &= C_{n,\alpha} PV \int_{\mathbb{R}^n} \frac{-w(y)}{|x^0 - y|^{n+\alpha}} dy \\
&= C_{n,\alpha} PV \left\{ \int_H \frac{-w(y)}{|x^0 - y|^{n+\alpha}} dy + \int_{\tilde{H}} \frac{-w(y)}{|x^0 - y|^{n+\alpha}} dy \right\} \\
&= C_{n,\alpha} PV \left\{ \int_H \frac{-w(y)}{|x^0 - y|^{n+\alpha}} dy + \int_H \frac{-w(\tilde{y})}{|x^0 - \tilde{y}|^{n+\alpha}} dy \right\} \\
&= C_{n,\alpha} PV \left\{ \int_H \frac{-w(y)}{|x^0 - y|^{n+\alpha}} dy + \int_H \frac{w(y)}{|x^0 - \tilde{y}|^{n+\alpha}} dy \right\} \\
&= C_{n,\alpha} PV \int_H w(y) \left\{ \frac{1}{|x^0 - \tilde{y}|^{n+\alpha}} - \frac{1}{|x^0 - y|^{n+\alpha}} \right\} dy \\
&\leq 0,
\end{aligned}
$$

since

$$\frac{1}{|x^0 - \tilde{y}|^{n+\alpha}} - \frac{1}{|x^0 - y|^{n+\alpha}} < 0.$$

This implies that

$$w(y) \equiv 0 \quad \text{in } H.$$

Then by the antisymmetry of w, we derive immediately that

$$w(x) = 0 \quad \text{almost everywhere in } \mathbb{R}^n.$$

This proves the theorem.

6.4.2 Narrow Region Principle

Theorem 6.4.2 *Let T be a hyperplane in $\mathbb{R}^n$. Without loss of generality, we may assume that*

$$T = \{x = (x_1, x') \in \mathbb{R}^n \mid x_1 = \lambda, \ \text{for some } \lambda \in \mathbb{R}\}.$$

Let

$$\tilde{x} = (2\lambda - x_1, x_2, \ldots, x_n),$$

$$H = \{x \in \mathbb{R}^n \mid x_1 < \lambda\}, \quad \tilde{H} = \{x \mid \tilde{x} \in H\}.$$

Let Ω be a bounded narrow region in H, such that it is contained in $\{x \mid \lambda - l < x_1 < \lambda\}$ with small l. Suppose that $w \in L_\alpha \cap C^{1,1}(\Omega)$ and is lower semi-continuous on $\bar{\Omega}$. If $c(x)$ is bounded from below in Ω and

$$\begin{cases} (-\triangle)^{\alpha/2} w(x) + c(x)w(x) \geq 0 & \text{in } \Omega, \\ w(x) \geq 0 & \text{in } H \backslash \Omega, \\ w(\tilde{x}) = -w(x) & \text{in } H, \end{cases} \tag{6.48}$$

then for sufficiently small l, we have

$$w(x) \geq 0 \text{ in } \Omega. \tag{6.49}$$

This conclusion holds for unbounded region Ω if we further assume that

$$\lim_{|x| \to \infty} w(x) \geq 0.$$

Proof. If (6.49) does not hold, then the lower semi-continuity of w on $\bar{\Omega}$ indicates that there exists an $x^0 \in \bar{\Omega}$ such that

$$w(x^0) = \min_{\bar{\Omega}} w < 0.$$

And one can further deduce from condition (6.48) that x^0 is in the interior of Ω.

Then it follows that

$$(-\triangle)^{\alpha/2}w(x^0) = C_{n,\alpha}PV\int_{\mathbb{R}^n}\frac{w(x^0)-w(y)}{|x^0-y|^{n+\alpha}}dy$$

$$= C_{n,\alpha}PV\left\{\int_H\frac{w(x^0)-w(y)}{|x^0-y|^{n+\alpha}}dy+\int_{\tilde{H}}\frac{w(x^0)-w(y)}{|x^0-y|^{n+\alpha}}dy\right\}$$

$$= C_{n,\alpha}PV\left\{\int_H\frac{w(x^0)-w(y)}{|x^0-y|^{n+\alpha}}dy+\int_H\frac{w(x^0)-w(\tilde{y})}{|x^0-\tilde{y}|^{n+\alpha}}dy\right\}$$

$$= C_{n,\alpha}PV\left\{\int_H\frac{w(x^0)-w(y)}{|x^0-y|^{n+\alpha}}dy+\int_H\frac{w(x^0)+w(y)}{|x^0-\tilde{y}|^{n+\alpha}}dy\right\}$$

$$\leq C_{n,\alpha}\int_H\left\{\frac{w(x^0)-w(y)}{|x^0-\tilde{y}|^{n+\alpha}}+\frac{w(x^0)+w(y)}{|x^0-\tilde{y}|^{n+\alpha}}\right\}dy$$

$$= C_{n,\alpha}\int_H\frac{2w(x^0)}{|x^0-\tilde{y}|^{n+\alpha}}dy.$$

To estimate the integral above, given that x^o lies in a narrow region close to the hyperplane $P = \{y \in \mathbb{R}^n \mid y_1 = \lambda\}$, as an example, we may first consider the extreme case where $x_1^o = \lambda$. Then it's easy to see that

$$\int_H\frac{1}{|x^o-\tilde{y}|^{n+\alpha}}dy = \infty.$$

This suggests that we may obtain values arbitrarily large by integrating on a domain that is sufficiently close to the hyperplane P.

$$\int_H\frac{1}{|x^o-\tilde{y}|^{n+\alpha}}dy \geq \int_{H\cap(B_1(x^o)\backslash B_l(x^o))}\frac{1}{|x^o-\tilde{y}|^{n+\alpha}}dy$$

$$\geq \frac{C}{l^\alpha}. \tag{6.50}$$

Since $c(x)$ is lower bounded in Ω, by choosing l sufficiently small, it holds that

$$(-\triangle)^{\alpha/2}w(x^0) + c(x^0)w(x^0) < 0.$$

This is a contradiction with condition (6.48). Therefore, (6.49) must be true.

6.4.3 Decay at Infinity

Theorem 6.4.3 *Let $H = \{x \in \mathbb{R}^n \mid x_1 < \lambda$ for some $\lambda \in \mathbb{R}\}$ and let Ω be an unbounded region in H. Assume*

$$\begin{cases} (-\triangle)^{\alpha/2}w(x) + c(x)w(x) \geq 0 & in\ \Omega, \\ w(x) \geq 0 & in\ H\backslash\Omega, \\ w(\tilde{x}) = -w(x) & in\ H, \end{cases} \tag{6.51}$$

with

$$\lim_{|x|\to\infty} |x|^\alpha c(x) \geq 0, \tag{6.52}$$

then there exists a constant $R_0 > 0$ for all solutions w, such that if

$$w(x^0) = \min_{\Omega} w(x) < 0, \tag{6.53}$$

then

$$|x^0| \leq R_0. \tag{6.54}$$

Proof. It follows from (6.51) and (6.53) that

$$(-\triangle)^{\alpha/2} w(x^0) = C_{n,\alpha} PV \int_{\mathbb{R}^n} \frac{w(x^0) - w(y)}{|x^0 - y|^{n+\alpha}} dy$$

$$= C_{n,\alpha} PV \left\{ \int_H \frac{w(x^0) - w(y)}{|x^0 - y|^{n+\alpha}} dy + \int_{\tilde{H}} \frac{w(x^0) - w(y)}{|x^0 - y|^{n+\alpha}} dy \right\}$$

$$= C_{n,\alpha} PV \left\{ \int_H \frac{w(x^0) - w(y)}{|x^0 - y|^{n+\alpha}} dy + \int_H \frac{w(x^0) - w(\tilde{y})}{|x^0 - \tilde{y}|^{n+\alpha}} dy \right\}$$

$$= C_{n,\alpha} PV \left\{ \int_H \frac{w(x^0) - w(y)}{|x^0 - y|^{n+\alpha}} dy + \int_H \frac{w(x^0) + w(y)}{|x^0 - \tilde{y}|^{n+\alpha}} dy \right\}$$

$$\leq C_{n,\alpha} \int_H \left\{ \frac{w(x^0) - w(y)}{|x^0 - \tilde{y}|^{n+\alpha}} + \frac{w(x^0) + w(y)}{|x^0 - \tilde{y}|^{n+\alpha}} \right\} dy$$

$$= C_{n,\alpha} \int_H \frac{2w(x^0)}{|x^0 - \tilde{y}|^{n+\alpha}} dy.$$

For each fixed λ, when $|x^0| \geq \lambda$, we have $B_{|x^0|}(x^1) \subset \tilde{H}$ with $x^1 = (3|x^0| + x_1^0, (x^0)')$, and it follows that

$$\int_H \frac{1}{|x^0 - \tilde{y}|^{n+\alpha}} dy \geq \int_{B_{|x^0|}(x^1)} \frac{1}{|x^0 - y|^{n+\alpha}} dy$$

$$\geq \int_{B_{|x^0|}(x^1)} \frac{1}{4^{n+\alpha}|x^0|^{n+\alpha}} dy$$

$$= \frac{\omega_n}{4^{n+\alpha}|x^0|^{\alpha}}.$$

Then we have

$$0 \leq (-\triangle)^{\alpha/2} w(x^0) + c(x^0) w(x^0)$$

$$\leq \left[\frac{2\omega_n C_{n,\alpha}}{4^{n+\alpha}|x^0|^{\alpha}} + c(x^0) \right] w(x^0).$$

Or equivalently,

$$\frac{2\omega_n C_{n,\alpha}}{4^{n+\alpha}|x^0|^{\alpha}} + c(x^0) \leq 0.$$

Now if $|x^0|$ is sufficiently large, this would contradict (6.52). Therefore, (6.54) holds. This verifies the theorem.

Remark 6.4.1 *From the proof, one can see that the inequality*

$$(-\triangle)^{\alpha/2} w(x) + c(x) w(x) \geq 0$$

and condition (6.52) are only required at points where w is negative.

6.5 Various Maximum Principles for the Fractional Laplacian in $\mathbb{R}^n_+$

6.5.1 A Maximum Principle for the Anti-Symmetric Functions

When dealing with Dirichlet problems on a upper half space $\mathbb{R}^n_+$, such as

$$\begin{cases} (-\triangle)^{\alpha/2}u = u^p(x), \; x \in \mathbb{R}^n_+, \\ u(x) = 0, \qquad\qquad x \in \mathbb{R}^n_- \equiv \{x \mid x_n \le 0\}, \end{cases} \tag{6.55}$$

we can move the plane perpendicular to x_n-axis upward to show that the solutions are monotone increasing along x_n-direction, hence derive a contradiction if we have certain decay condition on u, and this will lead to non-existence of solutions.

Set
$$T = \{x \in \mathbb{R}^n_+ \mid x_n = \lambda, \text{ for some } \lambda \in \mathbb{R}_+\}.$$

Let
$$\tilde{x} = (x_1, x_2, \ldots, 2\lambda - x_n)$$

be the reflection of x about the plane T. Denote

$$H = \{x \in \mathbb{R}^n_+ \mid 0 < x_n < \lambda\},$$

the region between the boundary $\partial\mathbb{R}^n_+$ and T, and

$$\tilde{H} = \{x \in \mathbb{R}^n_+ \mid \tilde{x} \in H\},$$

the reflection of H about the plane T. Again we compare the value of $u(\tilde{x})$ with $u(x)$. Let $w(x) = u(\tilde{x}) - u(x)$, then obviously w satisfies

$$(-\triangle)^{\alpha/2}w(x) + c(x)w(x) = 0, \; \; x \in H \tag{6.56}$$

for some $c(x)$ depending on $u(x)$.

The main difference between the whole space and the half space is that, in the whole space, equation (6.56) holds in the whole lower half part of T, i.e. in $H \cup \mathbb{R}^n_-$; while in the half space here, it only holds in H, or sometimes, only in a subregion of H. Hence the *narrow region principle*, the *decay at infinity*, and other *maximum principles* we derived in the previous section cannot be applied directly here, and we need to prove suitable versions of *maximum principles* that work in this situation.

We first start with a simple maximum principle for anti-symmetric functions.

Theorem 6.5.1 *Let Ω be a bounded domain in H. Assume that $w \in L_\alpha \cap C^{1,1}_{loc}(\Omega)$ and is lower semi-continuous on $\bar{\Omega}$. If*

$$\begin{cases} (-\triangle)^{\alpha/2}w(x) \ge 0 & \text{in } \Omega, \\ w(x) \ge 0 & \text{in } (H \backslash \Omega) \cup \mathbb{R}^n_-, \\ w(\tilde{x}) = -w(x) & \text{in } H \cup \mathbb{R}^n_-, \end{cases} \tag{6.57}$$

then

$$w(x) \geq 0 \ \text{in } \Omega. \tag{6.58}$$

Furthermore, if $w = 0$ at some point in Ω, then

$$w(x) = 0 \ \text{almost everywhere in } \mathbb{R}^n.$$

These conclusions hold for unbounded region Ω if we further assume that

$$\varliminf_{|x| \to \infty} w(x) \geq 0.$$

Proof. If (6.58) does not hold, then the lower semi-continuity of w on $\bar{\Omega}$ indicates that there exists an $x^o \in \bar{\Omega}$ such that

$$w(x^o) = \min_{\bar{\Omega}} w < 0. \tag{6.59}$$

And one can further deduce from condition (6.57) that x^o is in the interior of Ω.

It follows that

$$(-\triangle)^{\alpha/2} w(x^o)$$
$$= C_{n,\alpha} PV \int_{\mathbb{R}^n} \frac{w(x^o) - w(y)}{|x^o - y|^{n+\alpha}} dy$$
$$= C_{n,\alpha} PV \left\{ \int_{H \cup \tilde{H}} \frac{w(x^o) - w(y)}{|x^o - y|^{n+\alpha}} dy + \int_{\mathbb{R}^n_- \cup \{y_n > 2\lambda\}} \frac{w(x^o) - w(y)}{|x^o - y|^{n+\alpha}} dy \right\}$$
$$= I_1 + I_2.$$

Through elementary calculations, by (6.59) we have

$$I_1 = C_{n,\alpha} PV \left\{ \int_H \frac{w(x^o) - w(y)}{|x^o - y|^{n+\alpha}} dy + \int_{\tilde{H}} \frac{w(x^o) - w(y)}{|x^o - y|^{n+\alpha}} dy \right\}$$
$$= C_{n,\alpha} PV \left\{ \int_H \frac{w(x^o) - w(y)}{|x^o - y|^{n+\alpha}} dy + \int_H \frac{w(x^o) - w(\tilde{y})}{|x^o - \tilde{y}|^{n+\alpha}} dy \right\}$$
$$= C_{n,\alpha} PV \left\{ \int_H \frac{w(x^o) - w(y)}{|x^o - y|^{n+\alpha}} dy + \int_H \frac{w(x^o) + w(y)}{|x^o - \tilde{y}|^{n+\alpha}} dy \right\}$$
$$\leq C_{n,\alpha} \int_H \left\{ \frac{w(x^o) - w(y)}{|x^o - \tilde{y}|^{n+\alpha}} + \frac{w(x^o) + w(y)}{|x^o - \tilde{y}|^{n+\alpha}} \right\} dy$$
$$= C_{n,\alpha} \int_H \frac{2w(x^o)}{|x^o - \tilde{y}|^{n+\alpha}} dy$$
$$< 0,$$

and

$$I_2 = C_{n,\alpha} PV \left\{ \int_{\mathbb{R}^n_-} \frac{w(x^o) - w(y)}{|x^o - y|^{n+\alpha}} dy + \int_{\{y_n > 2\lambda\}} \frac{w(x^o) - w(y)}{|x^o - y|^{n+\alpha}} dy \right\}$$

$$= C_{n,\alpha} PV \left\{ \int_{\mathbb{R}^n_-} \frac{w(x^o) - w(y)}{|x^o - y|^{n+\alpha}} dy + \int_{\mathbb{R}^n_-} \frac{w(x^o) - w(\tilde{y})}{|x^o - \tilde{y}|^{n+\alpha}} dy \right\}$$

$$= C_{n,\alpha} PV \left\{ \int_{\mathbb{R}^n_-} \frac{w(x^o) - w(y)}{|x^o - y|^{n+\alpha}} dy + \int_{\mathbb{R}^n_-} \frac{w(x^o) + w(y)}{|x^o - \tilde{y}|^{n+\alpha}} dy \right\}$$

$$\leq C_{n,\alpha} \int_{\mathbb{R}^n_-} \left\{ \frac{w(x^o) - w(y)}{|x^o - \tilde{y}|^{n+\alpha}} + \frac{w(x^o) + w(y)}{|x^o - \tilde{y}|^{n+\alpha}} \right\} dy$$

$$= C_{n,\alpha} \int_{\mathbb{R}^n_-} \frac{2w(x^o)}{|x^o - \tilde{y}|^{n+\alpha}} dy$$

$$< 0.$$

Therefore,

$$(-\triangle)^{\alpha/2} w(x^o) = I_1 + I_2 < 0,$$

which contradicts inequality (6.57). This verifies (6.58).

Now we have shown that $w \geq 0$ in $H \cup \mathbb{R}^n_-$. If there is some point $x^o \in \Omega$, such that $w(x^o) = 0$, then

$$0 \leq (-\triangle)^{\alpha/2} w(x^o) = C_{n,\alpha} PV \left\{ \int_{H \cup \tilde{H}} \frac{-w(y)}{|x^0 - y|^{n+\alpha}} dy + \int_{\mathbb{R}^n_- \cup \{y_n > 2\lambda\}} \frac{-w(y)}{|x^0 - y|^{n+\alpha}} dy \right\}$$

$$= C_{n,\alpha} PV \int_H w(y) \left\{ \frac{1}{|x^0 - \tilde{y}|^{n+\alpha}} - \frac{1}{|x^0 - y|^{n+\alpha}} \right\} dy$$

$$+ C_{n,\alpha} PV \int_{\mathbb{R}^n_-} w(y) \left\{ \frac{1}{|x^0 - \tilde{y}|^{n+\alpha}} - \frac{1}{|x^0 - y|^{n+\alpha}} \right\} dy$$

$$\leq 0,$$

since

$$\frac{1}{|x^0 - \tilde{y}|^{n+\alpha}} - \frac{1}{|x^0 - y|^{n+\alpha}} < 0.$$

This implies that

$$w(y) \equiv 0 \quad \text{in } H \cup \mathbb{R}^n_-.$$

Then by the antisymmetry of w, we derive immediately that

$$w(x) = 0 \quad \text{almost everywhere in } \mathbb{R}^n.$$

This completes the proof.

6.5.2 Narrow Region Principle

Theorem 6.5.2 *Let Ω be a bounded narrow region in H, such that it is contained in $\{x \in H \mid \lambda - l < x_n < \lambda\}$ with small l. Assume that $w \in L_\alpha \cap C^{1,1}_{loc}(\Omega)$ and is lower semi-continuous on $\bar{\Omega}$. If $c(x)$ is bounded from below in Ω and*

$$\begin{cases} (-\triangle)^{\alpha/2}w(x) + c(x)w(x) \geq 0 & in\ \Omega, \\ w(x) \geq 0 & in\ (H\backslash\Omega) \cup \mathbb{R}^n_-, \\ w(\tilde{x}) = -w(x) & in\ H \cup \mathbb{R}^n_-, \end{cases} \qquad (6.60)$$

then for sufficiently small l, then we have

$$w(x) \geq 0\ in\ \Omega. \qquad (6.61)$$

Furthermore, if $w = 0$ at some point in Ω, then

$$w(x) = 0\ \ almost\ everywhere\ in\ \mathbb{R}^n.$$

These conclusions hold for unbounded region Ω if we further assume that

$$\lim_{|x|\to\infty} w(x) \geq 0.$$

Proof. If (6.61) does not hold, then the lower semi-continuity of w on $\bar{\Omega}$ indicates that there exists an $x^o \in \bar{\Omega}$ such that

$$w(x^o) = \min_{\bar{\Omega}} w < 0.$$

And one can further deduce from condition (6.60) that x^o is in the interior of Ω.

Through same calculations in Theorem 6.5.1, we have

$$\begin{aligned} &(-\triangle)^{\alpha/2}w(x^o) \\ &= C_{n,\alpha}PV \int_{\mathbb{R}^n} \frac{w(x^o) - w(y)}{|x^o - y|^{n+\alpha}}dy \\ &= C_{n,\alpha}PV \left\{ \int_{H \cup \tilde{H}} \frac{w(x^o) - w(y)}{|x^o - y|^{n+\alpha}}dy + \int_{\mathbb{R}^n \cup \{y_n > 2\lambda\}} \frac{w(x^o) - w(y)}{|x^o - y|^{n+\alpha}}dy \right\} \\ &\leq C_{n,\alpha} \int_{H \cup \mathbb{R}^n_-} \frac{2w(x^o)}{|x^o - \tilde{y}|^{n+\alpha}}dy. \end{aligned}$$

To estimate the integral above, given that x^o lies in a narrow region close to the hyperplane $P = \{y \in \mathbb{R}^n \mid y_n = \lambda\}$, as an example, we may first consider the extreme case where $x^o_n = \lambda$. Then it's easy to see that

$$\begin{aligned} \int_{H \cup \mathbb{R}^n_-} \frac{1}{|x^o - \tilde{y}|^{n+\alpha}}dy &= \int_{\tilde{H} \cup \{y_n > 2\lambda\}} \frac{1}{|x^o - y|^{n+\alpha}}dy \\ &> \int_{\tilde{H}} \frac{1}{|x^o - y|^{n+\alpha}}dy \\ &= \infty. \end{aligned}$$

This suggests that we may obtain values arbitrarily large by integrating on a domain that is sufficiently close to the hyperplane P. Indeed

$$\int_{\tilde{H} \cup \{y_n > 2\lambda\}} \frac{1}{|x^o - y|^{n+\alpha}} dy \geq \int_{\tilde{H} \cap (B_1(x^o) \setminus B_l(x^o))} \frac{1}{|x^o - y|^{n+\alpha}} dy$$

$$\geq C \int_{B_1(x^o) \setminus B_l(x^o)} \frac{1}{|x^o - y|^{n+\alpha}} dy$$

$$\geq \frac{C}{l^\alpha}. \tag{6.62}$$

Since $c(x)$ is lower bounded in Ω, by choosing l sufficiently small, it holds that

$$(-\triangle)^{\alpha/2} w(x^o) + c(x^o) w(x^o) < 0.$$

This is a contradiction with condition (6.60). Therefore, (6.61) must be true.

6.6 Applications to Equations Involving the Fractional Laplacian

6.6.1 Radial Symmetry of $(-\triangle)^{\alpha/2} u(x) = u^p(x)$, $x \in \mathbb{R}^n$

Theorem 6.6.1 *Assume that $u \in L_\alpha \cap C_{loc}^{1,1}$ and*

$$(-\triangle)^{\alpha/2} u(x) = u^p(x), \ x \in \mathbb{R}^n. \tag{6.63}$$

Then

(i) in the subcritical case $1 < p < \frac{n+\alpha}{n-\alpha}$, (6.63) has no positive solution;

(ii) in the critical case $p = \frac{n+\alpha}{n-\alpha}$, the positive solutions must be radially symmetric and monotone decreasing about some point in $\mathbb{R}^n$.

Proof. With no decay assumption on u at infinity, we are not able to carry out the *method of moving planes* on u directly. To circumvent this difficulty, we make a Kelvin transform.

Let x^0 be a point in $\mathbb{R}^n$, and

$$\bar{u}(x) = \frac{1}{|x - x^0|^{n-\alpha}} u \left(\frac{x - x^0}{|x - x^0|^2} + x^0 \right), \ x \in \mathbb{R}^n \setminus \{x^0\},$$

be the Kelvin transform of u centered at x^0. It is well-known that

$$(-\triangle)^{\alpha/2} \bar{u}(x) = \frac{1}{|x - x^0|^{n+\alpha}} \left((-\triangle)^{\alpha/2} u \right) \left(\frac{x - x^0}{|x - x^0|^2} + x^0 \right). \tag{6.64}$$

For the readers' convenience, we prove (6.64) in the appendix. By (6.63) we know

$$(-\triangle)^{\alpha/2} \bar{u}(x) = \frac{\bar{u}^p(x)}{|x - x^0|^\tau}, \ x \in \mathbb{R}^n \setminus \{x^0\} \tag{6.65}$$

with $\tau = n + \alpha - p(n - \alpha)$. Obviously, $\tau = 0$ in the critical case.

Choose any direction to be the x_1 direction. For $\lambda < x_1^0$, let

$$T_\lambda = \{x \in \mathbb{R}^n \mid x_1 = \lambda\}, \quad x^\lambda = (2\lambda - x_1, x'),$$

$$\bar{u}_\lambda(x) = \bar{u}(x^\lambda), \quad w_\lambda(x) = \bar{u}_\lambda(x) - \bar{u}(x),$$

and

$$\Sigma_\lambda = \{x \in \mathbb{R}^n \mid x_1 < \lambda\}, \quad \tilde{\Sigma}_\lambda = \{x^\lambda \mid x \in \Sigma_\lambda\}.$$

First, notice that, by the definition of w_λ, we have

$$\lim_{|x| \to \infty} w_\lambda(x) = 0.$$

Hence, if w_λ is negative somewhere in Σ_λ, then the negative minima of w_λ are attained in the interior of Σ_λ.

Let

$$\Sigma_\lambda^- = \{x \in \Sigma_\lambda \mid w_\lambda(x) < 0\}.$$

Then from (6.65), we have, for $x \in \Sigma_\lambda^- \backslash \{(x^0)^\lambda\}$,

$$\begin{aligned}
(-\triangle)^{\alpha/2} w_\lambda(x) &= \frac{\bar{u}_\lambda^p(x)}{|x^\lambda - x^0|^\tau} - \frac{\bar{u}^p(x)}{|x - x^0|^\tau} \\
&\geq \frac{\bar{u}_\lambda^p(x) - \bar{u}^p(x)}{|x - x^0|^\tau} \\
&\geq \frac{p\bar{u}^{p-1}(x) w_\lambda(x)}{|x - x^0|^\tau};
\end{aligned}$$

that is,

$$(-\triangle)^{\alpha/2} w_\lambda(x) + c(x) w_\lambda(x) \geq 0, \quad x \in \Sigma_\lambda^- \backslash \{(x^0)^\lambda\}, \tag{6.66}$$

with

$$c(x) = -\frac{p\bar{u}^{p-1}(x)}{|x - x^0|^\tau}. \tag{6.67}$$

The Subcritical Case

For $1 < p < \frac{n+\alpha}{n-\alpha}$, we show that (6.63) admits no positive solution.

Step 1. We show that, for λ sufficiently negative,

$$w_\lambda(x) \geq 0, \quad \text{in } \Sigma_\lambda \backslash \{(x^0)^\lambda\}. \tag{6.68}$$

This is done by using Theorem 6.4.3 (*decay at infinity*).

First, we claim that for λ sufficiently negative, there exists some $\epsilon > 0$ and $c_\lambda > 0$, such that

$$w_\lambda(x) \geq c_\lambda, \quad \forall x \in B_\epsilon((x^0)^\lambda) \backslash \{(x^0)^\lambda\}. \tag{6.69}$$

For the proof, please see the Lemma A.1.2 in the Appendix, or it can be derived from Theorem 2 in [LWX]. From (6.69), one can see that Σ_λ^- has no intersection with $B_\epsilon((x^0)^\lambda)$.

From (6.67), it is easy to verify that, for $|x|$ sufficiently large,

$$c(x) \sim \frac{1}{|x|^{2\alpha}}.$$ (6.70)

Hence $c(x)$ satisfies condition (6.52) in Theorem 6.4.3. Applying Theorem 6.4.3 to w_λ with

$$H = \Sigma_\lambda \backslash \{(x^0)^\lambda\} \text{ and } \Omega = \Sigma_\lambda^-$$

for any sufficiently small ϵ, we conclude that, there exists a $R_o > 0$ (independent of λ), such that if $\bar{x}$ is a negative minimum of w_λ in Σ_λ, then

$$|\bar{x}| \leq R_o.$$ (6.71)

Now for $\lambda \leq -R_o$, we must have

$$w_\lambda(x) \geq 0, \ \forall x \in \Sigma_\lambda \backslash \{(x^0)^\lambda\}.$$

This verifies (6.68).

Step 2. *Step 1* provides a starting point, from which we can now move the plane T_λ to the right as long as (6.68) holds to its limiting position.

Let

$$\lambda_0 = \sup\{\lambda < x_1^0 \mid w_\mu(x) \geq 0, \ \forall x \in \Sigma_\mu \backslash \{(x^0)^\mu\}, \mu \leq \lambda\}.$$

In this part, we show that

$$\lambda_0 = x_1^0$$

and

$$w_{\lambda_0}(x) \equiv 0, \quad x \in \Sigma_{\lambda_0}.$$ (6.72)

Suppose that

$$\lambda_0 < x_1^0,$$

we show that the plane T_λ can be moved further to the right. To be more rigorous, there exists some $\epsilon > 0$, such that for any $\lambda \in (\lambda_0, \lambda_0 + \epsilon)$, we have

$$w_\lambda(x) \geq 0, \quad x \in \Sigma_\lambda \backslash \{(x^0)^\lambda\}.$$ (6.73)

This is a contradiction with the definition of λ_0. Hence we must have

$$\lambda_0 = x_1^0.$$ (6.74)

Now we prove (6.73) by the combining use of *narrow region principle* and *decay at infinity*.

Again we need the fact (see Lemma A.1.3 in the Appendix or Theorem 2 in [LWX]) that there exists some $c_o > 0$ such that for sufficiently small η

$$w_{\lambda_0}(x) \geq c_o, \ \forall x \in B_\eta((x^0)^{\lambda_0}) \backslash \{(x^0)^{\lambda_0}\}.$$ (6.75)

As a result of (6.71), the negative minimum of w_λ cannot be attained outside of $B_{R_o}(0)$. Next we argue that it can neither be attained inside of $B_{R_o}(0)$. Specifically, we show that for λ sufficiently close to λ_0,

$$w_\lambda(x) \geq 0, \quad \forall x \in (\Sigma_\lambda \cap B_{R_o}(0)) \backslash \{(x^0)^\lambda\}. \tag{6.76}$$

Recall that in the *narrow region principle* (Theorem 6.4.2), we show that there is a small $\delta > 0$, such that for $\lambda \in [\lambda_0, \lambda_0 + \delta)$, if

$$w_\lambda(x) \geq 0, \quad \forall x \in \Sigma_{\lambda_0 - \delta} \backslash \{(x^0)^\lambda\}, \tag{6.77}$$

then

$$w_\lambda(x) \geq 0, \quad \forall x \in (\Sigma_\lambda \backslash \Sigma_{\lambda_0-\delta}) \backslash \{(x^0)^\lambda\}. \tag{6.78}$$

To obtain (6.78) in our case, in Theorem 6.4.2, we let

$$H = \Sigma_\lambda \text{ and the narrow region } \Omega = (\Sigma_\lambda^- \backslash \Sigma_{\lambda_0-\delta}),$$

while the lower bound of $c(x)$ can be seen from (6.70).

Then what's left is to show (6.77). Actually we only need

$$w_\lambda(x) \geq 0, \quad \forall x \in (\Sigma_{\lambda_0-\delta} \cap B_{R_o}(0)) \backslash \{(x^0)^\lambda\}. \tag{6.79}$$

In fact, when $\lambda_0 < x_1^0$, we have

$$w_{\lambda_0}(x) > 0, \quad x \in \Sigma_{\lambda_0} \backslash \{(x^0)^{\lambda_0}\}. \tag{6.80}$$

If not, there exists some $\hat{x}$ such that

$$w_{\lambda_0}(\hat{x}) = \min_{\Sigma_{\lambda_0}} w_{\lambda_0}(x) = 0.$$

It follows that

$$(-\triangle)^{\alpha/2} w_{\lambda_0}(\hat{x}) = C_{n,\alpha} PV \int_{\mathbb{R}^n} \frac{-w_{\lambda_0}(y)}{|\hat{x} - y|^{n+\alpha}} dy$$

$$= C_{n,\alpha} PV \left\{ \int_{\Sigma_{\lambda_0}} \frac{-w_{\lambda_0}(y)}{|\hat{x} - y|^{n+\alpha}} dy + \int_{\mathbb{R}^n \backslash \Sigma_{\lambda_0}} \frac{-w_{\lambda_0}(y)}{|\hat{x} - y|^{n+\alpha}} dy \right\}$$

$$= C_{n,\alpha} PV \left\{ \int_{\Sigma_{\lambda_0}} \frac{-w_{\lambda_0}(y)}{|\hat{x} - y|^{n+\alpha}} dy + \int_{\Sigma_{\lambda_0}} \frac{w_{\lambda_0}(y)}{|\hat{x} - y^\lambda|^{n+\alpha}} dy \right\}$$

$$= C_{n,\alpha} PV \int_{\Sigma_{\lambda_0}} \left(\frac{1}{|\hat{x} - y^\lambda|^{n+\alpha}} - \frac{1}{|\hat{x} - y|^{n+\alpha}} \right) w_{\lambda_0}(y) dy$$

$$\leq 0. \tag{6.81}$$

On the other hand

$$(-\triangle)^{\alpha/2} w_{\lambda_0}(\hat{x}) = \frac{\bar{u}_{\lambda_0}^p(\hat{x})}{|\hat{x}^{\lambda_0} - x^0|^\tau} - \frac{\bar{u}^p(\hat{x})}{|\hat{x} - x^0|^\tau} = \frac{\bar{u}^p(\hat{x})}{|\hat{x}^{\lambda_0} - x^0|^\tau} - \frac{\bar{u}^p(\hat{x})}{|\hat{x} - x^0|^\tau} > 0.$$

A contradiction with (6.81). This proves (6.80). It follows from (6.80) that there exists a constant $c_o > 0$, such that

$$w_{\lambda_0}(x) \geq c_o, \quad x \in \overline{\Sigma_{\lambda_0 - \delta} \cap B_{R_o}(0)} \setminus \{(x^0)^{\lambda_0}\}.$$

Since w_λ depends on λ continuously, there exists $\epsilon > 0$ and $\epsilon < \delta$, such that for all $\lambda \in (\lambda_0, \lambda_0 + \epsilon)$, we have

$$w_\lambda(x) \geq 0, \quad x \in \overline{\Sigma_{\lambda_0 - \delta} \cap B_{R_o}(0)} \setminus \{(x^0)^\lambda\}. \tag{6.82}$$

Combining (6.78), (6.71), and (6.82), we conclude that for all $\lambda \in (\lambda_0, \lambda_0 + \epsilon)$,

$$w_\lambda(x) \geq 0, \quad x \in \Sigma_\lambda \setminus \{(x^0)^\lambda\}. \tag{6.83}$$

This contradicts the definition of λ_0. Therefore, we must have

$$\lambda_0 = x_1^0 \text{ and } w_{\lambda_0} \geq 0 \ \forall \, x \in \Sigma_{\lambda_0}.$$

Similarly, one can move the plane T_λ from $+\infty$ to the left and show that

$$w_{\lambda_0} \leq 0 \ \forall \, x \in \Sigma_{\lambda_0}. \tag{6.84}$$

Now we have shown that

$$\lambda_0 = x_1^0 \text{ and } w_{\lambda_0}(x) \equiv 0, \quad x \in \Sigma_{\lambda_0}.$$

This completes *Step 2*.

So far, we have proved that $\bar{u}$ is symmetric about the plane $T_{x_1^0}$. Since the x_1 direction can be chosen arbitrarily, we have actually shown that $\bar{u}$ is radially symmetric about x^0.

For any two points $X^i \in \mathbb{R}^n$, $i = 1, 2$. Choose x^0 to be the midpoint: $x^0 = \frac{X^1 + X^2}{2}$. Since $\bar{u}$ is radially symmetric about x^0, so is u, hence $u(X^1) = u(X^2)$. This implies that u is constant. A positive constant function does not satisfy (6.63). This proves the non-existence of positive solutions for (6.63) when $1 < p < \frac{n+\alpha}{n-\alpha}$.

The Critical Case

Let $\bar{u}$ be the Kelvin transform of u centered at the origin, then

$$(-\triangle)^{\alpha/2} \bar{u}(x) = \bar{u}^p(x). \tag{6.85}$$

We show that either $\bar{u}$ is symmetric about the origin or u is symmetric about some point.

We still use the notation introduced in the subcritical case. Step 1 is entirely the same as that in the subcritical case, that is, we can show that for λ sufficiently negative,

$$w_\lambda(x) \geq 0, \ \forall \, x \in \Sigma_\lambda \setminus \{0^\lambda\}.$$

Hence we omit its proof.

For Step 2, let

$$\lambda_0 = \sup\{\lambda \le 0 \mid w_\mu(x) \ge 0, \ \forall x \in \Sigma_\mu \backslash \{0^\mu\}, \mu \le \lambda\}.$$

Case (i). $\lambda_0 < 0$. Similar to the subcritical case, one can show that

$$w_{\lambda_0}(x) \equiv 0, \ \forall x \in \Sigma_{\lambda_0}.$$

It follows that x^0 is not a singular point of $\bar{u}$ and hence

$$u(x) = O(\frac{1}{|x|^{n-\alpha}}) \ \text{when} \ |x| \to \infty.$$

This enables us to apply the method of moving plane to u directly and show that u is symmetric about some point in $\mathbb{R}^n$.

Case (ii). $\lambda_0 = 0$. Then by moving the planes from near $x_1 = +\infty$, we derive that $\bar{u}$ is symmetric about the origin, and so is u.

In either case, u is symmetric about some point in $\mathbb{R}^n$.

This completes the proof.

6.6.2 A Dirichlet Problem on a Half Space

We investigate a Dirichlet problem involving the fractional Laplacian on an upper half space

$$\mathbb{R}^n_+ = \{x = (x_1, \cdots, x_n) \mid x_n > 0\}.$$

Consider

$$\begin{cases} (-\triangle)^{\alpha/2}u = u^p(x), & x \in \mathbb{R}^n_+, \\ u(x) \equiv 0, & x \notin \mathbb{R}^n_+. \end{cases} \tag{6.86}$$

Theorem 6.6.2 *Assume that $0 < \alpha < 2$ and $u \in L_\alpha \cap C^{1,1}_{loc}$ is a nonnegative solution of problem (6.86). Then in the subcritical and critical case $1 < p \le \frac{n+\alpha}{n-\alpha}$, $u \equiv 0$.*

Proof. To prove this theorem, again we make a Kelvin transform. In order that $\mathbb{R}^n_+$ is invariant under the transform, we put the center x^o on the boundary $\partial \mathbb{R}^n_+$.

Let

$$v_{x^o}(x) = \frac{1}{|x - x^o|^{n-\alpha}} u \left(\frac{x - x^o}{|x - x^o|^2} + x^o \right).$$

Then

$$(-\triangle)^{\alpha/2} v_{x^o}(x) = \frac{v^p_{x^o}(x)}{|x - x^o|^\tau}, \quad x \in \mathbb{R}^n_+. \tag{6.87}$$

with $\tau = n + \alpha - p(n - \alpha)$. Obviously, $\tau = 0$ in the critical case.

The main ideas are as follows.

In the critical case $p = \frac{n+\alpha}{n-\alpha}$, we consider two possibilities.

(i) *There is a point $x^o \in \partial \mathbb{R}_+^n$, such that $v_{x^o}(x)$ is bounded near x^o.* In this situation, $u \leq \frac{C}{1+|x|^{n-\alpha}}$ has the needed asymptotic behavior near infinity, hence we move the planes in the direction of x_n-axis to show that the solution u is monotone increasing in x_n.

(ii) *For all $x^o \in \partial \mathbb{R}_+^n$, $v_{x^o}(x)$ are unbounded near x^o.* In this situation, we move the planes in $x_1, \cdots, x_{n-1}$ directions to show that, for every x^o, v_{x^o} is axially symmetric about the line that is parallel to x_n-axis and passing through x^o. This implies further that u depends on x_n only.

In the subcritical case, we only need to work on $v_{x^o}(x)$; and similar to the above possibility (ii), we show that for every x^o, v_{x^o} is axially symmetric about the line that is parallel to x_n-axis and passing through x^o, which implies again that u depends on x_n only.

In both cases, we will be able to derive contradictions.

The Critical Case

We consider two possibilities.

(i) *There is a point $x^o \in \partial \mathbb{R}_+^n$, such that $v_{x^o}(x)$ is bounded near x^o.* In this situation, from the symmetric expression

$$u(x) = \frac{1}{|x - x^o|^{n-\alpha}} v_{x^o}\left(\frac{x - x^o}{|x - x^o|^2} + x^o \right),$$

we see immediately that

$$u(x) \sim \frac{1}{|x|^{n-\alpha}}, \quad \text{near infinity.} \tag{6.88}$$

Consequently, by Theorem 3.1.2, we have

$$\text{either } u(x) > 0 \text{ or } u(x) \equiv 0, \quad \forall x \in \mathbb{R}_+^n.$$

Hence in the following, we may assume that $u > 0$ in $\mathbb{R}_+^n$.

Now we carry out the method of moving planes on the solution u along x_n direction.

Let

$$T_\lambda = \{x \in \mathbb{R}^n \mid x_n = \lambda\}, \quad \lambda > 0,$$

and

$$\Sigma_\lambda = \{x \in \mathbb{R}^n \mid 0 < x_n < \lambda\}.$$

Let

$$x^\lambda = (x_1, \cdots, x_{n-1}, 2\lambda - x_n)$$

be the reflection of x about the plane T_λ.

Denote $w_\lambda(x) = u(x^\lambda) - u(x)$, and

$$\Sigma_\lambda^- = \{x \in \Sigma_\lambda \mid w_\lambda(x) < 0\}.$$

Then

$$(-\triangle)^{\alpha/2} w_\lambda(x) + c(x) w_\lambda(x) \geq 0, \quad x \in \Sigma_\lambda^-, \tag{6.89}$$

with

$$c(x) = -p u^{p-1}(x). \tag{6.90}$$

From this and (6.88), we see that $c(x)$ is bounded from below in Σ_λ^-, and

$$\lim_{|x| \to \infty} w_\lambda(x) = 0 \text{ and } c(x) \sim \frac{1}{|x|^{2\alpha}} \text{ for } |x| \text{ large.} \tag{6.91}$$

It follows that we can apply the *narrow region principle* to conclude that for λ sufficiently small,

$$w_\lambda(x) \geq 0, \quad \forall x \in \Sigma_\lambda, \tag{6.92}$$

because Σ_λ is a narrow region.

Eq. (6.92) provides a starting point, from which we can move the plane T_λ upward as long as inequality (6.92) holds. Define

$$\lambda_o = \sup\{\lambda \mid w_\mu(x) \geq 0, x \in \Sigma_\mu; \mu \leq \lambda\}.$$

We show that

$$\lambda_o = \infty. \tag{6.93}$$

Otherwise, if $\lambda_o < \infty$, then by (6.91), combining the *Narrow Region Principle* and *Decay at Infinity* and going through the similar arguments as in the previous subsection, we are able to show that

$$w_{\lambda_o}(x) \equiv 0 \quad \text{in } \Sigma_{\lambda_o},$$

which implies

$$u(x_1, \cdots, x_{n-1}, 2\lambda_o) = u(x_1, \cdots, x_{n-1}, 0) = 0.$$

This is impossible, because we assume that $u > 0$ in $\mathbb{R}_+^n$.

Therefore, (6.93) must be valid. Consequently, the solution $u(x)$ is monotone increasing with respect to x_n. This contradicts (6.88). Therefore what left to be considered is

Possibility (ii): *For all $x^o \in \partial \mathbb{R}_+^n$, $v_{x^o}(x)$ are unbounded near x^o.*

In this situation, we carry out the method of moving planes on v_{x^o} along any direction in $\mathbb{R}^{n-1}$–the boundary of $\mathbb{R}_+^n$, and call it x_1 direction.

For a given real number λ, define

$$\hat{T}_\lambda = \{x \in \mathbb{R}^n \mid x_1 = \lambda\},$$

$$\hat{\Sigma}_\lambda = \{x = (x_1, \cdots, x_n) \in \mathbb{R}_+^n \mid x_1 < \lambda\}$$

and let

$$x^\lambda = (2\lambda - x_1, x_2, \cdots, x_n).$$

Let $w_\lambda(x) = v_{x^o}(x^\lambda) - v_{x^o}(x)$ and

$$\hat{\Sigma}_\lambda^- = \{x \in \hat{\Sigma}_\lambda \mid w_\lambda(x) < 0\}.$$

Then

$$(-\triangle)^{\alpha/2} w_\lambda(x) + c(x)w_\lambda(x) \geq 0, \quad x \in \hat{\Sigma}_\lambda^-,$$

with

$$c(x) = -pv_{x^o}^{p-1}(x).$$

By the asymptotic behavior

$$v_{x^o}(x) \sim \frac{1}{|x|^{n-\alpha}}, \quad \text{for } |x| \text{ large,}$$

we derive

$$\lim_{|x| \to \infty} w_\lambda(x) = 0 \text{ and } c(x) \sim \frac{1}{|x|^{2\alpha}} \text{ for } |x| \text{ large.}$$

These guarantee that we can apply the *narrow region principle* and *decay at infinity* to show the following:

(i) For λ sufficiently negative,

$$w_\lambda(x) \geq 0, \quad x \in \Sigma_\lambda.$$

(ii) Define

$$\lambda_o = \sup\{\lambda \mid w_\mu(x) \geq 0, x \in \Sigma_\mu, \mu \leq \lambda < x_1^o\},$$

where x_1^o is the first component of x^o. Then if $\lambda_o < x_1^o$, we must have

$$w_{\lambda_o}(x) \equiv 0, \quad x \in \Sigma_{\lambda_o},$$

that is

$$v_{x^o}(x^{\lambda_o}) \equiv v_{x^o}(x), \quad x \in \Sigma_{\lambda_o}.$$

This is impossible, because by our assumption, v_{x^o} is unbounded near x^o, while in fact it is bounded near $(x^o)^{\lambda_o}$. Therefore, we must have

$$\lambda_o = x_1^o.$$

Based on this, and by moving the plane $\hat{T}_\lambda$ from near $x_1 = +\infty$ to the left to its limiting position, we show that v_{x^o} is symmetric about the plane $\hat{T}_{x_1^o}$. Since x_1 direction can be chosen arbitrarily, we conclude that v_{x^o} is axially symmetric about the line parallel to x_n axis and passing through x^o. Because x^o is any point on $\partial\mathbb{R}_+^n$, we deduce that the original solution u is independent of the first $n-1$ variables, i.e., $u = u(x_n)$.

To finally derive a contradiction, we need two results from [CFY].

Proposition 6.6.1 *(Theorem 4.1 in [CFY]) Assume that $u \in L_\alpha$ is a locally bounded positive solution of*

$$\begin{cases} (-\triangle)^{\alpha/2}u(x) = u^p(x), & x \in \mathbb{R}^n_+, \\ u(x) = 0, & x \notin \mathbb{R}^n_+. \end{cases}$$

Then it is also a solution of

$$u(x) = \int_{\mathbb{R}^n_+} G_\infty(x,y)u^p(y)dy;$$

and vice versa. Here $G_\infty(x,y)$ is the Green's function of the corresponding problem:

$$G_\infty(x,y) = \frac{A_{n,\alpha}}{s^{\frac{n-\alpha}{2}}}\left[1 - B\frac{1}{(t+s)^{\frac{n-2}{2}}}\int_0^{\frac{s}{t}}\frac{(s-tb)^{\frac{n-2}{2}}}{b^{\alpha/2}(1+b)}db\right],$$

with $s = |x-y|^2$ and $t = 4x_ny_n$.

Proposition 6.6.2 *If $u = u(x_n) > 0$, then*

$$\int_{\mathbb{R}^n_+} G_\infty(x,y)u^p(y)dy = \infty.$$

(See the proof between pages 23 and 27 in [CFY].)

Now these two propositions imply that if $u = u(x_n)$ is a positive solution of problem (6.86), then

$$u(x) = \int_{\mathbb{R}^n_+} G_\infty(x,y)u^p(y)dy = \infty,$$

which is obviously impossible. This completes the proof in the critical case.

The Subcritical Case

Recall that

$$(-\triangle)^{\alpha/2}v_{x^o}(x) = \frac{v_{x^o}^p(x)}{|x-x^o|^\tau}, \quad x \in \mathbb{R}^n_+. \tag{6.94}$$

Similar to possibility (ii) in the critical case, we apply the method of moving planes to v_{x^o} along any direction in $\mathbb{R}^{n-1}$, and call it x_1 direction. Due to the monotonicity of the term $\frac{1}{|x-x^o|^\tau}$ in equation (6.94) with $\tau > 0$, through a similar argument, we can derive that v_{x^o} is axially symmetric about the line parallel to x_n axis and passing through x^o, and hence the original solution u is independent of the first $n-1$ variables, i.e. $u = u(x_n)$, which leads to a contradiction as in the critical case.

This completes the proof of the theorem.

6.6.3 The Nonlinear Schrödinger Equation

We study positive solutions for the nonlinear Schrödinger equation with fractional diffusion

$$(-\triangle)^{\alpha/2}u + u = u^p, \ x \in \mathbb{R}^n. \tag{6.95}$$

Theorem 6.6.3 *Assume that $u \in L_\alpha \cap C_{loc}^{1,1}$ is a positive solution of (6.95) with $1 < p < \infty$. If*

$$\lim_{|x|\to\infty} u(x) = a < \left(\frac{1}{p}\right)^{\frac{1}{p-1}}, \tag{6.96}$$

then u must be radially symmetric and monotone decreasing about some point in $\mathbb{R}^n$.

Proof. Because of the presence of the term u in the equation, if one makes a Kelvin transform, the coefficients in the resulting equation do not possess the monotonicity needed in the method of moving planes. Hence we directly work on the original equation (6.95).

Let $T_\lambda, \Sigma_\lambda, x^\lambda$, and u_λ be defined as in the previous section. And let $w_\lambda(x) = u_\lambda(x) - u(x)$. Then at the points where w_λ is negative, it is easy to verify that

$$(-\triangle)^{\alpha/2}w_\lambda + \left(1 - pu^{p-1}\right)w_\lambda(x) \geq 0. \tag{6.97}$$

Step 1. We apply Theorem 6.4.3 (*decay at infinity*) to show that for sufficiently negative λ, it holds

$$w_\lambda(x) \geq 0, \ x \in \Sigma_\lambda. \tag{6.98}$$

Here in (6.97), our $c(x) = \left(1 - pu^{p-1}(x)\right)$.

First, by our assumption that $\lim_{|x|\to\infty} u(x) = a$, we have, for each fixed λ,

$$\lim_{|x|\to\infty} w_\lambda(x) = 0. \tag{6.99}$$

Hence if (6.98) is violated, then a negative minimum of w_λ is attained at some point, say at x^o.

By condition (6.96), we have,

$$c(x) \geq 0, \ \text{for } |x| \text{ sufficiently large,}$$

and hence assumption (6.52) in Theorem 6.4.3 is satisfied. Consequently, there exists R_o (independent of λ), such that

$$|x^o| \leq R_o. \tag{6.100}$$

It follows that, for $\lambda < -R_o$, we must have

$$w_\lambda(x) \geq 0, \ x \in \Sigma_\lambda.$$

Step 2. Step 1 provides a starting point, from which we can now move the plane T_λ to the right as long as (6.98) holds to its limiting position.

Let
$$\lambda_0 = \sup\{\lambda \mid w_\mu(x) \geq 0, \ \forall x \in \Sigma_\mu, \mu \leq \lambda\}.$$
It follows from (6.99) that $\lambda_0 < \infty$.

We show that
$$w_{\lambda_0}(x) \equiv 0, \quad x \in \Sigma_{\lambda_0}. \tag{6.101}$$

Suppose in the contrary,
$$w_{\lambda_0}(x) \geq 0 \text{ and } w_{\lambda_0}(x) \not\equiv 0, \ \text{ in } \Sigma_{\lambda_0},$$
we must have
$$w_{\lambda_0}(x) > 0 \text{ in } \Sigma_{\lambda_0}. \tag{6.102}$$
In fact, if (6.102) is violated, then there exists a point $\hat{x} \in \Sigma_{\lambda_0}$, such that
$$w_{\lambda_0}(\hat{x}) = \min_{\Sigma_{\lambda_0}} w_{\lambda_0} = 0.$$
Consequently, similar to (6.81), we have
$$(-\triangle)^{\alpha/2} w_{\lambda_0}(\hat{x}) = C_{n,\alpha} PV \int_{\Sigma_{\lambda_0}} \left(\frac{1}{|\hat{x} - y^{\lambda_0}|^{n+\alpha}} - \frac{1}{|\hat{x} - y|^{n+\alpha}} \right) w_{\lambda_0}(y) dy$$
$$< 0.$$

This contradicts (6.97). Hence (6.102) holds.

Then we show that the plane T_λ can be moved further to the right. To be more rigorous, there exists some $\epsilon > 0$, such that for any $\lambda \in (\lambda_0, \lambda_0 + \epsilon)$, we have
$$w_\lambda(x) \geq 0, \quad x \in \Sigma_\lambda. \tag{6.103}$$
This is a contradiction with the definition of λ_0. Therefore (6.101) must be valid.

Under our assumptions, we have
$$\lim_{|x| \to \infty} w_\lambda(x) = 0 \text{ and } c(x) \text{ is bounded from below.}$$

Then combining the *narrow region principle* and the *decay at infinity*, through a similar argument as in the previous section, we derive (6.103). This completes the proof of the theorem.

6.6.4 More General Nonlinearities on a Bounded Domain

Consider
$$\begin{cases} (-\triangle)^{\alpha/2} u(x) = f(u(x)), & x \in B_1(0), \\ u(x) = 0, & x \notin B_1(0). \end{cases} \tag{6.104}$$

We prove

Theorem 6.6.4 *Assume that $u \in L_\alpha \cap C_{loc}^{1,1}(B_1(0))$ is a positive solution of (6.104) with $f(\cdot)$ being Lipschitz continuous. Then u must be radially symmetric and monotone decreasing about the origin.*

Proof. Let $T_\lambda, x^\lambda, u_\lambda$, and w_λ be as defined in the previous section. Let

$$\Sigma_\lambda = \{x \in B_1(0) \mid x_1 < \lambda\}.$$

Then it is easy to verify that

$$(-\triangle)^{\alpha/2} w_\lambda(x) + c_\lambda(x) w_\lambda(x) = 0, \quad x \in \Sigma_\lambda,$$

where

$$c_\lambda(x) = -\frac{f(u(x)) - f(u_\lambda(x))}{u(x) - u_\lambda(x)}.$$

Our Lipschitz continuity assumption on f guarantees that $c_\lambda(x)$ is uniformly bounded from below. Now we can apply Theorem 6.4.2 (*narrow region principle*) to conclude that for $\lambda > -1$ and sufficiently close to -1,

$$w_\lambda(x) \geq 0, \quad x \in \Sigma_\lambda; \tag{6.105}$$

because Σ_λ is a narrow region for such λ.

Define

$$\lambda_0 = \sup\{\lambda \leq 0 \mid w_\mu(x) \geq 0, x \in \Sigma_\mu, \mu \leq \lambda\}.$$

Then we must have $\lambda_0 = 0$. Otherwise, we can use the *narrow region principle* and similar arguments as in the previous section to show that we would be able to move the plane T_λ further to the right to contradict the definition of λ_0. Therefore

$$w_0(x) \geq 0, \quad x \in \Sigma_0;$$

or more apparently,

$$u(-x_1, x_2, \cdots, x_n) \leq u(x_1, x_2, \cdots, x_n), \quad 0 < x_1 < 1. \tag{6.106}$$

Since the x_1-direction can be chosen arbitrarily, (6.106) implies u is radially symmetric about the origin. The monotonicity is a consequence of the fact that (6.105) holds for all $-1 < \lambda \leq 0$. This completes the proof of the theorem.

6.7 Various Maximum Principles for a Fully Nonlinear Fractional Order Operator

Let

$$F_\alpha(u(x)) = C_{n,\alpha} \, PV \int_{\mathbb{R}^n} \frac{G(u(x) - u(z))}{|x - z|^{n+\alpha}} dz. \tag{6.107}$$

This is a fully nonlinear fractional order operator, and in particular, when $G(\cdot)$ a linear function, it becomes the fractional Laplacian, and when $G(t) = |t|^{p-2}t$ and $\alpha = ps$, it becomes the fractional p-Laplacian (see [CL9]).

To ensure the integral on the right hand side converges, G needs to be at least Lipschitz continuous at 0, and we require that $u \in C_{loc}^{1,1} \cap L_{G,\alpha}$ with

$$L_{G,\alpha} = \{G(u) \in L_{loc}^1 \mid \int_{\mathbb{R}^n} \frac{|G(1 + u(x))|}{1 + |x|^{n+\alpha}} dx < \infty\}.$$

6.7.1 The Limit of F_α as $\alpha \to 2$

One significance in considering such an operator is indicated in the following.

Theorem 6.7.1 *Assume that $u \in C^2_{loc} \cap L_{G,\alpha}$, $G(\cdot)$ is second order differentiable, G' is bounded and $G(0) = 0$. Then*

$$\lim_{\alpha \to 2} F_\alpha(u(x)) = a(-\triangle u)(x) + b|\triangledown u(x)|^2, \tag{6.108}$$

where a and b are constant multiples of $G'(0)$ and $G''(0)$ respectively.

Proof. We use the fact $C_{n,\alpha} = c_n(2-\alpha)$ with some constant c_n depending on n.

It follows that

$$F_\alpha(u(x))$$
$$= c_n(2-\alpha)PV \int_{B_\epsilon(x)} \frac{G(u(x) - u(y))}{|x-y|^{n+\alpha}} dy$$
$$+ c_n(2-\alpha) \int_{\mathbb{R}^n \setminus B_\epsilon(x)} \frac{G(u(x) - u(y))}{|x-y|^{n+\alpha}} dy$$
$$= I_1 + I_2. \tag{6.109}$$

First fix ϵ and let $\alpha \to 2$. Then obviously

$$I_2 \to 0. \tag{6.110}$$

To estimate I_1, we apply Taylor expansion to G near 0 and to u near x.

$$G(u(x) - u(y))$$
$$= G'(0)(u(x) - u(y)) + \frac{1}{2}G''(0)(u(x) - u(y))^2 + o(\epsilon)(u(x) - u(y))^2$$
$$= -G'(0)[\triangledown u(x) \cdot z + \frac{1}{2}u_{ij}(x)z_i z_j] + \frac{1}{2}G''(0)(u_i(x)u_j(x)z_i z_j) + o(\epsilon)|z|^2.$$

Here we write $z = x - y$ to avoid lengthy expressions and adapt the summation convention that

$$u_{ij}z_i z_j = \sum_{i,j} u_{ij}z_i z_j,$$

and

$$o(\epsilon) \to 0, \quad \text{as } \epsilon \to 0.$$

Evaluate I_1 in four separate parts:

$$I_1 = I_{11} + I_{12} + I_{13} + I_{14}. \tag{6.111}$$

Due to the symmetry, we have

$$I_{11} = -c_n(2-\alpha)G'(0)PV \int_{B_\epsilon(0)} \frac{\triangledown u(x) \cdot z}{|z|^{n+\alpha}} dz = 0. \tag{6.112}$$

$$I_{12} = -c_n(2-\alpha)G'(0)\frac{u_{ij}(x)}{2}\int_{B_\epsilon(0)}\frac{z_iz_j}{|z|^{n+\alpha}}dz$$

$$= -c_n(2-\alpha)G'(0)\frac{u_{ii}(x)}{2}\int_{B_\epsilon(0)}\frac{z_i^2}{|z|^{n+\alpha}}dz$$

$$= -c_n(2-\alpha)G'(0)\frac{\triangle u(x)}{2n}\int_{B_\epsilon(0)}\frac{1}{|z|^{n+\alpha-2}}dz$$

$$= a(-\triangle u(x))\epsilon^{2-\alpha}$$

$$\rightarrow a(-\triangle u(x)), \text{ as } \alpha\rightarrow 2. \tag{6.113}$$

$$I_{13} = 1/2c_n(2-\alpha)G''(0)u_i(x)u_j(x)\int_{B_\epsilon(0)}\frac{z_iz_j}{|z|^{n+\alpha}}dz$$

$$= 1/2c_n(2-\alpha)G''(0)u_i^2(x)\int_{B_\epsilon(0)}\frac{z_i^2}{|z|^{n+\alpha}}dz$$

$$= 1/2c_n(2-\alpha)G''(0)\frac{|\bigtriangledown u(x)|^2}{n}\int_{B_\epsilon(0)}\frac{1}{|z|^{n+\alpha-2}}dz$$

$$= b(|\bigtriangledown u(x)|^2)\epsilon^{2-\alpha}$$

$$\rightarrow b(|\bigtriangledown u(x)|^2), \text{ as } \alpha\rightarrow 2. \tag{6.114}$$

Then we let $\epsilon\rightarrow 0$ and obtain

$$I_{14} = o(\epsilon)\rightarrow 0. \tag{6.115}$$

Combining (6.109) through (6.115), we prove the theorem. $\square$

Throughout the section, we assume that $G \in C^1(\mathbb{R})$,

$$G(0) = 0, \text{ and } G'(t) \geq c_0 > 0 \ \forall t \in \mathbb{R}. \tag{6.116}$$

6.7.2 A Maximum Principle

Theorem 6.7.2 (The Simple Maximum Principle) *Let Ω be a bounded domain in $\mathbb{R}^n$. Assume that $u \in C_{loc}^{1,1}(\Omega) \cap L_{G,\alpha}$ is lower semi-continuous on $\bar{\Omega}$, and satisfies*

$$\begin{cases} F_\alpha(u(x)) \geq 0, x \in \Omega, \\ u(x) \geq 0, \qquad x \in \Omega^c. \end{cases} \tag{6.117}$$

Then

$$u(x) \geq 0, \quad x \in \Omega. \tag{6.118}$$

The same conclusion holds for unbounded domains Ω if we further assume that

$$\liminf_{|x|\rightarrow\infty} u(x) \geq 0.$$

Proof. Suppose (6.118) is violated, then since u is lower semi-continuous on $\bar{\Omega}$, there exists some x^o in Ω such that

$$u(x^o) = \min_{\Omega} u < 0.$$

It follows from (6.116) that

$$PV \int_{\mathbb{R}^n} \frac{G(u(x^o) - u(z))}{|x^o - z|^{n+\alpha}} dz = PV \int_{\mathbb{R}^n} \frac{G'(\psi(z))[u(x^o) - u(z)]}{|x^o - z|^{n+\alpha}} dz$$

$$\leq PV c_0 \int_{\mathbb{R}^n} \frac{u(x^o) - u(z)}{|x^o - z|^{n+\alpha}} dz$$

$$< 0.$$

This contradicts (6.117) and hence proves the theorem.   □

Let

$$w_\lambda(x) = u(x^\lambda) - u(x).$$

For simplicity of notation, we denote w_λ by w and Σ_λ by Σ.

6.7.3 Maximum Principle for Anti-Symmetric Functions

Theorem 6.7.3 (Maximum Principle for Anti-Symmetric Functions)
Let Ω be a bounded domain in Σ. Assume that $u \in L_{G,\alpha} \cap C^{1,1}_{loc}(\Omega)$ and is lower semi-continuous on $\bar{\Omega}$. If

$$\begin{cases} F_\alpha(u_\lambda(x)) - F_\alpha(u(x)) \geq 0 & in \ \Omega, \\ w(x) \geq 0 & in \ \Sigma \backslash \Omega, \\ w(x^\lambda) = -w(x) & in \ \Sigma, \end{cases}$$

then

$$w(x) \geq 0 \ in \ \Omega.$$

Furthermore, if $w = 0$ at some point in Ω, then

$$w(x) = 0 \quad almost \ everywhere \ in \ \mathbb{R}^n.$$

These conclusions hold for unbounded regions Ω if we further assume that

$$\varliminf_{|x| \to \infty} w(x) \geq 0.$$

Proof. Suppose otherwise, then there exists a point x in Ω, such that

$$w(x) = \min_{\Omega} w < 0. \tag{6.119}$$

$$F_\alpha(u_\lambda(x)) - F_\alpha(u(x))$$
$$= C_{n,\alpha} PV \int_{\mathbb{R}^n} \frac{G(u_\lambda(x) - u_\lambda(y)) - G(u(x) - u(y))}{|x-y|^{n+\alpha}} dy$$
$$= \cdots \int_\Sigma \frac{G(u_\lambda(x) - u_\lambda(y)) - G(u(x) - u(y))}{|x-y|^{n+\alpha}} dy$$
$$+ \cdots \int_\Sigma \frac{G(u_\lambda(x) - u(y)) - G(u(x) - u_\lambda(y))}{|x-y^\lambda|^{n+\alpha}} dy$$
$$= \cdots \int_\Sigma (\frac{1}{|x-y|^{n+\alpha}} - \frac{1}{|x-y^\lambda|^{n+\alpha}})[G(u_\lambda(x) - u_\lambda(y)) - G(u(x) - u(y))] dy$$
$$+ \cdots \int_\Sigma \frac{G(u_\lambda(x) - u_\lambda(y)) - G(u(x) - u(y)) + G(u_\lambda(x) - u(y)) - G(u(x) - u_\lambda(y))}{|x-y^\lambda|^{n+\alpha}} dy$$
$$= \cdots (I_1 + I_2). \tag{6.120}$$

Notice that

$$G(u_\lambda(x) - u_\lambda(y)) - G(u(x) - u(y)) = G'(\xi)[w(x) - w(y)] \le 0,$$

and

$$\frac{1}{|x-y|^{n+\alpha}} - \frac{1}{|x-y^\lambda|^{n+\alpha}} > 0, \quad \forall x, y \in \Sigma_\lambda.$$

We have

$$I_1 \le 0.$$

To estimate I_2, we regroup the terms:

$$I_2 = \int_\Sigma \frac{[G(u_\lambda(x) - u_\lambda(y)) - G(u(x) - u_\lambda(y))] + [G(u_\lambda(x) - u(y)) - G(u(x) - u(y))]}{|x-y^\lambda|^{n+\alpha}} dy$$
$$= \int_\Sigma \frac{G'(\xi)w(x) + G'(\eta)w(x)}{|x-y^\lambda|^{n+\alpha}} dy$$
$$\le 2c_0 w(x) \int_\Sigma \frac{1}{|x-y^\lambda|^{n+\alpha}} dy < 0.$$

This is a contradiction with the equation and hence we must have

$$w_\lambda(x) \ge 0, \quad \forall x \in \Omega. \tag{6.121}$$

Remark 6.7.1 *Note inequality*

$$F_\alpha(u_\lambda(x)) - F_\alpha(u(x)) < 0 \tag{6.122}$$

holds at a negative minimum x of w, and it is a key ingredient in obtaining various maximum principles, *which will be used several times later.*

Suppose $w = 0$ at some point in Ω, say $w(x^o) = 0$, then we must have

$$w(x) = 0 \text{ almost everywhere in } \mathbb{R}^n. \tag{6.123}$$

To this end, we re-estimate I_1 and I_2 at $x = x^o$. From the previous argument, we have

$$I_1 = C_{n,\alpha} PV \int_\Sigma (\frac{1}{|x-y|^{n+\alpha}} - \frac{1}{|x-y^\lambda|^{n+\alpha}}) G'(\xi)(-w(y)) dy$$

$$\leq -C_0 \int_\Sigma (\frac{1}{|x-y|^{n+\alpha}} - \frac{1}{|x-y^\lambda|^{n+\alpha}}) w(y) dy.$$

If $w \not\equiv 0$ in Σ, then we have

$$I_1 < 0 \quad (\text{maybe } -\infty).$$

Obviously, we also have $I_2 = 0$, hence

$$F_\alpha(u_\lambda(x^o)) - F_\alpha(u(x^o)) < 0,$$

a contradiction. Therefore, we must have

$$w(y) \equiv 0, \quad \forall y \in \Sigma,$$

and consequently, by the antisymmetry of w, we arrive at (6.123).

If Ω is an unbounded region, then under the condition

$$\varliminf_{|x| \to \infty} w(x) \geq 0,$$

any negative minimum of w in Ω is attained at some point $x^o \in \Omega$, and similar to the above arguments, we can deduce all the same conclusions. □

6.7.4 Narrow Region Principle

Theorem 6.7.4 (Narrow Region Principle) *Let Ω be a bounded narrow region in Σ, such that it is contained in*

$$\{x| \lambda - \delta < x_1 < \lambda\}$$

with small δ. Suppose that $u \in L_{G,\alpha} \cap C^{1,1}_{loc}(\Omega)$ and is lower semi-continuous on $\bar{\Omega}$. If $c(x)$ is bounded from below in Ω and

$$\begin{cases} F_\alpha(u_\lambda(x)) - F_\alpha(u(x)) + c(x)w(x) \geq 0 & in\ \Omega, \\ w(x) \geq 0 & in\ \Sigma \backslash \Omega, \\ w(x^\lambda) = -w(x) & in\ \Sigma, \end{cases}$$

then for sufficiently small δ, we have

$$w(x) \geq 0\ in\ \Omega.$$

Furthermore, if $w = 0$ at some point in Ω, then

$$w(x) \equiv 0\ \ \mathbb{R}^n.$$

These conclusions hold for unbounded region Ω if we further assume that

$$\varliminf_{|x| \to \infty} w(x) \geq 0.$$

Proof. Suppose in the contrary, there exists an $x^0 \in \Omega$, such that

$$w(x^0) = \min_{\Omega} w < 0.$$

Then by the *key inequality* (6.122), we deduce

$$F_\alpha(u_\lambda(x^0)) - F_\alpha(u(x^0)) + c(x^0)w(x^0)$$
$$\leq w(x^0) \left[2C_{n,\alpha}c_0 \int_\Sigma \frac{1}{|x^0 - y^\lambda|^{n+\alpha}} dy + c(x^0) \right]. \qquad (6.124)$$

Since $c(x)$ is bounded from below, to derive a contradiction, it suffices to show that the integral in the above brackets can be arbitrarily large as δ becomes sufficiently small. To see this, let

$$D = \{y \mid \delta < y_1 - x_1^0 < 1, \ |y' - (x^0)'| < 1\},$$

$$s = y_1 - x_1^0, \quad \tau = |y' - (x^0)'|,$$

and ω_{n-2} be the area of $(n-2)$-dimensional unit sphere. Here we write $x = (x_1, x')$.

Then we have

$$\int_\Sigma \frac{1}{|x^0 - \tilde{y}|^{n+\alpha}} dy \geq \int_D \frac{1}{|x^0 - y|^{n+\alpha}} dy$$
$$= \int_\delta^1 \int_0^1 \frac{\omega_{n-2}\tau^{n-2}d\tau}{(s^2 + \tau^2)^{\frac{n+\alpha}{2}}} ds$$
$$= \int_\delta^1 \int_0^{\frac{1}{s}} \frac{\omega_{n-2}(st)^{n-2}s dt}{s^{n+\alpha}(1 + t^2)^{\frac{n+\alpha}{2}}} ds$$
$$= \int_\delta^1 \frac{1}{s^{1+\alpha}} \int_0^{\frac{1}{s}} \frac{\omega_{n-2}t^{n-2}dt}{(1 + t^2)^{\frac{n+\alpha}{2}}} ds$$
$$\geq \int_\delta^1 \frac{1}{s^{1+\alpha}} \int_0^1 \frac{\omega_{n-2}t^{n-2}dt}{(1 + t^2)^{\frac{n+\alpha}{2}}} ds$$
$$\geq C \int_\delta^1 \frac{1}{s^{1+\alpha}} ds \geq \frac{c}{\delta^\alpha}. \qquad (6.125)$$

Combining (6.124) with (6.125), we arrive at

$$F_\alpha(u_\lambda(x^0)) - F_\alpha(u(x^0)) + c(x^0)w(x^0)$$
$$\leq w(x^0) \left[2C_{n,\alpha}c_0 \frac{c}{\delta^\alpha} + c(x^0) \right].$$

Notice that $w(x^0) < 0$, so for sufficiently small δ we have

$$F_\alpha(u_\lambda(x^0)) - F_\alpha(u(x^0)) + c(x^0)w(x^0) < 0.$$

This contradicts the equation and hence proves $w(x) \geq 0$ in Ω. The proof of the rest of the theorem is similar to that of Theorem 6.7.3. $\square$

6.7.5 Decay at Infinity

Theorem 6.7.5 (Decay at Infinity) *Let Ω be an unbounded region in Σ. Assume $u \in L_{G,\alpha} \cap C_{loc}^{1,1}(\Omega)$ is a solution of*

$$
\begin{cases}
F_\alpha(u_\lambda(x)) - F_\alpha(u(x)) + c(x)w(x) \geq 0 & \text{in } \Omega, \\
w(x) \geq 0 & \text{in } \Sigma \backslash \Omega, \\
w(x^\lambda) = -w(x) & \text{in } \Sigma,
\end{cases}
$$

with

$$
\lim_{|x| \to \infty} |x|^\alpha c(x) \geq 0,
$$

then there exists a constant $R_0 > 0$ (depending on $c(x)$, but independent of w), such that if

$$
w(x^0) = \min_\Omega w(x) < 0,
$$

then

$$
|x^0| \leq R_0.
$$

Proof. Assume that $x^0 \in \Omega$, such that

$$
w(x^0) = \min_\Omega w < 0.
$$

Again by the *key inequality* (6.122), we deduce

$$
F_\alpha(u_\lambda(x^0)) - F_\alpha(u(x^0)) + c(x^0)w(x^0)
$$
$$
\leq w(x^0) \left[2C_{n,\alpha}c_0 \int_\Sigma \frac{1}{|x^0 - y^\lambda|^{n+\alpha}} dy + c(x^0) \right]. \tag{6.126}
$$

We now estimate the above integral. Let $\Sigma^c = \mathbb{R}^n \backslash \Sigma$. Choose a point x^1 in Σ^c, such as $x^1 = (3|x^0| + x_1^0, (x^0)')$, then $B_{|x^0|}(x^1) \subset \Sigma^c$. It follows that

$$
\int_\Sigma \frac{1}{|x^0 - y^\lambda|^{n+\alpha}} dy = \int_{\Sigma^c} \frac{1}{|x^0 - y|^{n+\alpha}} dy
$$
$$
\geq \int_{B_{|x^0|}(x^1)} \frac{1}{|x^0 - y|^{n+\alpha}} dy
$$
$$
\geq \int_{B_{|x^0|}(x^1)} \frac{1}{4^{n+\alpha}|x^0|^{n+\alpha}} dy
$$
$$
= \frac{\omega_n}{4^{n+\alpha}|x^0|^\alpha}.
$$

Then from the equation and (6.126), we have

$$
0 \leq F_\alpha(u_\lambda(x^0)) - F_\alpha(u(x^0)) + c(x^0)w(x^0)
$$
$$
\leq \left[\frac{2\omega_n C_{n,\alpha}c_0}{4^{n+\alpha}|x^0|^\alpha} + c(x^0) \right] w(x^0).
$$

Or equivalently,

$$\frac{2\omega_n C_{n,\alpha} c_0}{4^{n+\alpha}|x^0|^\alpha} + c(x^0) \leq 0.$$

Now if $|x^0|$ is sufficiently large, this contradicts the decay assumption on $c(x)$.

$\square$

6.8 Applications to Fully Nonlinear Fractional Order Equations

We consider nonlinear equations involving fully nonlinear nonlocal operators

$$F_\alpha(u) = f(x, u) \tag{6.127}$$

where

$$F_\alpha(u(x)) = C_{n,\alpha} \lim_{\epsilon \to 0} \int_{\mathbb{R}^n \setminus B_\epsilon(x)} \frac{G(u(x) - u(z))}{|x - z|^{n+\alpha}} dz$$

$$= C_{n,\alpha} PV \int_{\mathbb{R}^n} \frac{G(u(x) - u(z))}{|x - z|^{n+\alpha}} dz, \tag{6.128}$$

where PV stands for the Cauchy principle value. This operator was introduced by Caffarelli and Silvestre in [CaS].

In order for the integral to make sense, we require that

$$u \in C_{loc}^{1,1} \cap L_{G,\alpha}$$

with

$$L_{G,\alpha} = \{G(u) \in L_{loc}^1 \mid \int_{\mathbb{R}^n} \frac{|G(1 + u(x))|}{1 + |x|^{n+\alpha}} dx < \infty\},$$

and G being at least locally Lipschitz continuous, and $G(0) = 0$.

In the special case when $G(\cdot)$ is an identity map, F_α becomes the fractional Laplacian $(-\triangle)^{\alpha/2}$.

6.8.1 Symmetry of Solutions in a Unit Ball

Consider

$$\begin{cases} F_\alpha(u(x)) = f(u(x)), & x \in B_1(0), \\ u(x) = 0, & x \notin B_1(0). \end{cases} \tag{6.129}$$

Theorem 6.8.1 *Assume that $u \in L_{G,\alpha} \cap C_{loc}^{1,1}(B_1(0))$ is a positive solution of (6.129) with $f(\cdot)$ being Lipschitz continuous. Then u must be radially symmetric and monotone decreasing about the origin.*

Proof. Let $T_\lambda, x^\lambda, u_\lambda$, and w_λ be as defined in the previous section. Let

$$\Sigma_\lambda = \{x \in B_1(0) \mid x_1 < \lambda\}.$$

Then it is easy to verify that

$$F_\alpha(u_\lambda(x)) - F_\alpha(u(x)) + c_\lambda(x)w_\lambda(x) = 0, \quad x \in \Sigma_\lambda,$$

where

$$c_\lambda(x) = \frac{f(u(x)) - f(u_\lambda(x))}{u_\lambda(x) - u(x)}.$$

Our Lipschitz continuity assumption on f guarantees that $c_\lambda(x)$ is uniformly bounded from below. Now we can apply Theorem 6.7.4 (*narrow region principle*) to conclude that for $\lambda > -1$ and sufficiently close to -1,

$$w_\lambda(x) \geq 0, \quad x \in \Sigma_\lambda, \tag{6.130}$$

because Σ_λ is a narrow region for such λ.

Define

$$\lambda_0 = \sup\{\lambda \leq 0 \mid w_\mu(x) \geq 0, x \in \Sigma_\mu, \mu \leq \lambda\}.$$

Then we must have

$$\lambda_0 = 0.$$

Otherwise, suppose that $\lambda_0 < 0$, we show that the plane can be moved to the right a little more and inequality (6.130) is still valid. More precisely, there exists a small $\epsilon > 0$, such that for all $\lambda \in [\lambda_0, \lambda_0 + \epsilon)$, inequality (6.130) holds, which contradicts the definition of λ_0.

First, since $w_{\lambda_0}(x)$ is not identically zero, from the strong *maximum principle* (Theorem 6.7.3), we have

$$w_{\lambda_0}(x) > 0, \quad \forall x \in \Sigma_{\lambda_0}.$$

Thus for any $\delta > 0$,

$$w_{\lambda_0}(x) > c_\delta > 0, \quad \forall x \in \Sigma_{\lambda_0 - \delta}.$$

By the continuity of w_λ with respect to λ, there exists an $\epsilon > 0$, such that

$$w_\lambda(x) \geq 0, \quad \forall x \in \Sigma_{\lambda_0 - \delta}, \; \forall \lambda \in [\lambda_0, \lambda_0 + \epsilon). \tag{6.131}$$

In the *narrow region principle* (Theorem 6.7.4), let

$$\Sigma = \Sigma_\lambda \text{ and the narrow region } \Omega = \Sigma_\lambda \backslash \Sigma_{\lambda_0 - \delta},$$

then we have

$$w_\lambda(x) \geq 0, \quad \forall x \in \Sigma_\lambda \backslash \Sigma_{\lambda_0 - \delta}.$$

This together with (6.131) implies

$$w_\lambda(x) \geq 0, \quad \forall x \in \Sigma_\lambda, \; \forall \lambda \in [\lambda_0, \lambda_0 + \epsilon).$$

This contradicts the definition of λ_0. Therefore, we must have $\lambda_0 = 0$. As a result,

$$w_0(x) \geq 0, \quad x \in \Sigma_0;$$

or more apparently,

$$u(-x_1, x_2, \cdots, x_n) \leq u(x_1, x_2, \cdots, x_n), \quad 0 < x_1 < 1. \tag{6.132}$$

Since the x_1-direction can be chosen arbitrarily, (6.132) implies u is radially symmetric about the origin. The monotonicity is a consequence of the fact that

$$w_\lambda(x) > 0, \quad \forall x \in \Sigma_\lambda$$

holds for all $-1 < \lambda < 0$. $\square$

6.8.2 Symmetry of Solutions in the Whole Space $\mathbb{R}^n$

Let $T_\lambda, x^\lambda, u_\lambda$, and w_λ be as defined in the previous section. Let

$$\Sigma_\lambda = \{x \in \mathbb{R}^n \mid x_1 < \lambda\}.$$

Theorem 6.8.2 *Assume that $u \in C_{loc}^{1,1} \cap L_{G,\alpha}$ is a positive solution of*

$$F_\alpha(u(x)) = g(u(x)), \quad x \in \mathbb{R}^n. \tag{6.133}$$

Suppose, for some $\gamma > 0$,

$$u(x) = o(\frac{1}{|x|^\gamma}), \quad as \ |x| \to \infty, \tag{6.134}$$

and

$$g'(s) \leq s^q, \quad with \ q\gamma \geq \alpha. \tag{6.135}$$

Then u must be radially symmetric about some point in $\mathbb{R}^n$.

Proof. We carry out the proof in two steps. To begin with, we show that for λ sufficiently negative,

$$w_\lambda(x) \geq 0, \quad \forall x \in \Sigma_\lambda \tag{6.136}$$

with an application of the *decay at infinity* (Theorem 6.7.5).

Next, we move the plane T_λ along the x_1-axis to the right as long as inequality (6.136) holds. The plane will eventually stop at some limiting position $\lambda = \lambda_o$. Then we claim that

$$w_{\lambda_o}(x) \equiv 0, \quad \forall x \in \Sigma_{\lambda_o}.$$

The symmetry and monotone decreasing properties of solution u about T_{λ_o} follow naturally from the proof. Also, because of the arbitrariness of the x_1-axis, we conclude that u must be radially symmetric and monotone about some point.

Step 1. Start moving the plane T_λ along the x_1-axis from near $-\infty$ to the right.

By the *mean value theorem* it is easy to see that

$$F_\alpha(u_\lambda(x)) - F_\alpha(u(x)) = g(u_\lambda(x)) - g(u(x)) = g'(\psi_\lambda(x))w_\lambda(x),$$

where $\psi_\lambda(x)$ is between $u_\lambda(x)$ and $u(x)$. By the *decay at infinity* argument (Theorem 6.7.5), it suffices to check the decay rate of $g'(\psi_\lambda(x))$, and to be more precise, only at the points $\tilde{x}$ where w_λ is negative. Since

$$u_\lambda(\tilde{x}) < u(\tilde{x}),$$

we have

$$0 \le u_\lambda(\tilde{x}) \le \psi_\lambda(\tilde{x}) \le u(\tilde{x}).$$

The decay assumptions (6.134) and (6.135) instantly yield that

$$g'(\psi_\lambda(\tilde{x})) = o\left(\frac{1}{|\tilde{x}|^\alpha}\right).$$

Consequently, there exists a $R_0 > 0$, such that, if x^o is a negative minimum of w_λ in Σ_λ, then

$$|x^o| \le R_0. \tag{6.137}$$

Now take $\Omega = \Sigma_\lambda$, then by Theorem 6.7.5, it's easy to conclude that for $\lambda \le -R_0$, we must have (6.136). This completes the preparation for the moving of planes.

Step 2. Keep moving the plane to the limiting position T_{λ_0} as long as (6.136) holds.

Let

$$\lambda_0 = \sup\{\lambda \mid w_\mu(x) \ge 0, \ \forall x \in \Sigma_\mu, \mu \le \lambda\}.$$

In this part, we show that

$$w_{\lambda_0}(x) \equiv 0, \quad x \in \Sigma_{\lambda_0}. \tag{6.138}$$

Otherwise, the plane T_{λ_0} can still be moved to the right. More rigorously, there exists a $\delta_0 > 0$ such that for all $0 < \delta < \delta_0$, we have

$$w_{\lambda_0+\delta}(x) \ge 0, \quad \forall x \in \Sigma_{\lambda_0+\delta}. \tag{6.139}$$

This contradicts the definition of λ_0, and hence (6.138) must be true. Below we prove (6.139).

Suppose (6.138) is false, then w_{λ_0} is positive somewhere in Σ_{λ_0}, and the *strong maximum principle for anti-symmetric functions* (Theorem 6.7.3) implies

$$w_{\lambda_0}(x) > 0, \ x \in \Sigma_{\lambda_0}.$$

It follows that for any small positive number σ,

$$w_{\lambda_0}(x) \geq c_o > 0, \quad x \in \overline{\Sigma_{\lambda_0-\sigma} \cap B_{R_0}(0)}, \tag{6.140}$$

where R_0 is defined in *Step 1*. Since w_λ depends on λ continuously, for δ small positive, we have

$$w_{\lambda_0+\delta}(x) \geq 0, \quad x \in \overline{\Sigma_{\lambda_0-\sigma} \cap B_{R_0}(0)}.$$

Suppose (6.139) is false, then there exists some $x^o \in \Sigma_{\lambda_0+\delta}$, such that

$$w_{\lambda_0+\delta}(x^o) = \min_{\Sigma_{\lambda_0+\delta}} w_{\lambda_0+\delta} < 0.$$

From the *decay at infinity theorem* and (6.139), we must have

$$x^o \in (\Sigma_{\lambda_0+\delta} \backslash \Sigma_{\lambda_0-\sigma}) \cap B_{R_0}(0). \tag{6.141}$$

Notice that $(\Sigma_{\lambda_0+\delta} \backslash \Sigma_{\lambda_0-\sigma}) \cap B_{R_0}(0)$ is a narrow region for sufficiently small σ and δ. By the *narrow region principle* (Theorem 6.7.4), $w_{\lambda_0+\delta}$ cannot attain its negative minimum here. This contradicts (6.141), and hence (6.139) holds.   □

6.8.3 Non-Existence of Solutions on a Half Space

We investigate a Dirichlet problem on an upper half space

$$\mathbb{R}^n_+ = \{x = (x_1, \cdots, x_n) \mid x_n > 0\}.$$

Consider

$$\begin{cases} F_\alpha(u(x)) = h(u(x)), & x \in \mathbb{R}^n_+, \\ u(x) \equiv 0, & x \notin \mathbb{R}^n_+. \end{cases} \tag{6.142}$$

Theorem 6.8.3 *Assume that $u \in L_{G,\alpha} \cap C^{1,1}_{loc}$ is a nonnegative solution of problem (6.142). Suppose*

$$\lim_{|x|\to\infty} u(x) = 0, \tag{6.143}$$

and

$$h'(s) \leq s^q, \quad \text{with } q\gamma \geq \alpha; \quad h(0) = 0.$$

Then $u \equiv 0$.

Proof. Based on the assumption (6.143) and $h(0) = 0$, from the proof of Theorem 6.7.2, one can see that

$$\text{either } u(x) > 0 \text{ or } u(x) \equiv 0, \quad \forall x \in \mathbb{R}^n_+.$$

Hence in the following, we may assume that $u > 0$ in $\mathbb{R}^n_+$.

Now we carry out the method of moving planes on the solution u along x_n direction.

Let

$$T_\lambda = \{x \in \mathbb{R}^n \mid x_n = \lambda\}, \ \ \lambda > 0,$$

and

$$\Sigma_\lambda = \{x \in \mathbb{R}^n \mid 0 < x_n < \lambda\}.$$

Let

$$x^\lambda = (x_1, \cdots, x_{n-1}, 2\lambda - x_n)$$

be the reflection of x about the plane T_λ. Denote $w_\lambda(x) = u(x^\lambda) - u(x)$.

The key ingredient in this proof is the *narrow region principle* (Theorem 6.7.4). To see that it still applies in this situation, we only need to take

$$\Sigma = \Sigma_\lambda \cup \mathbb{R}^n_-,$$

where

$$\mathbb{R}^n_- = \{x \in \mathbb{R}^n \mid x_n \leq 0\}.$$

Step 1. For λ sufficiently small, since Σ_λ is a narrow region, we have immediately

$$w_\lambda(x) \geq 0, \ \ \forall\, x \in \Sigma_\lambda. \tag{6.144}$$

Step 2. Inequality (6.144) provides a starting point, from which we can move the plane T_λ upward as long as inequality (6.144) holds. Define

$$\lambda_o = \sup\{\lambda \mid w_\mu(x) \geq 0, x \in \Sigma_\mu, \mu \leq \lambda\}.$$

We show that

$$\lambda_o = \infty. \tag{6.145}$$

Otherwise, if $\lambda_o < \infty$, then by (6.144), combining the *narrow region principle* and *decay at infinity* and going through the similar arguments as in the previous subsection, we are able to show that

$$w_{\lambda_o}(x) \equiv 0 \ \ \text{in } \Sigma_{\lambda_o},$$

which implies

$$u(x_1, \cdots, x_{n-1}, 2\lambda_o) = u(x_1, \cdots, x_{n-1}, 0) = 0.$$

This is impossible, because we assume that $u > 0$ in $\mathbb{R}^n_+$.

Therefore, (6.145) must be valid. Consequently, the solution $u(x)$ is monotone increasing with respect to x_n. This contradicts (6.143).

6.9 Various Maximum Principles for a Fractional System in $\mathbb{R}^n$

6.9.1 Decay at Infinity

Let

$$\Sigma_\lambda = \{x \in \mathbb{R}^n \mid x_1 < \lambda\}.$$

Theorem 6.9.1 (Decay at infinity) *For $0 < \alpha, \beta < 2$, assume that $U \in L_\alpha(\mathbb{R}^n) \cap C^{1,1}_{loc}(\Omega)$, $V \in L_\beta(\mathbb{R}^n) \cap C^{1,1}_{loc}(\Omega)$, and U, V are lower semi-continuous on $\bar{\Omega}$. If*

$$\begin{cases} (-\triangle)^{\alpha/2}U(x) + c_1(x)V(x) \geq 0, \, x \in \Omega, \\ (-\triangle)^{\beta/2}V(x) + c_2(x)U(x) \geq 0, \, x \in \Omega, \\ U(x), V(x) \geq 0, & x \in \Sigma_\lambda \backslash \Omega, \\ U(x^\lambda) = -U(x), & x \in \Sigma_\lambda, \\ V(x^\lambda) = -V(x), & x \in \Sigma_\lambda, \end{cases} \qquad (6.146)$$

with

$$c_1(x) \sim o(\frac{1}{|x|^\alpha}), \quad c_2(x) \sim o(\frac{1}{|x|^\beta}), \quad for \ |x| \ large, \qquad (6.147)$$

and

$$c_i(x) < 0, \quad i = 1, 2,$$

then there exists a constant $R > 0$ (depending on $c_i(x)$, but is independent of U, V) such that if

$$U(\tilde{x}) = \min_\Omega U(x) < 0, \quad V(\bar{x}) = \min_\Omega V(x) < 0,$$

then at least one of $\tilde{x}$ and $\bar{x}$ satisfies

$$|x| \leq R. \qquad (6.148)$$

Proof. By the defining integral of the fractional Laplacian, we have

$$(-\triangle)^{\alpha/2}U(\tilde{x}) = C_{n,\alpha} PV \int_{\mathbb{R}^n} \frac{U(\tilde{x}) - U(y)}{|\tilde{x} - y|^{n+\alpha}} dy$$

$$= C_{n,\alpha} PV \left\{ \int_{\Sigma_\lambda} \frac{U(\tilde{x}) - U(y)}{|\tilde{x} - y|^{n+\alpha}} dy + \int_{\mathbb{R}^n \backslash \Sigma_\lambda} \frac{U(\tilde{x}) - U(y)}{|\tilde{x} - y|^{n+\alpha}} dy \right\}$$

$$= C_{n,\alpha} PV \left\{ \int_{\Sigma_\lambda} \frac{U(\tilde{x}) - U(y)}{|\tilde{x} - y|^{n+\alpha}} dy + \int_{\Sigma_\lambda} \frac{U(\tilde{x}) - U(y^\lambda)}{|\tilde{x} - y^\lambda|^{n+\alpha}} dy \right\}$$

$$= C_{n,\alpha} PV \left\{ \int_{\Sigma_\lambda} \frac{U(\tilde{x}) - U(y)}{|\tilde{x} - y|^{n+\alpha}} dy + \int_{\Sigma_\lambda} \frac{U(\tilde{x}) + U(y)}{|\tilde{x} - y^\lambda|^{n+\alpha}} dy \right\}$$

$$\leq C_{n,\alpha} \int_{\Sigma_\lambda} \left\{ \frac{U(\tilde{x}) - U(y)}{|\tilde{x} - y^\lambda|^{n+\alpha}} + \frac{U(\tilde{x}) + U(y)}{|\tilde{x} - y^\lambda|^{n+\alpha}} \right\} dy$$

$$= C_{n,\alpha} \int_{\Sigma_\lambda} \frac{2U(\tilde{x})}{|\tilde{x} - y^\lambda|^{n+\alpha}} dy.$$

For each fixed λ, there exists a $C > 0$ such that for $\tilde{x} \in \Sigma_\lambda$ and $|\tilde{x}|$ sufficiently large,

$$\int_{\Sigma_\lambda} \frac{1}{|\tilde{x} - y^\lambda|^{n+\alpha}} dy \geq \int_{(B_{3|\tilde{x}|}(\tilde{x}) \backslash B_{2|\tilde{x}|}(\tilde{x})) \cap \tilde{\Sigma}_\lambda} \frac{1}{|\tilde{x} - y|^{n+\alpha}} dy$$

$$\sim \frac{C}{|\tilde{x}|^\alpha}. \qquad (6.149)$$

Hence

$$(-\triangle)^{\alpha/2}U(\tilde{x}) \leq \frac{CU(\tilde{x})}{|\tilde{x}|^\alpha} < 0. \tag{6.150}$$

Similar to (6.149), we can derive that

$$(-\triangle)^{\beta/2}V(\bar{x}) \leq \frac{CV(\bar{x})}{|\bar{x}|^\beta} < 0. \tag{6.151}$$

Combining (6.146), (6.150), and (6.151), for λ sufficiently negative, we have

$$\begin{aligned}
0 &\leq (-\triangle)^{\beta/2}V(\bar{x}) + c_2(\bar{x})U(\bar{x}) \\
&\leq \frac{CV(\bar{x})}{|\bar{x}|^\beta} + c_2(\bar{x})U(\bar{x}) \\
&\leq C\left(\frac{V(\bar{x})}{|\bar{x}|^\beta} - c_2(\bar{x})c_1(\tilde{x})|\tilde{x}|^\alpha V(\tilde{x})\right) \\
&\leq C\left(\frac{V(\bar{x})}{|\bar{x}|^\beta} - c_2(\bar{x})c_1(\tilde{x})|\tilde{x}|^\alpha V(\bar{x})\right) \\
&\leq \frac{CV(\bar{x})}{|\bar{x}|^\beta}(1 - c_1(\tilde{x})|\tilde{x}|^\alpha c_2(\bar{x})|\bar{x}|^\beta) \\
&< 0.
\end{aligned}$$

The last inequality follows from assumption (6.147). This contradiction shows that (6.148) must be true for at least one of $\tilde{x}$ and $\bar{x}$.   □

6.9.2 Narrow Region Principle

Theorem 6.9.2 (Narrow Region Principle) *Let Ω be a bounded narrow region in Σ_λ, such that it is contained in $\{x \mid \lambda - l < x_1 < \lambda\}$ with small l. For $0 < \alpha, \beta < 2$, assume that $U \in L_\alpha(\mathbb{R}^n) \cap C_{loc}^{1,1}(\Omega)$, $V \in L_\beta(\mathbb{R}^n) \cap C_{loc}^{1,1}(\Omega)$, and U, V are lower semi-continuous on $\bar{\Omega}$. If $c_i(x) < 0$, $i = 1, 2$, are bounded from below in Ω and*

$$\begin{cases}
(-\triangle)^{\alpha/2}U(x) + c_1(x)V(x) \geq 0, & x \in \Omega, \\
(-\triangle)^{\beta/2}V(x) + c_2(x)U(x) \geq 0, & x \in \Omega, \\
U(x), V(x) \geq 0, & x \in \Sigma_\lambda \backslash \Omega, \\
U(x^\lambda) = -U(x), & x \in \Sigma_\lambda, \\
V(x^\lambda) = -V(x), & x \in \Sigma_\lambda,
\end{cases} \tag{6.152}$$

then for sufficiently small l, we have

$$U(x), V(x) \geq 0 \text{ in } \Omega. \tag{6.153}$$

If Ω is unbounded, the conclusion still holds under the condition that

$$\varliminf_{|x|\to\infty} U(x), V(x) \geq 0.$$

Further, if either $U(x)$ or $V(x)$ attains 0 somewhere in Ω, then

$$U(x) = V(x) \equiv 0, \quad x \in \mathbb{R}^n. \tag{6.154}$$

Proof. If (6.153) does not hold, then the lower semi-continuity of U on $\bar{\Omega}$ guarantees that there exists some $\tilde{x} \in \bar{\Omega}$ such that

$$U(\tilde{x}) = \min_{\bar{\Omega}} U < 0.$$

And one can further deduce from condition (6.152) that $\tilde{x}$ is in the interior of Ω.

Similar to the previous argument, we have

$$(-\triangle)^{\alpha/2} U(\tilde{x}) = C_{n,\alpha} PV \int_{\mathbb{R}^n} \frac{U(\tilde{x}) - U(y)}{|\tilde{x} - y|^{n+\alpha}} dy$$

$$\leq C_{n,\alpha} \int_{\Sigma_\lambda} \frac{2U(\tilde{x})}{|\tilde{x} - y^\lambda|^{n+\alpha}} dy.$$

Let $D = B_{2l}(\tilde{x}) \cap \tilde{\Sigma}_\lambda$. Then we have

$$\int_{\Sigma_\lambda} \frac{1}{|\tilde{x} - y^\lambda|^{n+\alpha}} dy \geq \int_D \frac{1}{|\tilde{x} - y|^{n+\alpha}} dy$$

$$\geq C \int_{B_{2l}(\tilde{x})} \frac{1}{|\tilde{x} - y|^{n+\alpha}} dy$$

$$\geq \frac{C}{l^\alpha}. \tag{6.155}$$

Thus,

$$(-\triangle)^{\alpha/2} U(\tilde{x}) \leq \frac{CU(\tilde{x})}{l^\alpha} < 0. \tag{6.156}$$

Together with (6.152), we have

$$U(\tilde{x}) \geq -c_1(\tilde{x}) l^\alpha V(\tilde{x}). \tag{6.157}$$

From (6.157), we know that there exists an $\bar{x}$ such that

$$V(\bar{x}) = \min_{\Omega} V(x) < 0.$$

Similar to (6.156), we can derive that

$$(-\triangle)^{\beta/2} V(\bar{x}) \leq \frac{CV(\bar{x})}{l^\beta} < 0.$$

Together (6.157), for l sufficiently small, we have

$$0 \leq (-\triangle)^{\beta/2} V(\bar{x}) + c_2(\bar{x}) U(\bar{x})$$

$$\leq \frac{CV(\bar{x})}{l^\beta} + c_2(\bar{x}) U(\tilde{x})$$

$$\leq C\left(\frac{V(\bar{x})}{l^\beta} - c_2(\bar{x}) c_1(\tilde{x}) l^\alpha V(\tilde{x}) \right)$$

$$\leq C\left(\frac{V(\bar{x})}{l^\beta} - c_2(\bar{x}) c_1(\tilde{x}) l^\alpha V(\bar{x}) \right)$$

$$\leq \frac{CV(\bar{x})}{l^\beta} (1 - c_1(\tilde{x}) c_2(\bar{x}) l^{\alpha+\beta})$$

$$< 0.$$

This contradiction shows that (6.153) must be true.

To prove (6.154), we suppose that there exists an $\eta \in \Omega$ such that

$$V(\eta) = 0.$$

Then

$$(-\triangle)^{\beta/2} V(\eta)$$

$$= C_{n,\beta} PV \int_{\mathbb{R}^n} \frac{-V(y)}{|\eta - y|^{n+\beta}} dy$$

$$= C_{n,\beta} PV \left\{ \int_{\Sigma_\lambda} \frac{-V(y)}{|\eta - y|^{n+\beta}} dy + \int_{\Sigma_\lambda} \frac{-V(y^\lambda)}{|\eta - y^\lambda|^{n+\beta}} dy \right\}$$

$$= C_{n,\beta} PV \int_{\Sigma_\lambda} \left(\frac{1}{|\eta - y^\lambda|^{n+\beta}} - \frac{1}{|\eta - y|^{n+\beta}} \right) V(y)\, dy. \qquad (6.158)$$

If $V(x) \not\equiv 0$, then (6.158) implies that

$$(-\triangle)^{\beta/2} V(\eta) < 0.$$

Together with (6.152), it shows that

$$U(\eta) < 0.$$

This is a contradiction with (6.153). Hence $V(x)$ must be identically 0 in Σ_λ. Since

$$V(x^\lambda) = -V(x), \quad x \in \Sigma_\lambda,$$

it shows that

$$V(x) \equiv 0, \quad x \in \mathbb{R}^n.$$

Again from (6.152), we know that

$$U(x) \leq 0, \quad x \in \Sigma_\lambda.$$

Since we already know that

$$U(x) \geq 0, \quad x \in \Sigma_\lambda,$$

it must hold that

$$U(x) = 0, x \in \Sigma_\lambda.$$

Together with $U(x^\lambda) = -U(x)$, we arrive at

$$U(x) \equiv 0, \quad x \in \mathbb{R}^n.$$

Similarly, one can show that if $U(x)$ attains 0 at one point in Σ_λ, then both $U(x)$ and $V(x)$ are identically 0 in $\mathbb{R}^n$.

This completes the proof.

6.10 Applications to Fractional Systems

We consider

$$\begin{cases} (-\triangle)^{\alpha/2}u(x) = f(v(x)), \ x \in \mathbb{R}^n, \\ (-\triangle)^{\beta/2}v(x) = g(u(x)), \ x \in \mathbb{R}^n, \\ u, v \geq 0, \qquad\qquad\quad x \in \mathbb{R}^n. \end{cases} \tag{6.159}$$

Assume that for $r \geq 0$, f, g are nonnegative continuous functions satisfying:

(a) $f(r)$ and $g(r)$ are non-decreasing about r,

(b) $\frac{f(r)}{r^p}$, $\frac{g(r)}{r^q}$ are bounded near $r = 0$ and non-increasing with $p = \frac{n+\alpha}{n-\beta}$ and $q = \frac{n+\beta}{n-\alpha}$.

Theorem 6.10.1 *If u and v are nonnegative solutions for (6.159), then one of the following statements must be true:*

- *u and v are constant,*
- *u and v are radially symmetric about some point in $\mathbb{R}^n$,*
- *$f(v) = C_1 v^{\frac{n+\alpha}{n-\beta}}$ and $g(u) = C_2 u^{\frac{n+\beta}{n-\alpha}}$.*

In particular, from Theorem 6.10.1, we have

Theorem 6.10.2 *Assume f and g satisfy the conditions in Theorem 6.10.1. If u and v are nonnegative solutions for (6.159), then when $\alpha = \beta$,*

- *either u and v are constant,*
- *or $u(x) = C_1 \left(\frac{c}{c^2 + |x - x_0|^2} \right)^{\frac{n-\alpha}{2}}$ and $v(x) = C_2 \left(\frac{c}{c^2 + |x - x_0|^2} \right)^{\frac{n-\alpha}{2}}$.*

6.10.1 Proof of Theorem 6.10.1

To carry out the *method of moving planes*, we need to know the behavior of the solutions at infinity. To illustrate the idea, we first prove Theorem 6.10.1 with additional growth assumptions on solutions at infinity. Then one will see that Theorem 6.10.1 still holds after the removal of these assumptions.

Radial Symmetry under Decay Assumption

Proof. Assume that for $|x|$ large, $b(p-1) \geq \alpha$ and $a(q-1) > \beta$ (or $b(p-1) > \alpha$ and $a(q-1) \geq \beta$), we have

$$u(x) \sim \frac{1}{|x|^a}, \quad v(x) \sim \frac{1}{|x|^b}.$$

Choose an arbitrary direction for the x_1-axis. Let

$$u_\lambda(x) = u(x^\lambda),$$

$$\tilde{U}_\lambda(x) = u_\lambda(x) - u(x), \quad \tilde{V}_\lambda(x) = v_\lambda(x) - v(x).$$

Step 1. Start moving the plane T_λ from $-\infty$ to the right along the x_1-axis. We show that for λ sufficiently negative,

$$\tilde{U}_\lambda(x), \tilde{V}_\lambda(x) \geq 0, \quad x \in \Sigma_\lambda.$$

For a fixed λ, by the decay rate, we know that

$$u_\lambda(x), \, u(x) \to 0, \quad |x| \to \infty.$$

Hence

$$\tilde{U}_\lambda(x) \to 0, \quad |x| \to \infty. \tag{6.160}$$

Similarly, one can show that for $x \in \Sigma_\lambda$,

$$\tilde{V}_\lambda(x) \to 0, \quad |x| \to \infty.$$

If

$$\Sigma_{\tilde{U}_\lambda}^- = \{x \in \Sigma_\lambda \mid \tilde{U}_\lambda(x) < 0\} \neq \emptyset,$$

then by (6.160), we know that there must exist some $\tilde{x} \in \Sigma_\lambda$ such that

$$\tilde{U}_\lambda(\tilde{x}) = \min_{\Sigma_\lambda} \tilde{U}_\lambda < 0.$$

On one hand, by (6.150),

$$(-\triangle)^{\alpha/2} \tilde{U}_\lambda(\tilde{x}) \leq \frac{C \tilde{U}_\lambda(\tilde{x})}{|\tilde{x}|^\alpha} < 0. \tag{6.161}$$

On the other hand, by (6.159),

$$(-\triangle)^{\alpha/2} \tilde{U}_\lambda(\tilde{x}) = f(v_\lambda(\tilde{x})) - f(v(\tilde{x})).$$

Therefore, by the monotonicity of f, we have

$$\tilde{V}_\lambda(\tilde{x}) < 0.$$

This implies that there exists some $\bar{x} \in \Sigma_\lambda$ such that

$$\tilde{V}_\lambda(\bar{x}) = \min_{\Sigma_\lambda} \tilde{V}_\lambda < 0.$$

By the *mean value theorem*, we have

$$(-\triangle)^{\alpha/2} \tilde{U}_\lambda(\tilde{x}) \tag{6.162}$$
$$= f(v_\lambda(\tilde{x})) - f(v(\tilde{x}))$$
$$= \frac{f(v_\lambda(\tilde{x}))}{v_\lambda^p(\tilde{x})} v_\lambda^p(\tilde{x}) - \frac{f(v(\tilde{x}))}{v^p(\tilde{x})} v^p(\tilde{x})$$
$$\geq \frac{f(v(\tilde{x}))}{v^p(\tilde{x})} [v_\lambda^p(\tilde{x}) - v^p(\tilde{x})]$$
$$= \frac{f(v(\tilde{x}))}{v^p(\tilde{x})} p \xi^{p-1}(\tilde{x}) \tilde{V}_\lambda(\tilde{x}), \qquad \xi \in [v_\lambda(\tilde{x}), v(\tilde{x})]$$
$$\geq \frac{f(v(\tilde{x}))}{v^p(\tilde{x})} p v^{p-1}(\tilde{x}) \tilde{V}_\lambda(\tilde{x}). \tag{6.163}$$

Through a similar argument, one can show that

$$\tilde{U}_\lambda(\bar{x}) < 0,$$

and

$$(-\triangle)^{\beta/2}\tilde{V}_\lambda(\bar{x}) \geq \frac{g(u(\bar{x}))}{u^q(\bar{x})}qu^{q-1}(\bar{x})\tilde{U}_\lambda(\bar{x}),$$

Let

$$c_1(x) = -\frac{f(v(x))}{v^p(x)}pv^{p-1}(x),$$

and

$$c_2(x) = -\frac{g(u(x))}{u^q(x)}qu^{q-1}(x).$$

From assumption *(b)*, we know that for $|\tilde{x}|$, $|\bar{x}|$ large,

$$c_1(\tilde{x}) \sim \frac{1}{|\tilde{x}|^{b(p-1)}} \sim O(\frac{1}{|\tilde{x}|^\alpha}),$$

$$c_2(\bar{x}) \sim \frac{1}{|\bar{x}|^{a(q-1)}} \sim O(\frac{1}{|\bar{x}|^\beta}).$$

By Theorem 6.9.1 *(decay at infinity)*, for λ sufficiently negative (or $|\lambda| < R$ in Theorem 6.9.1), one of $\tilde{U}_\lambda(x)$ and $\tilde{V}_\lambda(x)$ must be nonnegative in Σ_λ. Without loss of generality, we assume that

$$\tilde{U}_\lambda(x) \geq 0, \quad x \in \Sigma_\lambda. \qquad (6.164)$$

To prove (6.164) also holds for $\tilde{V}_\lambda(x)$, we argue by contradiction.

If $\tilde{V}_\lambda(x)$ is negative somewhere in Σ_λ, then there must exist some $\bar{x} \in \Sigma_\lambda$ such that

$$\tilde{V}_\lambda(\bar{x}) = \min_{\Sigma_\lambda} \tilde{V}_\lambda < 0.$$

From previous arguments, we know that

$$0 > \frac{C\tilde{V}_\lambda(\bar{x})}{|\bar{x}|^\beta} \geq (-\triangle)^{\beta/2}\tilde{V}_\lambda(\bar{x})$$

$$\geq \frac{g(u(\bar{x}))}{u^q(\bar{x})}qu^{q-1}(\bar{x})\tilde{U}_\lambda(\bar{x}) \geq 0.$$

A contradiction.

This completes step 1.

Step 2. Continue to move the plane T_λ until the limiting position

$$\lambda_o = \sup\{\lambda \mid U_\rho(x), V_\rho(x) \geq 0, \, x \in \Sigma_\rho, \forall \rho \leq \lambda\}.$$

Obviously,

$$\lambda_o < \infty.$$

Otherwise, for any $\lambda > 0$,

$$u(0^\lambda) \geq u(0) \geq 0.$$

At same time, we know

$$u(0^\lambda) \sim \frac{1}{|0^\lambda|^a} \to 0, \quad \lambda \to \infty.$$

This makes a contradiction.

Then we show that

$$U_{\lambda_o}, V_{\lambda_o} \equiv 0 \tag{6.165}$$

through a contradiction argument. The main idea is that if (6.165) is violated, then we can keep moving the plane to the right. In other words, there exists an $\varepsilon > 0$ small such that for any $\lambda \in (\lambda_o, \lambda_o + \varepsilon)$, it holds that

$$U_\lambda, V_\lambda \geq 0, \quad x \in \Sigma_\lambda.$$

This contradicts the definition of λ_o. Here we omit the proof since a similar proof will be given for the next case where no decay assumptions are imposed on the solutions.

Due to the arbitrariness of the x_1 direction, we conclude that u, v are symmetric about some point in $\mathbb{R}^n$. $\square$

Radial Symmetry without Decay Assumption

Without any decay assumption on u and v, we first consider their Kelvin transforms. For any $x^o \in \mathbb{R}^n$, let $\bar{u}$ and $\bar{v}$ be the Kelvin transform of u and v respectively:

$$\bar{u} = \frac{1}{|x - x^o|^{n-\alpha}} u\left(\frac{x - x^o}{|x - x^o|^2} + x^o\right), \tag{6.166}$$

$$\bar{v} = \frac{1}{|x - x^o|^{n-\beta}} v\left(\frac{x - x^o}{|x - x^o|^2} + x^o\right). \tag{6.167}$$

Without loss of generality, let $x^o = 0$, then

$$\bar{u} = \frac{1}{|x|^{n-\alpha}} u\left(\frac{x}{|x|^2}\right),$$

$$\bar{v} = \frac{1}{|x|^{n-\beta}} v\left(\frac{x}{|x|^2}\right). \tag{6.168}$$

By (6.159) and (6.168),

$$(-\triangle)^{\alpha/2}\bar{u}(x) = \frac{1}{|x|^{n+\alpha}} [(-\triangle)^{\alpha/2}u]\left(\frac{x}{|x|^2}\right)$$

$$= \frac{1}{|x|^{n+\alpha}} f(|x|^{n-\beta}\bar{v}(x)). \tag{6.169}$$

Similarly,

$$(-\triangle)^{\beta/2}\bar{v}(x) = \frac{1}{|x|^{n+\beta}}g(|x|^{n-\alpha}\bar{u}(x)).$$

Let

$$U_\lambda(x) = \bar{u}_\lambda(x) - \bar{u}(x), \quad V_\lambda(x) = \bar{v}_\lambda(x) - \bar{v}(x).$$

Then

$$(-\triangle)^{\alpha/2}U_\lambda(x) = \frac{f(|x^\lambda|^{n-\beta}\bar{v}_\lambda(x))}{|x^\lambda|^{n+\alpha}} - \frac{f(|x|^{n-\beta}\bar{v}(x))}{|x|^{n+\alpha}}, \tag{6.170}$$

$$(-\triangle)^{\beta/2}V_\lambda(x) = \frac{g(|x^\lambda|^{n-\alpha}\bar{u}_\lambda(x))}{|x^\lambda|^{n+\beta}} - \frac{g(|x|^{n-\alpha}\bar{u}(x))}{|x|^{n+\beta}}. \tag{6.171}$$

Step 1. Start moving the plane T_λ from $-\infty$ to the right along the x_1-axis.
We show that for λ sufficiently negative,

$$U_\lambda(x), V_\lambda(x) \geq 0, x \in \Sigma_\lambda \backslash \{0^\lambda\}.$$

We note that for λ sufficiently negative, there exists a positive constant C such that

$$U_\lambda(x), V_\lambda(x) \geq C > 0, x \in B_\varepsilon(0^\lambda)\backslash\{0^\lambda\}.$$

The proof appears in the appendix.
   Meanwhile, by definition, for λ fixed,

$$U_\lambda(x), V_\lambda(x) \to 0, \text{ as } |x| \to \infty.$$

Hence if

$$\Sigma_{U_\lambda}^- = \{x \in \Sigma_\lambda \mid U_\lambda(x) < 0\} \neq \emptyset,$$

there must exist a point $\tilde{x}$ such that

$$U_\lambda(\tilde{x}) = \min_{\Sigma_\lambda} U_\lambda < 0.$$

Through arguments similar to those in Theorem 6.9.1, we obtain

$$(-\triangle)^{\alpha/2}U_\lambda(\tilde{x}) \leq \frac{CU_\lambda(\tilde{x})}{|\tilde{x}|^\alpha} < 0. \tag{6.172}$$

We claim that

$$V_\lambda(\tilde{x}) < 0. \tag{6.173}$$

Indeed, otherwise, we have $V_\lambda(\tilde{x}) \geq 0$. From (6.170),

$$(-\triangle)^{\alpha/2}U_\lambda(\tilde{x})$$
$$= \frac{f(|\tilde{x}^\lambda|^{n-\beta}\bar{v}_\lambda(\tilde{x}))}{|\tilde{x}^\lambda|^{n+\alpha}} - \frac{f(|\tilde{x}|^{n-\beta}\bar{v}(\tilde{x}))}{|\tilde{x}|^{n+\alpha}}$$
$$= \frac{f(|\tilde{x}^\lambda|^{n-\beta}\bar{v}_\lambda(\tilde{x}))}{[|\tilde{x}^\lambda|^{n-\beta}\bar{v}_\lambda(\tilde{x})]^p}\bar{v}_\lambda^p(\tilde{x}) - \frac{f(|\tilde{x}|^{n-\beta}\bar{v}_\lambda(\tilde{x}))}{[|\tilde{x}|^{n-\beta}\bar{v}_\lambda(\tilde{x})]^p}\bar{v}_\lambda^p(\tilde{x})$$
$$+ \frac{f(|\tilde{x}|^{n-\beta}\bar{v}_\lambda(\tilde{x}))}{|\tilde{x}|^{n+\alpha}} - \frac{f(|\tilde{x}|^{n-\beta}\bar{v}(\tilde{x}))}{|\tilde{x}|^{n+\alpha}} \tag{6.174}$$
$$\geq 0.$$

This is a contradiction with (6.172), and it proves (6.173).

From (6.173), we can see that there exists a point $\bar{x}$ such that

$$V_\lambda(\bar{x}) = \min_{\Sigma_\lambda} V_\lambda(x) < 0,$$

and

$$
\begin{aligned}
(-\triangle)^{\alpha/2} U_\lambda(\tilde{x}) &= \frac{f(|\tilde{x}^\lambda|^{n-\beta}\bar{v}_\lambda(\tilde{x}))}{|\tilde{x}^\lambda|^{n+\alpha}} - \frac{f(|\tilde{x}|^{n-\beta}\bar{v}(\tilde{x}))}{|\tilde{x}|^{n+\alpha}} \\
&= \frac{f(|\tilde{x}^\lambda|^{n-\beta}\bar{v}_\lambda(\tilde{x}))}{[|\tilde{x}^\lambda|^{n-\beta}\bar{v}_\lambda(\tilde{x})]^p} \bar{v}_\lambda^p(\tilde{x}) - \frac{f(|\tilde{x}|^{n-\beta}\bar{v}(\tilde{x}))}{[|\tilde{x}|^{n-\beta}\bar{v}(\tilde{x})]^p} \bar{v}^p(\tilde{x}) \\
&\geq \frac{f(|\tilde{x}|^{n-\beta}\bar{v}(\tilde{x}))}{[|\tilde{x}|^{n-\beta}\bar{v}(\tilde{x})]^p} [\bar{v}_\lambda^p(\tilde{x}) - \bar{v}^p(\tilde{x})] \\
&= \frac{f(|\tilde{x}|^{n-\beta}\bar{v}(\tilde{x}))}{[|\tilde{x}|^{n-\beta}\bar{v}(\tilde{x})]^p} p\xi^{p-1} V_\lambda(\tilde{x}), \qquad \xi \in [\bar{v}_\lambda(\tilde{x}), \bar{v}(\tilde{x})] \\
&\geq \frac{f(|\tilde{x}|^{n-\beta}\bar{v}(\tilde{x}))}{[|\tilde{x}|^{n-\beta}\bar{v}(\tilde{x})]^p} p\bar{v}^{p-1} V_\lambda(\tilde{x}) \\
&= \frac{pf(v(\frac{\tilde{x}}{|\tilde{x}|^2}))}{|\tilde{x}|^{\alpha+\beta}v(\frac{\tilde{x}}{|\tilde{x}|^2})} V_\lambda(\tilde{x}).
\end{aligned}
\tag{6.175}
$$

Similar to (6.173), we can show that

$$U_\lambda(\bar{x}) < 0.$$

Then we have

$$
\begin{aligned}
(-\triangle)^{\beta/2} V_\lambda(\bar{x}) &= \frac{g(|\bar{x}^\lambda|^{n-\alpha}\bar{u}_\lambda(\tilde{x}))}{|\bar{x}^\lambda|^{n+\beta}} - \frac{g(|\bar{x}|^{n-\alpha}\bar{u}(\tilde{x}))}{|\bar{x}|^{n+\beta}} \\
&\geq \frac{qg(u(\frac{\tilde{x}}{|\tilde{x}|^2}))}{|\bar{x}|^{\alpha+\beta}u(\frac{\tilde{x}}{|\tilde{x}|^2})} U_\lambda(\bar{x}).
\end{aligned}
$$

Let

$$c_1(x) = -\frac{pf(v(\frac{x}{|x|^2}))}{|x|^{\alpha+\beta}v(\frac{x}{|x|^2})},$$

and

$$c_2(x) = -\frac{qg(u(\frac{x}{|x|^2}))}{|x|^{\alpha+\beta}u(\frac{x}{|x|^2})}.$$

From Theorem 6.9.1 (*decay at infinity*), for λ sufficiently negative (or $\lambda < -R$ in Theorem 6.9.1), one of $U_\lambda(x)$ and $V_\lambda(x)$ must be positive in $\Sigma_\lambda\backslash\{0^\lambda\}$. Without loss of generality, we assume that

$$U_\lambda(x) \geq 0, \quad x \in \Sigma_\lambda\backslash\{0^\lambda\}. \tag{6.176}$$

To prove (6.176) also holds for $V_\lambda(x)$, we argue by contradiction.

If $V_\lambda(x)$ is negative somewhere in $\Sigma_\lambda\backslash\{0^\lambda\}$, then there must exist some $\bar{x} \in \Sigma_\lambda$ such that

$$V_\lambda(\bar{x}) = \min_{\Sigma_\lambda} V_\lambda < 0.$$

From previous arguments, we know that

$$0 > \frac{CV_\lambda(\bar{x})}{|\bar{x}|^\beta} \geq (-\triangle)^{\beta/2}V_\lambda(\bar{x})$$

$$\geq \frac{qg(u(\frac{\bar{x}}{|\bar{x}|^2}))}{|\bar{x}|^{\alpha+\beta}u(\frac{\bar{x}}{|\bar{x}|^2})}U_\lambda(\bar{x}) \geq 0.$$

A contradiction. This completes step 1.

Step 2. Keeping moving the plane T_λ until the limiting position. Let

$$\lambda_o = \sup\{\lambda \leq 0 \mid U_\rho(x), V_\rho(x) \geq 0, x \in \Sigma_\rho\backslash\{0^\rho\}, \forall \rho \leq \lambda\}.$$

We claim that

$$U_{\lambda_o}(x), V_{\lambda_o}(x) \equiv 0, \quad x \in \Sigma_{\lambda_o}\backslash\{0^{\lambda_o}\}. \tag{6.177}$$

We prove the above identity in two cases.

Case i. $\lambda_o < 0$.
By the definition of λ_o,

$$U_{\lambda_o}(x), V_{\lambda_o}(x) \geq 0, \quad x \in \Sigma_{\lambda_o}\backslash\{0^{\lambda_o}\}. \tag{6.178}$$

We have either

$$U_{\lambda_o}(x) = V_{\lambda_o}(x) \equiv 0, \quad x \in \Sigma_{\lambda_o}\backslash\{0^{\lambda_o}\}, \tag{6.179}$$

or

$$U_{\lambda_o}(x), V_{\lambda_o}(x) > 0, \quad x \in \Sigma_{\lambda_o}\backslash\{0^{\lambda_o}\}. \tag{6.180}$$

To see this, we suppose there exists some $\tilde{x} \in \Sigma_{\lambda_o}$ such that

$$U_{\lambda_o}(\tilde{x}) = \min_{\Sigma_{\lambda_o}} U_{\lambda_o} = 0,$$

then it must be true that

$$U_{\lambda_o}(x) \equiv 0, \quad x \in \Sigma_{\lambda_o}\backslash\{0^{\lambda_o}\}. \tag{6.181}$$

Otherwise,

$$(-\triangle)^{\alpha/2}U_{\lambda_o}(\tilde{x}) = C_{n,\alpha}PV\int_{\mathbb{R}^n}\frac{-U_{\lambda_o}(y)}{|x^0 - y|^{n+\alpha}}dy$$

$$< 0. \tag{6.182}$$

On the other hand,

$$(-\triangle)^{\alpha/2} U_{\lambda_o}(\tilde{x})$$

$$= \frac{f(|\tilde{x}^{\lambda_o}|^{n-\beta}\bar{v}_{\lambda_o}(\tilde{x}))}{|\tilde{x}^{\lambda_o}|^{n+\alpha}} - \frac{f(|\tilde{x}|^{n-\beta}\bar{v}(\tilde{x}))}{|\tilde{x}|^{n+\alpha}}$$

$$= \frac{f(|\tilde{x}^{\lambda_o}|^{n-\beta}\bar{v}_{\lambda_o}(\tilde{x}))}{[|\tilde{x}^{\lambda_o}|^{n-\beta}\bar{v}_{\lambda_o}(\tilde{x})]^p} \bar{v}^p_{\lambda_o}(\tilde{x}) - \frac{f(|\tilde{x}|^{n-\beta}\bar{v}_{\lambda_o}(\tilde{x}))}{[|\tilde{x}|^{n-\beta}\bar{v}_{\lambda_o}(\tilde{x})]^p} \bar{v}^p_{\lambda_o}(\tilde{x})$$

$$+ \frac{f(|\tilde{x}|^{n-\beta}\bar{v}_{\lambda_o}(\tilde{x}))}{|\tilde{x}|^{n+\alpha}} - \frac{f(|\tilde{x}|^{n-\beta}\bar{v}(\tilde{x}))}{|\tilde{x}|^{n+\alpha}}$$

$$\geq 0.$$

A contradiction with (6.182). This proves (6.181).

When $U_{\lambda_o}(x) \equiv 0$, since

$$U_{\lambda_o}(x) = -U_{\lambda_o}(x^{\lambda_o}),$$

we have

$$U_{\lambda_o}(x) \equiv 0, \quad x \in \mathbb{R}^n.$$

Hence

$$(-\triangle)^{\alpha/2} U_{\lambda_o}(x) = 0.$$

Together with (6.174), one can deduce that

$$\bar{v}_{\lambda_o}(x) \leq \bar{v}(x), \quad x \in \Sigma_{\lambda_o}\setminus\{0^{\lambda_o}\}.$$

By the definition of λ_o,

$$\bar{v}_{\lambda_o}(x) \geq \bar{v}(x), \quad x \in \Sigma_{\lambda_o}\setminus\{0^{\lambda_o}\}.$$

Thus

$$V_{\lambda_o}(x) = \bar{v}_{\lambda_o}(x) - \bar{v}(x) \equiv 0, \quad x \in \Sigma_{\lambda_o}\setminus\{0^{\lambda_o}\},$$

and

$$V_{\lambda_o}(x) \equiv 0, \quad x \in \mathbb{R}^n. \tag{6.183}$$

Similarly, if $V_{\lambda_o}(x) = 0$ somewhere, then we can show that

$$V_{\lambda_o}(x) = U_{\lambda_o}(x) \equiv 0, \quad x \in \mathbb{R}^n.$$

Further, we can prove that

$$\frac{f(t)}{t^{\frac{n+\alpha}{n-\beta}}} = C, \tag{6.184}$$

and

$$\frac{g(t)}{t^{\frac{n+\beta}{n-\alpha}}} = C. \tag{6.185}$$

We give the detailed proofs in Lemma A.1.6 in the appendix.

When

$$U_{\lambda_o}(x), V_{\lambda_o}(x) > 0, \quad x \in \Sigma_{\lambda_o}\setminus\{0^{\lambda_o}\}, \tag{6.186}$$

we are able to keep moving the plane T_λ. To be precise, for some $\varepsilon > 0$ small such $\lambda_o + \varepsilon < 0$, it holds that

$$U_\lambda(x), V_\lambda(x) \geq 0, \quad x \in \Sigma_\lambda \backslash \{0^\lambda\}, \ \forall \lambda \in (\lambda_o, \lambda_o + \varepsilon). \tag{6.187}$$

This is a contradiction with the definition of λ_o. Therefore,

$$U_{\lambda_o}(x), V_{\lambda_o}(x) > 0$$

must not happen.

From (6.186), we have

$$U_{\lambda_o}(x), \ V_{\lambda_o}(x) > 0, \quad x \in (\Sigma_{\lambda_o} \backslash \{0^{\lambda_o}\}) \cap B_R(0).$$

Later, in Lemma A.1.5, we show that

$$U_{\lambda_o}(x), \ V_{\lambda_o}(x) \geq C > 0, \quad x \in B_\varepsilon(0^{\lambda_o}) \backslash \{0^{\lambda_o}\}.$$

Together with the above bounded-away-from-0 result, we derive that for $\delta > 0$ small there exists some C that

$$U_{\lambda_o}(x), \ V_{\lambda_o}(x) \geq C > 0, \quad x \in (\Sigma_{\lambda_o - \delta} \backslash \{0^{\lambda_o}\}) \cap B_R(0). \tag{6.188}$$

For $\delta, \varepsilon \ll |\lambda_o|$,

$$0^\lambda \in (\Sigma_{\lambda_o - \delta} \backslash \{0^{\lambda_o}\}) \cap B_R(0).$$

Since $U_\lambda(x), V_\lambda(x)$ depend on λ continuously, it holds

$$U_\lambda(x), \ V_\lambda(x) \geq 0, \quad x \in (\Sigma_{\lambda_o - \delta} \backslash \{0^{\lambda_o}\}) \cap B_R(0). \tag{6.189}$$

By *decay at infinity*, we know that if

$$U_\lambda(\tilde{x}) = \min_{\Sigma_\lambda} U_\lambda < 0,$$

then there exists a R large such that

$$|\tilde{x}| \leq R.$$

Hence $\tilde{x} \in (\Sigma_\lambda \backslash \Sigma_{\lambda_o - \delta}) \cap B_R(0)$.

Similar to (6.173), we can show that

$$V_\lambda(\tilde{x}) < 0.$$

Therefore, there exists some $\bar{x}$ such that

$$V_\lambda(\bar{x}) = \min_{\Sigma_\lambda} V_\lambda < 0.$$

- If $\bar{x} \in B_R^c \cap \Sigma_\lambda$, then

$$0 > \frac{C V_\lambda(\bar{x})}{|\bar{x}|^\beta} \geq (-\triangle)^{\beta/2} V_\lambda(\bar{x}) \geq \frac{qg(u(\frac{\bar{x}}{|\bar{x}|^2}))}{|\bar{x}|^{\alpha+\beta} u(\frac{\bar{x}}{|\bar{x}|^2})} U_\lambda(\bar{x})$$

$$\geq \frac{qg(u(\frac{\bar{x}}{|\bar{x}|^2}))}{|\bar{x}|^{\alpha+\beta} u(\frac{\bar{x}}{|\bar{x}|^2})} U_\lambda(\tilde{x}). \tag{6.190}$$

Meanwhile, at $\tilde{x}$, it holds

$$\frac{U_\lambda(\tilde{x})}{(\delta+\varepsilon)^\alpha} \geq (-\triangle)^{\alpha/2}U_\lambda(\tilde{x}) \geq \frac{pf(v(\frac{\tilde{x}}{|\tilde{x}|^2}))}{|\tilde{x}|^{\alpha+\beta}v(\frac{\tilde{x}}{|\tilde{x}|^2})}V_\lambda(\tilde{x})$$

$$\geq \frac{pf(v(\frac{\tilde{x}}{|\tilde{x}|^2}))}{|\tilde{x}|^{\alpha+\beta}v(\frac{\tilde{x}}{|\tilde{x}|^2})}V_\lambda(\bar{x}). \qquad (6.191)$$

By (6.190) and (6.191), we have

$$1 \leq \frac{qg(u(\frac{\bar{x}}{|\bar{x}|^2}))}{|\bar{x}|^{\alpha+\beta}u(\frac{\bar{x}}{|\bar{x}|^2})}(\delta+\varepsilon)^\alpha \frac{pf(v(\frac{\tilde{x}}{|\tilde{x}|^2}))}{|\tilde{x}|^{\alpha+\beta}v(\frac{\tilde{x}}{|\tilde{x}|^2})}. \qquad (6.192)$$

For a fixed $\lambda_o < 0$, when ε is sufficiently small, we have $\lambda < \lambda_o + \varepsilon < \frac{\lambda_o}{2}$. Since $\tilde{x} \in \Sigma_\lambda$, it holds that $|\tilde{x}| > -\frac{\lambda_o}{2}$. Therefore $\frac{pf(v(\frac{\tilde{x}}{|\tilde{x}|^2}))}{|\tilde{x}|^{\alpha+\beta}v(\frac{\tilde{x}}{|\tilde{x}|^2})}$ is bounded. Notice that $\frac{qg(u(\frac{\bar{x}}{|\bar{x}|^2}))}{|\bar{x}|^{\alpha+\beta}u(\frac{\bar{x}}{|\bar{x}|^2})}$ is also bounded for $|\bar{x}| > R$. This shows that (6.192) must not be true for δ sufficiently small.

It thus implies that $\bar{x} \in B_R^c \cap \Sigma_\lambda$ must not happen.

- If $\bar{x} \in (\Sigma_\lambda \backslash \Sigma_{\lambda_o-\delta}) \cap B_R(0)$, then by (6.155),

$$(-\triangle)^{\alpha/2}U_\lambda(\tilde{x}) \leq \frac{U_\lambda(\tilde{x})}{(\delta+\varepsilon)^\alpha} < 0.$$

Similarly, one can show that

$$(-\triangle)^{\beta/2}V_\lambda(\bar{x}) \leq \frac{V_\lambda(\bar{x})}{(\delta+\varepsilon)^\beta} < 0.$$

Since

$$\tilde{x}, \bar{x} \in (\Sigma_\lambda \backslash \Sigma_{\lambda_o-\delta}) \cap B_R(0),$$

we can choose $\delta, \varepsilon \ll \frac{|\lambda_o|}{2}$ such that $|\tilde{x}|, |\bar{x}| > \frac{2|\lambda_o|}{3}$. Then for some C, it holds that

$$v(\frac{\tilde{x}}{|\tilde{x}|^2}), \; u(\frac{\bar{x}}{|\bar{x}|^2}) \leq C,$$

and

$$\frac{1}{(\delta+\varepsilon)^{\alpha+\beta}} > C + 2 > c_1(\tilde{x})c_2(\bar{x}).$$

It then follows from Theorem 6.9.2 (*narrow region principle*) that

$$U_\lambda(x), \; V_\lambda(x) \geq 0, \quad x \in (\Sigma_\lambda \backslash \Sigma_{\lambda_o-\delta}) \cap B_R(0). \qquad (6.193)$$

This shows that neither $U_\lambda(x)$ nor $V_\lambda(x)$ has negative minimum in $\Sigma_\lambda \backslash \{0^\lambda\}$. Therefore,

$$U_\lambda(x), \; V_\lambda(x) \geq 0, \quad \Sigma_\lambda \backslash \{0^\lambda\}.$$

This completes the proof of (6.187).

From

$$U_{\lambda_o}(x),\, V_{\lambda_o}(x) \geq 0, \quad \Sigma_{\lambda_o} \backslash \{0^{\lambda_o}\},$$

it follows that

$$u(x) = O(\frac{1}{|x|^{n-\alpha}}),\; v(x) = O(\frac{1}{|x|^{n-\beta}}), \quad |x| \to \infty.$$

Through an identical argument to that in the previous section, we can show that $u(x)$, $v(x)$ are radially symmetric about some point.

Case ii. $\lambda_o = 0$.

In this case, we can move the plane from $+\infty$ to the left and show that

$$U_0(x), V_0(x) \leq 0,\, x \in \Sigma_0.$$

Thus,

$$U_0(x), V_0(x) \equiv 0,\, x \in \Sigma_0.$$

For a more general Kelvin transform as in (6.166) and (6.167), through a similar argument as above one can show that

$$\lambda_o = x_1^o,$$

and

$$U_{\lambda_o}(x),\, V_{\lambda_o}(x) \equiv 0, \quad x \in \Sigma_{\lambda_o}.$$

For any $x^1, x^2 \in \mathbb{R}^n$, let their midpoint be the center of the Kelvin transform:

$$x^o = \frac{x^1 + x^2}{2}.$$

Let

$$y^1 = \frac{x^1 - x^o}{|x^1 - x^o|^2} + x^o, \quad y^2 = \frac{x^2 - x^o}{|x^2 - x^o|^2} + x^o.$$

Then

$$y^2 = (y^1)^{\lambda_o},$$

and

$$\bar{u}(y^1) = \bar{u}(y^2), \quad \bar{v}(y^1) = \bar{v}(y^2).$$

It thus implies that

$$u(x^1) = u(x^2), \quad v(x^1) = v(x^2).$$

Since x^1, x^2 are arbitrary, u and v must be constant.

Combining *Case i* and *Case ii*, we complete the proof. □

6.10.2 Proof of Theorem 6.10.2

From Theorem 6.10.1, we can see that when $\alpha = \beta$, if u and v are nonnegative solutions for (6.159), then

- either u and v are constant,
- or $f(v) = C_1 v^{\frac{n+\alpha}{n-\alpha}}$ and $g(u) = C_2 u^{\frac{n+\alpha}{n-\alpha}}$.

By Theorem 2 in [ZCCY], we know that when $\alpha = \beta$, system (6.159) is equivalent to

$$\begin{cases} u(x) = \int_{\mathbb{R}^n} \frac{C_1 v^{\frac{n+\alpha}{n-\alpha}}(y)}{|x-y|^{n-\alpha}} dy \\ v(x) = \int_{\mathbb{R}^n} \frac{C_2 u^{\frac{n+\alpha}{n-\alpha}}(y)}{|x-y|^{n-\alpha}} dy. \end{cases}$$

From the results in [CLO2], we have

$$u(x) = C_1 \left(\frac{c}{c^2 + |x - x_0|^2} \right)^{\frac{n-\alpha}{2}}, \quad v(x) = C_2 \left(\frac{c}{c^2 + |x - x_0|^2} \right)^{\frac{n-\alpha}{2}}.$$

This proves the theorem.

7

Method of Moving Planes in Integral Forms

7.1 Introduction

In Chapter 5, applying Liouville theorems, we established the equivalence between the pseudo-differential equations and the integral equations. In order to study the properties of solutions of the latter, such as symmetry, monotonicity and non-existence, in this chapter, we introduce the *method of moving planes in integral forms*, which has been proven to be a powerful tool in investigating integral equations. This method was introduced when the first author and Li and Ou considered an open problem posed by Lieb – showing that all positive solutions of the integral equation

$$u(x) = \int_{\mathbb{R}^n} \frac{1}{|x-y|^{n-\alpha}} u^{\frac{n+\alpha}{n-\alpha}}(y) dy$$

are radially symmetric and then classifying them. This problem was originated from the well-known Hardy-Littlewood-Sobolev inequality as described in the following.

Let $0 < \alpha < n$ and let $s, r > 1$ such that $\frac{1}{r} + \frac{1}{s} = \frac{n+\alpha}{n}$. The Hardy-Littlewood-Sobolev inequality states that:

$$\int_{\mathbb{R}^n} \int_{\mathbb{R}^n} f(x) \frac{1}{|x-y|^{n-\alpha}} g(y) dx dy \leq C(n, s, \alpha) \|f\|_r \|g\|_s \qquad (7.1)$$

for any $f \in L^r(\mathbb{R}^n)$ and $g \in L^s(\mathbb{R}^n)$, where $\|f\|_r$ is the $L^r(\mathbb{R}^n)$ norm of f.

To find the best constant $C = C(n, s, \alpha)$ in the inequality, one usually maximize the functional

$$J(f, g) = \int_{\mathbb{R}^n} \int_{\mathbb{R}^n} f(x) \frac{1}{|x-y|^{n-\alpha}} g(y) dx dy \qquad (7.2)$$

under the constraints

$$\|f\|_r = \|g\|_s = 1. \qquad (7.3)$$

Let (f, g) be a maximizer, or more generally, a critical point of (7.2) under the constraints (7.3). Making a substitution

$$u = \lambda_1 f^{r-1}, \quad v = \lambda_2 g^{s-1}, \quad p = \frac{1}{r-1}, \quad q = \frac{1}{s-1},$$

and by a proper choice of constants λ_1 and λ_2, one can derive that (u, v) satisfies the following system of integral equations in $\mathbb{R}^n$:

$$\begin{cases} u(x) = \int_{\mathbb{R}^n} \frac{1}{|x-y|^{n-\alpha}} v^q(y) dy \\ v(x) = \int_{\mathbb{R}^n} \frac{1}{|x-y|^{n-\alpha}} u^p(y) dy \end{cases} \tag{7.4}$$

with

$$\frac{1}{q+1} + \frac{1}{p+1} = \frac{n-\alpha}{n}.$$

This integral system is closely related to the system of partial differential (or pseudo-differential) equations

$$\begin{cases} (-\triangle)^{\alpha/2} u = v^q, \quad u > 0, \text{ in } \mathbb{R}^n, \\ (-\triangle)^{\alpha/2} v = u^p, \quad v > 0, \text{ in } \mathbb{R}^n. \end{cases} \tag{7.5}$$

For all values of $0 < \alpha < n$, one can show that, if (u, v) is a pair of solutions of (7.4), then it is also a pair of solutions for (7.5). The converse is also true in many cases. So far as we know:

(a) In the case $0 < \alpha < 2$, using the method introduced in Chapter 5, one can show that every pair of locally bounded positive solutions of (7.5) also satisfies integral system (7.4).

(b) When α is any even number between 0 and n, it is shown in [CL7] that locally bounded positive solutions are also solutions of the integral system.

(c) For all real values of α between 0 and n, the two systems are equivalent in the sense of weak solutions in $H^{\alpha/2}(\mathbb{R}^n)$.

When $\alpha = 2$, system (7.5) becomes the Lane-Emden system

$$\begin{cases} -\triangle u = v^q, \quad u > 0, \text{ in } \mathbb{R}^n, \\ -\triangle v = u^p, \quad v > 0, \text{ in } \mathbb{R}^n. \end{cases}$$

The well-known Lane-Emden conjecture states that, in the subcritical case

$$\frac{1}{q+1} + \frac{1}{p+1} > \frac{n-2}{n},$$

the above system admits no positive solution. This conjecture has attracted much attention from the mathematical community, and a series of partial results have been obtained. The study of the corresponding integral systems will definitely shed some light on the complete solution of this conjecture.

In the special case where $p = q = \frac{n+\alpha}{n-\alpha}$, and $u(x) = v(x)$, the system becomes:

$$u(x) = \int_{\mathbb{R}^n} \frac{1}{|x-y|^{n-\alpha}} u^{\frac{n+\alpha}{n-\alpha}}(y) \, dy, \quad u > 0, \text{ in } \mathbb{R}^n. \tag{7.6}$$

And the corresponding PDE is the well-known family of semi-linear equations

$$(-\triangle)^{\alpha/2}u = u^{\frac{n+\alpha}{n-\alpha}}, \quad u > 0, \text{ in } \mathbb{R}^n. \tag{7.7}$$

In particular, when $n \geq 3$ and $\alpha = 2$, equation (7.7) becomes

$$-\triangle u = u^{\frac{n+2}{n-2}}, \quad u > 0, \text{ in } \mathbb{R}^n. \tag{7.8}$$

The classification of the solutions of (7.8) provides an important ingredient in the study of the well-known Yamabe problem and the prescribing scalar curvature problem on Riemannian manifolds. It is also essential in deriving a priori estimates in many related nonlinear elliptic equations.

Solutions to (7.8) were studied by Gidas, Ni, and Nirenberg [GNN]. They proved that all the positive solutions of (7.8) with reasonable behavior at infinity

$$u(x) = O(\frac{1}{|x|^{n-2}}) \tag{7.9}$$

are radially symmetric and therefore assume the form of

$$C(\frac{t}{t^2 + |x - x_o|^2})^{(n-2)/2}$$

with some positive constants C and t.

Later, in their elegant paper [CGS], Caffarelli, Gidas, and Spruck removed the growth condition (7.9) and obtained the same result. Then Chen and Li [CL4], and Li [Li] simplified their proof. Later, Lin [Lin] obtained the similar classification result for (7.7) with $\alpha = 4$, and Wei and Xu [WX] extended it to the case where α is any even number between 0 and n.

Obviously, for other real values of α between 0 and n, equation (7.7) is also of practical interest and importance. For instance, it arises as the Euler-Lagrange equation of the functional

$$I(u) = \int_{\mathbb{R}^n} |(-\triangle)^{\frac{\alpha}{4}}u|^2 dx \Big/ \left(\int_{\mathbb{R}^n} |u|^{\frac{2n}{n-\alpha}} dx\right)^{\frac{n-\alpha}{n}}.$$

The classification of the solutions would provide the best constant in the inequality of the critical Sobolev imbedding from $H^{\frac{\alpha}{2}}(\mathbb{R}^n)$ to $L^{\frac{2n}{n-\alpha}}(\mathbb{R}^n)$:

$$\left(\int_{\mathbb{R}^n} |u|^{\frac{2n}{n-\alpha}} dx\right)^{\frac{n-\alpha}{n}} \leq C \int_{\mathbb{R}^n} |(-\triangle)^{\frac{\alpha}{4}}u|^2 dx.$$

Moreover, these fractional powers of Laplacian in (7.7) have numerous applications and are very interesting in its own right.

In [L], Lieb classified all the maximizers of the functional (7.2) under the constraint (7.3) in the special case where $p = q = \frac{n+\alpha}{n-\alpha}$ and thus obtained the best constants in the HLS inequalities in that case. He then posed the classification of all the critical points of the functional – the solutions of integral equation (7.6) as an open problem.

This open problem was solved in [CLO]:

Proposition 7.1.1 *All solutions of partial differential equation (7.7) satisfy the integral equation (7.6), and vice versa. Every positive solution $u(x) \in L_{loc}^{\frac{2n}{n-\alpha}}(\mathbb{R}^n)$ of (7.6) is radially symmetric and decreasing about some point x_o and therefore assumes the form of*

$$C(\frac{t}{t^2 + |x - x_o|^2})^{(n-\alpha)/2} \tag{7.10}$$

with some positive constants C and t.

To prove Proposition 7.1.1, a new idea – the *method of moving planes in integral forms* was introduced.

As is known to people with experience in the *method of moving planes*, each problem has its own difficulty. For partial differential equations, the local properties of the differential operators are used extensively. For integral equation (7.6), the lack of knowledge of the local properties prevents us from using many known results, such as maximum principles. However, by exploring various special features possessed by the integral equation in its global form, and through estimating some specific integral norms, we are still able to establish the symmetry of solutions.

This new approach is quite different from the traditional ones for PDEs. Since its introduction, there have seen many applications to various systems of integral equations as well as to PDEs.

In Section 7.2, we use two simple examples in $\mathbb{R}^n$ to illustrate the idea. In Section 7.3, we prove the non-existence of positives solutions for the Dirichlet problem of a semi-linear equation involving the fractional Laplacian or higher integer power of the Laplacian. In this situation, the kernel of the integral equation – the Green's function has a more complex form, and we show how to carry out the *method of moving planes in integral forms* for more general kernels. In Section 7.4, we consider a system of m equations with more general nonlinearity. We use the *method of moving planes* to establish the equivalence between the PDE and the integral systems in the critical case, then we prove the non-existence of positive solutions in the subcritical case. In Section 7.5, we list a variety of applications of this method.

7.2 Illustration of MMP in Integral Forms

7.2.1 The Idea of Method of Moving Planes in Integral Forms

To briefly illustrate the difference between the usual *method of moving planes* and the *method of moving planes in integral forms*, we recall the basics of this method. Still take the Euclidean space $\mathbb{R}^n$ for an example. Let u be a positive solution of a certain equation. We want to prove that it is symmetric and monotone in a given direction. First we choose an arbitrary direction for the x_1 axis. For any real number λ, let

$$T_\lambda = \{x = (x_1, x_2, \cdots, x_n) \in \mathbb{R}^n \mid x_1 = \lambda\}.$$

This is a plane perpendicular to x_1-axis and the plane that we will move with. Let Σ_λ denote the region to the left of the plane, i.e.,

$$\Sigma_\lambda = \{x \in \mathbb{R}^n \mid x_1 < \lambda\}.$$

Let

$$x^\lambda = (2\lambda - x_1, x_2, \cdots, x_n)$$

be the reflection of the point $x = (x_1, \cdots, x_n)$ about the plane T_λ (see Figure 6.1).

We compare the values of the solution u at point x and x^λ, and we want to show that u is symmetric about some plane T_{λ_o}. Let

$$w_\lambda(x) = u(x^\lambda) - u(x),$$

then it is equivalent to show that, there exists some λ_o, such that

$$w_{\lambda_o}(x) \equiv 0 \, , \, \forall x \in \Sigma_{\lambda_o}.$$

To this end, we briefly go through the following two steps.

Step 1. We show that for λ sufficiently negative, we have

$$w_\lambda(x) \geq 0 \, , \, \forall x \in \Sigma_\lambda. \tag{7.11}$$

This is plausible because we usually assume that u approaches zero near infinity and that u is positive.

Then we are able to start off from this neighborhood of $x_1 = -\infty$, and move the plane T_λ along the x_1 direction to the right as long as the inequality (7.11) holds.

Step 2. We continuously move the plane this way up to its limiting position. More precisely, define

$$\lambda_o = \sup\{\lambda \mid w_\mu(x) \geq 0, \forall x \in \Sigma_\mu, \mu \leq \lambda\}.$$

We prove that $w_{\lambda_o}(x) \equiv 0$ for all $x \in \Sigma_{\lambda_o}$, or equivalently, u is symmetric about the plane T_{λ_o}. This is usually done by a contradiction argument. We show that if $w_{\lambda_o}(x) \not\equiv 0$, then there exists $\lambda > \lambda_o$, such that (7.11) holds, and this contradicts with the definition of λ_o.

From the above explanation, one can see that the key to the *method of moving planes* is to establish inequality (7.11), and for partial differential equations, *maximum principles* are powerful tools for this task. While for integral equations, we use an entirely different idea. We estimate a certain norm of w_λ on the set

$$\Sigma_\lambda^- = \{x \in \Sigma_\lambda \mid w_\lambda(x) < 0\}$$

where inequality (7.11) is violated. We then show that this norm must be zero, and hence Σ_λ^- is almost empty. This idea is illustrated by examples in the next two subsections.

7.2.2 Symmetry of Solutions for the Single Equation

In this subsection, we consider the single integral equation

$$u(x) = \int_{\mathbb{R}^n} \frac{1}{|x-y|^{n-\alpha}} u^{\frac{n+\alpha}{n-\alpha}}(y)\, dy, \quad u > 0, \text{ in } \mathbb{R}^n. \tag{7.12}$$

Using the *method of moving planes in integral forms*, we prove

Theorem 7.2.1 *Assume $0 < \alpha < n$. Let $u \in L_{loc}^{\frac{2n}{n-\alpha}}(\mathbb{R}^n)$ be a positive solution of (7.12). Then it is radially symmetric and monotone decreasing about some point in $\mathbb{R}^n$.*

To better illustrate the idea, we first prove Theorem 7.2.1 under a stronger assumption that $u \in L^{\frac{2n}{n-\alpha}}(\mathbb{R}^n)$. Then we show how the idea of this proof can be extended under a weaker condition that $u \in L_{loc}^{\frac{2n}{n-\alpha}}$.

As before, define

$$\Sigma_\lambda = \{x = (x_1, \ldots, x_n) \mid x_1 < \lambda\},$$

let $x^\lambda = (2\lambda - x_1, x_2, \ldots, x_n)$ and $u_\lambda(x) = u(x^\lambda)$ (see the previous Figure 6.1.)

We need the following two lemmas.

Lemma 7.2.1 (An Equivalent Form of the Hardy-Littlewood-Sobolev Inequality) *Let $g \in L^{\frac{np}{n+\alpha p}}(\mathbb{R}^n)$ for $\frac{n}{n-\alpha} < p < \infty$. Define*

$$Tg(x) = \int_{\mathbb{R}^n} \frac{1}{|x-y|^{n-\alpha}} g(y)\, dy.$$

Then

$$\|Tg\|_p \le C(n, p, \alpha)\|g\|_{\frac{np}{n+\alpha p}}. \tag{7.13}$$

Proof. From the original HLS inequality (7.1), we have

$$\int_{\mathbb{R}^n} f(x)Tg(x)dx \le C\|f\|_r\|g\|_s$$

with

$$\frac{1}{r} + \frac{1}{s} = \frac{n+\alpha}{n}.$$

Consequently, for $\frac{1}{p} + \frac{1}{r} = 1$, the duality between L^p and L^r implies

$$\|Tg\|_p \le C\|g\|_s.$$

Then it is easy to calculate that $s = \frac{np}{n+\alpha p}$.

Lemma 7.2.2 *Write $\tau = \frac{n+\alpha}{n-\alpha}$. For any solution $u(x)$ of (7.12), we have*

$$u(x) - u_\lambda(x) = \int_{\Sigma_\lambda} \Big(\frac{1}{|x-y|^{n-\alpha}} - \frac{1}{|x^\lambda - y|^{n-\alpha}}\Big)(u^\tau(y) - u_\lambda^\tau(y))dy. \quad (7.14)$$

It is also true for $v(x) = \frac{1}{|x|^{n-\alpha}}u\big(\frac{x}{|x|^2}\big)$, the Kelvin type transform of $u(x)$, for any $x \neq 0$.

Proof. Since $|x - y^\lambda| = |x^\lambda - y|$, we have

$$u(x) = \int_{\Sigma_\lambda} \frac{1}{|x-y|^{n-\alpha}}u^\tau(y)dy + \int_{\Sigma_\lambda} \frac{1}{|x^\lambda - y|^{n-\alpha}}u_\lambda^\tau(y)dy,$$

$$u(x^\lambda) = \int_{\Sigma_\lambda} \frac{1}{|x^\lambda - y|^{n-\alpha}}u^\tau(y)dy + \int_{\Sigma_\lambda} \frac{1}{|x-y|^{n-\alpha}}u_\lambda^\tau(y)dy.$$

This implies (7.14). The conclusion for $v(x)$ follows similarly since it satisfies the same integral equation (7.6) for $x \neq 0$. $\square$

Proof of Theorem 7.2.1 (*Under Stronger Assumption $u \in L^{\frac{2n}{n-\alpha}}(\mathbb{R}^n)$*).

Let $w_\lambda(x) = u_\lambda(x) - u(x)$. As indicated in the previous subsections, we will first show that, for λ sufficiently negative,

$$w_\lambda(x) \ge 0, \quad \forall x \in \Sigma_\lambda. \quad (7.15)$$

Here, and in the following, when we write an inequality or equality for an L^p function, we mean that it holds almost everywhere in the given set.

Then we can start moving the plane $T_\lambda = \partial \Sigma_\lambda$ from near $x_1 = -\infty$ to the right, as long as inequality (7.15) holds. Let

$$\lambda_o = \sup\{\lambda \mid w_\mu(x) \geq 0, \ \forall x \in \Sigma_\mu, \ \mu \leq \lambda\}.$$

We will show that

$$\lambda_o < \infty, \quad \text{and } w_{\lambda_o}(x) \equiv 0.$$

Unlike traditional method of moving planes, here we do not have any differential equations and the corresponding maximum principles for w_λ. Instead, we will exploit some global properties of the integral equations and estimate some L^q norm of w_λ.

Step 1. Start moving the plane from near $x_1 = -\infty$.

Define

$$\Sigma_\lambda^- = \{x \in \Sigma_\lambda \mid w_\lambda(x) < 0\}. \tag{7.16}$$

This is the set where inequality (7.15) is violated. We show that for sufficiently negative values of λ, Σ_λ^- must be measure zero. By Lemma 7.2.2, it is easy to verify that

$$
\begin{aligned}
u(x) - u_\lambda(x) &\leq \int_{\Sigma_\lambda^-} \left(\frac{1}{|x-y|^{n-\alpha}} - \frac{1}{|x^\lambda - y|^{n-\alpha}} \right) [u^\tau(y) - u_\lambda^\tau(y)] dy \\
&\leq \int_{\Sigma_\lambda^-} \frac{1}{|x-y|^{n-\alpha}} [u^\tau(y) - u_\lambda^\tau(y)] dy \\
&= \tau \int_{\Sigma_\lambda^-} \frac{1}{|x-y|^{n-\alpha}} \psi_\lambda^{\tau-1}(y)[u(y) - u_\lambda(y)] dy \\
&\leq \tau \int_{\Sigma_\lambda^-} \frac{1}{|x-y|^{n-\alpha}} u^{\tau-1}(y)[u(y) - u_\lambda(y)] dy,
\end{aligned}
$$

where $\psi_\lambda(x)$ is valued between $u_\lambda(x)$ and $u(x)$ by the *mean value theorem*, and since on Σ_λ^-, $u_\lambda(x) < u(x)$, we have $\psi_\lambda(x) \leq u(x)$.

We first apply Hardy-Littlewood-Sobolev inequality (7.13) (its equivalent form) to the above to obtain, for any $q > \frac{n}{n-\alpha}$:

$$\|w_\lambda\|_{L^q(\Sigma_\lambda^-)} \leq C \|u^{\tau-1} w_\lambda\|_{L^{nq/(n+\alpha q)}(\Sigma_\lambda^-)}.$$

Then use the generalized Hölder inequality

$$\|f \cdot g\|_{L^t} \leq \|f\|_{L^r} \|g\|_{L^s}, \quad \text{with } \frac{1}{t} = \frac{1}{r} + \frac{1}{s},$$

to the last integral in the above. Choosing

$$t = \frac{nq}{n + \alpha q}, \quad s = q, \quad \text{and} \quad r = \frac{n}{\alpha},$$

we arrive at

$$\|w_\lambda\|_{L^q(\Sigma_\lambda^-)} \le C \|u^{\tau-1}\|_{L^{\frac{n}{\alpha}}(\Sigma_\lambda^-)} \|w_\lambda\|_{L^q(\Sigma_\lambda^-)}$$

$$= C \left\{ \int_{\Sigma_\lambda^-} u^{\tau+1}(y)\, dy \right\}^{\frac{\alpha}{n}} \|w_\lambda\|_{L^q(\Sigma_\lambda^-)}. \tag{7.17}$$

Due to the condition that $u \in L^{\tau+1}(\mathbb{R}^n)$, we can choose N sufficiently large, such that for $\lambda \le -N$, we have

$$C \left\{ \int_{\Sigma_\lambda^-} u^{\tau+1}(y)\, dy \right\}^{\frac{\alpha}{n}} \le \frac{1}{2}.$$

Now (7.17) implies that $\|w_\lambda\|_{L^q(\Sigma_\lambda^-)} = 0$, and therefore Σ_λ^- must be measure zero. This verifies (7.15) and hence completes *Step 1*.

Step 2. We now move the plane $T_\lambda = \{x \mid x_1 = \lambda\}$ to the right as long as (7.15) holds. Let λ_o be as defined before, then we must have $\lambda_o < \infty$. This can be seen by applying a similar argument as in *Step 1* from λ near $+\infty$.

Now we show that

$$w_{\lambda_o}(x) \equiv 0, \quad \forall x \in \Sigma_{\lambda_o}. \tag{7.18}$$

Otherwise, we have $w_{\lambda_o}(x) \ge 0$, but $w_{\lambda_o}(x) \not\equiv 0$ on Σ_{λ_o}; we show that the plane can be moved further to the right. More precisely, there exists an ϵ depending on n, α, and the solution $u(x)$ itself such that $w_\lambda(x) \ge 0$ on Σ_λ for all λ in $[\lambda_o, \lambda_o + \epsilon)$.

We will use (7.17) again and show that condition $u \in L^{\tau+1}(\mathbb{R}^n)$ ensures one to choose ϵ sufficiently small, so that for all λ in $[\lambda_o, \lambda_o + \epsilon)$,

$$C \left\{ \int_{\Sigma_\lambda^-} u^{\tau+1}(y)\, dy \right\}^{\frac{\alpha}{n}} \le \frac{1}{2}. \tag{7.19}$$

Now by (7.17) and (7.19), we have $\|w_\lambda\|_{L^q(\Sigma_\lambda^-)} = 0$, and therefore Σ_λ^- must be measure zero. Hence for these values of $\lambda > \lambda_o$, we have

$$w_\lambda(x) \ge 0, \quad a.e. \ \forall x \in \Sigma_\lambda.$$

This contradicts with the definition of λ_o. Therefore, (7.18) must hold.

Now what's left is to prove inequality (7.19).

For any small $\eta > 0$, we can choose R sufficiently large, so that

$$\left(\int_{\mathbb{R}^n \setminus B_R(0)} u^{\tau+1}(y) dy \right)^{\frac{\alpha}{n}} < \eta. \tag{7.20}$$

Fix this R, we show that the measure of $\Sigma_\lambda^- \cap B_R(0)$ is sufficiently small for λ close to λ_o. By Lemma 7.2.2, we have in fact $w_{\lambda_o}(x) > 0$ in the interior of Σ_{λ_o}. For any $\delta > 0$, let

$$E_\delta = \{x \in \Sigma_{\lambda_o} \cap B_R(0) \mid w_{\lambda_o}(x) > \delta\} \quad \text{and} \quad F_\delta = (\Sigma_{\lambda_o} \cap B_R(0)) \backslash E_\delta.$$

Then obviously

$$\lim_{\delta \to 0} \mu(F_\delta) = 0.$$

For $\lambda > \lambda_o$, let

$$D_\lambda = (\Sigma_\lambda \backslash \Sigma_{\lambda_o}) \cap B_R(0).$$

Then it is easy to see that

$$\left(\Sigma_\lambda^- \cap B_R(0)\right) \subset \left(\Sigma_\lambda^- \cap E_\delta\right) \cup F_\delta \cup D_\lambda. \tag{7.21}$$

Apparently, the measure of D_λ gets smaller as λ approaches λ_o. We show that the measure of $\Sigma_\lambda^- \cap E_\delta$ can also be sufficiently small as λ goes to λ_o. In fact, for any $x \in \Sigma_\lambda^- \cap E_\delta$, we have

$$w_\lambda(x) = u(x^\lambda) - u(x^{\lambda_o}) + u(x^{\lambda_o}) - u(x) < 0.$$

Hence

$$u(x^{\lambda_o}) - u(x^\lambda) > w_{\lambda_o}(x) > \delta.$$

It follows that

$$\left(\Sigma_\lambda^- \cap E_\delta\right) \subset G_\delta \equiv \{x \in B_R(0) \mid u(x^{\lambda_o}) - u(x^\lambda) > \delta\}. \tag{7.22}$$

While by the well-known Chebyshev inequality, we have

$$\mu(G_\delta) \leq \frac{1}{\delta^{\tau+1}} \int_{G_\delta} |u(x^{\lambda_o}) - u(x^\lambda)|^{\tau+1} dx$$

$$\leq \frac{1}{\delta^{\tau+1}} \int_{B_R(0)} |u(x^{\lambda_o}) - u(x^\lambda)|^{\tau+1} dx.$$

For each fixed δ, as λ goes to λ_o, the right hand side of the above inequality can be as small as we wish. Therefore by (7.22) and (7.21), the measure of $\Sigma_\lambda^- \cap B_R(0)$ can also be sufficiently small. Combining this with (7.20), we arrive at (7.19). This completes the proof of the theorem. $\square$

Remark 7.2.1 *Actually, one can show that the solution $u(x)$ is continuous, and hence $w_\lambda(x)$ is continuous in both x and λ. Then the proof for inequality (7.19) is much simpler.*

Outline of Proof of Theorem 7.2.1 (*Under Weaker Assumption That u Is only Locally $L^{\tau+1}$*).

Since we do not assume any integrability condition of $u(x)$ near infinity, we are not able to carry out the *method of moving planes* directly on $u(x)$. To overcome this difficulty, we consider $v(x)$, the Kelvin type transform of $u(x)$. It is easy to verify that $v(x)$ satisfies the same equation (7.12), but has a possible singularity at origin, where we need to pay some particular attention. Since u is locally $L^{\tau+1}$, it is easy to see that $v(x)$ has no singularity at infinity, more precisely, for any domain Ω that is a positive distance away from the origin,

$$\int_\Omega v^{\tau+1}(y)\, dy < \infty. \tag{7.23}$$

We compare $v(x)$ and $v_\lambda(x)$ on $\Sigma_\lambda\backslash\{0^\lambda\}$, where 0^λ is the reflection of the origin 0 about the plane T_λ. The proof also consists of two steps. In step 1, we show that there exists an $N > 0$ such that for $\lambda \leq -N$, we have

$$v(x) \leq v_\lambda(x), \quad \forall x \in \Sigma_\lambda\backslash\{0^\lambda\}. \tag{7.24}$$

Thus we can start moving the plane continuously from $\lambda \leq -N$ to the right as long as (7.24) holds. If the plane stops at $x_1 = \lambda_o$ for some $\lambda_o < 0$, then $v(x)$ must be symmetric and monotone about the plane $x_1 = \lambda_o$. This implies that $v(x)$ has no singularity at the origin and $u(x)$ has no singularity at infinity. In this case, we can carry on the moving planes on $u(x)$ directly to obtain the radial symmetry and monotonicity. Otherwise, we can move the plane all the way to $x_1 = 0$, which is shown in step 2. Since the direction of x_1 can be chosen arbitrarily, we deduce that $v(x)$ must be radially symmetric and decreasing about the origin, so does $u(x)$. These arguments are entirely similar to the proof under the stronger assumption. What we need to do is to replace u there by v and Σ_λ there by $\Sigma_\lambda\backslash\{0^\lambda\}$.

7.2.3 Symmetry of Solutions for Systems

As another example, we consider the system of integral equations in $\mathbb{R}^n$

$$\begin{cases} u(x) = \int_{\mathbb{R}^n} \frac{1}{|x-y|^{n-\alpha}} v^q(y) dy \\ v(x) = \int_{\mathbb{R}^n} \frac{1}{|x-y|^{n-\alpha}} u^p(y) dy \end{cases} \tag{7.25}$$

in the critical case, that is

$$\frac{1}{q+1} + \frac{1}{p+1} = \frac{n-\alpha}{n}.$$

We prove

Theorem 7.2.2 *Assume $0 < \alpha < n$. Let (u, v) be a pair of positive solutions of (7.25) and $p, q \geq 1$. Assume that $u \in L^{p+1}(\mathbb{R}^n)$ and $v \in L^{q+1}(\mathbb{R}^n)$. Then u and v are radially symmetric and decreasing about some point x_o in $\mathbb{R}^n$.*

Define Σ_λ, T_λ, x^λ, and $u_\lambda(x)$ as before, and write $v_\lambda(x) = v(x^\lambda)$. Similar to the proof of Lemma 7.2.2, one can derive

Lemma 7.2.3 *For any pair of solutions $(u(x), v(x))$ of (7.25), we have*

$$u(x) - u_\lambda(x) = \int_{\Sigma_\lambda} \left(\frac{1}{|x-y|^{n-\alpha}} - \frac{1}{|x^\lambda - y|^{n-\alpha}} \right) (v^q(y) - v_\lambda^q(y)) \, dy, \quad (7.26)$$

and

$$v(x) - v_\lambda(x) = \int_{\Sigma_\lambda} \left(\frac{1}{|x-y|^{n-\alpha}} - \frac{1}{|x^\lambda - y|^{n-\alpha}} \right) (u^p(y) - u_\lambda^p(y)) \, dy.$$

Proof. To prove Theorem 7.2.2, we compare $u(x)$ with $u_\lambda(x)$ and $v(x)$ with $v_\lambda(x)$ on Σ_λ. The proof consists of two steps. In Step 1, we show that there exists an $N > 0$ such that for $\lambda \le -N$, we have

$$u(x) \le u_\lambda(x) \text{ and } v(x) \le v_\lambda(x), \quad \forall x \in \Sigma_\lambda. \quad (7.27)$$

Thus we can start moving the plane continuously from $\lambda \le -N$ to the right as long as (7.27) holds. In Step 2, we show that the plane must stop at $x_1 = \lambda_o$ for some $\lambda_o < \infty$, and $u(x)$ and $v(x)$ must be symmetric and monotone about the plane T_{λ_o}. Since the direction of x_1 can be chosen arbitrarily, we deduce that $u(x)$ and $v(x)$ must be radially symmetric and decreasing about some point (the same point).

Step 1. Define

$$\Sigma_\lambda^u = \{ x \in \Sigma_\lambda \mid u_\lambda(x) < u(x) \},$$

and

$$\Sigma_\lambda^v = \{ x \in \Sigma_\lambda \mid v_\lambda(x) < v(x) \},$$

We show that for sufficiently negative values of λ, both Σ_λ^u and Σ_λ^v must be measure zero. By Lemma 7.2.3, it is easy to verify that

$$u(x) - u_\lambda(x) \le C \int_{\Sigma_\lambda^v} \frac{1}{|x-y|^{n-\alpha}} [v^{q-1}(v - v_\lambda)](y) dy.$$

It follows from the Hardy-Littlewood-Sobolev inequality that

$$\|u_\lambda - u\|_{L^{p+1}(\Sigma_\lambda^u)} \le C \|v^{q-1}(v_\lambda - v)\|_{L^{(q+1)/q}(\Sigma_\lambda^v)}.$$

Then by the Hölder inequality,

$$\|u_\lambda - u\|_{L^{p+1}(\Sigma_\lambda^u)} \le C \|v\|_{L^{q+1}(\Sigma_\lambda^v)}^{q-1} \|v_\lambda - v\|_{L^{q+1}(\Sigma_\lambda^v)}. \quad (7.28)$$

Similarly, one can show that

$$\|v_\lambda - v\|_{L^{q+1}(\Sigma_\lambda^v)} \le C \|u\|_{L^{p+1}(\Sigma_\lambda^u)}^{p-1} \|u_\lambda - u\|_{L^{p+1}(\Sigma_\lambda^u)}. \quad (7.29)$$

Combining (7.28) and (7.29), we arrive at

$$\|u_\lambda - u\|_{L^{p+1}(\Sigma_\lambda^u)} \le C \|v\|_{L^{q+1}(\Sigma_\lambda)}^{q-1} \|u\|_{L^{p+1}(\Sigma_\lambda)}^{p-1} \|u_\lambda - u\|_{L^{p+1}(\Sigma_\lambda^u)}. \quad (7.30)$$

By the integrability condition $u \in L^{p+1}(\mathbb{R}^n)$ and $v \in L^{q+1}(\mathbb{R}^n)$, we can choose N sufficiently large, such that for $\lambda \leq -N$, we have

$$C\|v\|_{L^{q+1}(\Sigma_\lambda)}^{q-1}\|u\|_{L^{p+1}(\Sigma_\lambda)}^{p-1} \leq \frac{1}{2}.$$

Now (7.30) implies that $\|u_\lambda - u\|_{L^{p+1}(\Sigma_\lambda^u)} = 0$, and therefore Σ_λ^u must be measure zero. Similarly, one can show that Σ_λ^v is measure zero. Therefore (7.27) holds. This completes Step 1.

Step 2. We now move the plane $x_1 = \lambda$ to the right as long as (7.27) holds. Define

$$\lambda_o = \sup\{\lambda \mid u(x) \leq u_\mu(x) \text{ and } v(x) \leq v_\mu(x), \ \forall x \in \Sigma_\mu, \ \mu \leq \lambda\}.$$

Then one can see that

$$\lambda_o < +\infty.$$

Otherwise, one can apply a similar argument as in *Step 1*, but place the plane T_λ near $+\infty$ to derive a contradiction.

Next, we show that both u and v must be symmetric about T_{λ_o} by using a contradiction argument. Suppose that on Σ_{λ_o}, we have

$$u(x) \leq u_{\lambda_o}(x) \text{ and } v(x) \leq v_{\lambda_o}(x), \text{ but } u(x) \not\equiv u_{\lambda_o}(x) \text{ or } v(x) \not\equiv v_{\lambda_o}(x);$$

we show that the plane can be moved further to the right. More precisely, there exists an ϵ depending on n, α, and the solution $(u(x), v(x))$ itself such that

$$u(x) \leq u_\lambda(x) \text{ and } v(x) \leq v_\lambda(x) \text{ on } \Sigma_\lambda \text{ for all } \lambda \text{ in } [\lambda_o, \lambda_o + \epsilon). \tag{7.31}$$

In the case

$$v(x) \not\equiv v_{\lambda_o}(x) \text{ on } \Sigma_{\lambda_o},$$

by (7.26), we have in fact $u(x) < u_{\lambda_o}(x)$ in the interior of Σ_{λ_o}. Let

$$\tilde{\Sigma}_{\lambda_o}^u = \{x \in \Sigma_{\lambda_o} \mid u(x) \geq u_{\lambda_o}(x)\}, \quad \text{and} \quad \tilde{\Sigma}_{\lambda_o}^v = \{x \in \Sigma_{\lambda_o} \mid v(x) \geq v_{\lambda_o}(x)\}.$$

Then obviously, $\tilde{\Sigma}_{\lambda_o}^u$ has measure zero, and $\lim_{\lambda \to \lambda_o} \Sigma_\lambda^u \subset \tilde{\Sigma}_{\lambda_o}^u$. The same is true for that of v. From (7.28) and (7.29), we deduce,

$$\|u_\lambda - u\|_{L^{p+1}(\Sigma_\lambda^u)} \leq C\|v\|_{L^{q+1}(\Sigma_\lambda^v)}^{q-1}\|u\|_{L^{p+1}(\Sigma_\lambda^u)}^{p-1}\|u_\lambda - u\|_{L^{p+1}(\Sigma_\lambda^u)}. \tag{7.32}$$

Again the integrability conditions $u \in L^{p+1}(\mathbb{R}^n)$ and $v \in L^{q+1}(\mathbb{R}^n)$ ensure that one can choose ϵ sufficiently small, so that for all λ in $[\lambda_o, \lambda_o + \epsilon)$,

$$C\|v\|_{L^{q+1}(\Sigma_\lambda^v)}^{q-1}\|u\|_{L^{p+1}(\Sigma_\lambda^u)}^{p-1} \leq \frac{1}{2}.$$

Now by (7.32), we have $\|u_\lambda - u\|_{L^{p+1}(\Sigma_\lambda^u)} = 0$, and hence Σ_λ^u must be measure zero. Similarly, Σ_λ^v must also be measure zero. This verifies (7.31), and therefore completes the proof of the theorem. $\square$

7.3 Integral Equations Involving General Green's Functions

Let
$$\mathbb{R}^n_+ = \{x = (x_1, \cdots, x_n) \in \mathbb{R}^n \mid x_n > 0\}$$
be the upper half Euclidean space.

Consider the Dirichlet problem for poly-harmonic operators

$$\begin{cases} (-\triangle)^m u = u^p, & x \in \mathbb{R}^n_+, \\ u = \frac{\partial u}{\partial x_n} = \cdots = \frac{\partial^{m-1} u}{\partial x_n^{m-1}} = 0, & x \in \partial \mathbb{R}^n_+, \end{cases} \qquad (7.33)$$

where m is any positive integer, $2m < n$, and $1 < p \le \frac{n+2m}{n-2m}$.

Also consider the Dirichlet problem for the fractional Laplacian

$$\begin{cases} (-\triangle)^m u = u^p, & x \in \mathbb{R}^n_+, \\ u = 0, & x \in (\mathbb{R}^n_+)^c. \end{cases} \qquad (7.34)$$

Here m is any real number between 0 and 1.

It can be shown that both problem (7.33) and (7.34) are equivalent to the following integral equation

$$u(x) = \int_{\mathbb{R}^n_+} G_+(x, y) u^p(y) dy, \qquad (7.35)$$

where $G_+(x, y)$ is the corresponding Green's function. In this section, we will show how the *method of moving planes in integral forms* can be used to investigate integral equations involving general Green's functions.

For each fixed $y \in \mathbb{R}^n_+$, the Green's function is the solution of

$$\begin{cases} (-\triangle)^m G_+(x, y) = \delta(x - y) & \text{in } \mathbb{R}^n_+, \\ G_+ = \frac{\partial G_+}{\partial x_n} = \cdots = \frac{\partial^{m-1} G_+}{\partial x_n^{m-1}} = 0 & \text{on } \partial \mathbb{R}^n_+; \text{ or} \\ G_+(x, y) = 0 \text{ (in the case } 0 < m < 1) \ x \text{ or } y \notin \mathbb{R}^n_+. \end{cases} \qquad (7.36)$$

Let
$$s(x, y) = |x - y|^2 \text{ and } t = 4 x_n y_n.$$

Then for $x, y \in \mathbb{R}^n_+$, $x \neq y$, we have the following representation

$$G_+(x, y) = \frac{C_{n,m}}{|x - y|^{n-2m}} \int_0^{t/s} \frac{z^{m-1}}{(z+1)^{\frac{n}{2}}} dz,$$

where $C_{n,m}$ is a positive constant depending on m and n.

In the following, when studying integral equation (7.35), we assume that m is any real number between 0 and $n/2$. We will establish

Theorem 7.3.1 *Assume $\frac{n}{n-2m} < p \le \frac{n+2m}{n-2m}$. If $u \in L_{loc}^{\frac{n(p-1)}{2m}}(\mathbb{R}^n_+)$ is a non-negative solution of (7.35), then $u(x) \equiv 0$.*

To prove the theorem, we will first employ the *method of moving planes in integral forms* to show that in both critical and subcritical cases, the positive solutions must depend on x_n only. Then we will derive a contradiction with the finiteness of the integral on the right hand side of (7.35).

7.3.1 Properties of Green's Functions

In this section, we introduce some properties of the Green's function $G_+(x,y)$ on the upper half space $\mathbb{R}^n_+$, which will be the key ingredients in carrying out the *method of moving planes*.

We prepare to move the plane perpendicular to x_n axis from the boundary upward to show that the solutions are monotone increasing with respect to x_n. To this end, let λ be a positive real number and let the moving plane be

$$T_\lambda = \{x \in \mathbb{R}^n_+ \mid x_n = \lambda\}.$$

We denote Σ_λ the region between the plane $x_n = 0$ and the plane $x_n = \lambda$. That is

$$\Sigma_\lambda = \{x = (x_1, \cdots, x_{n-1}, x_n) \in \mathbb{R}^n_+ \mid 0 < x_n < \lambda\}.$$

Let

$$x^\lambda = (x_1, \cdots, x_{n-1}, 2\lambda - x_n)$$

be the reflection of the point $x = (x_1, \cdots, x_{n-1}, x_n)$ about the plane T_λ,

$$\Sigma_\lambda^C = \mathbb{R}^n_+ \setminus \Sigma_\lambda,$$

the complement of Σ_λ,

$$u_\lambda(x) = u(x^\lambda) \text{ and } w_\lambda(x) = u_\lambda(x) - u(x).$$

Before applying the method of moving planes, we state some properties of the Green's function, which were established independently in [FZ] and [LZ].

Lemma 7.3.1 *(i) For any*

$$x, y \in \Sigma_\lambda, \quad x \neq y,$$

we have

$$G_+(x^\lambda, y^\lambda) > max\{G_+(x^\lambda, y), G_+(x, y^\lambda)\} \tag{7.37}$$

and

$$G_+(x^\lambda, y^\lambda) - G_+(x, y) > |G_+(x^\lambda, y) - G_+(x, y^\lambda)|. \tag{7.38}$$

(ii) For any

$$x \in \Sigma_\lambda, \ y \in \Sigma_\lambda^C,$$

it holds

$$G_+(x^\lambda, y) > G_+(x, y). \tag{7.39}$$

Proof. Since $x, y \in \Sigma_\lambda$, it is easy to verify that

$$s(x^\lambda, y^\lambda) < s(x, y^\lambda) \quad \text{and} \quad t(x^\lambda, y^\lambda) > t(x, y^\lambda). \tag{7.40}$$

Moreover we have

$$t(x^\lambda, y^\lambda) > t(x, y). \tag{7.41}$$

Consider

$$G_+(x, y) = C_{n,m} H(s, t) = C_{n,m} s^{m-\frac{n}{2}} \int_0^{\frac{t}{s}} \frac{z^{m-1}}{(z+1)^{\frac{n}{2}}} \, dz$$

$$= C_{n,m} \int_0^t \frac{z^{m-1}}{(z+s)^{\frac{n}{2}}} \, dz.$$

Then for $s, t > 0$

$$\frac{\partial H}{\partial s} = -\frac{n}{2} \int_0^t \frac{z^{m-1}}{(z+s)^{\frac{n}{2}+1}} < 0, \tag{7.42}$$

$$\frac{\partial H}{\partial t} = \frac{t^{m-1}}{(t+s)^{\frac{n}{2}}} > 0 \tag{7.43}$$

and

$$\frac{\partial^2 H}{\partial t \partial s} = -\frac{n}{2} \frac{t^{m-1}}{(t+s)^{\frac{n}{2}+1}} < 0. \tag{7.44}$$

(i) From (7.40), (7.42), and (7.43), we arrive at (7.37).
While by (7.42) and (7.44), we have

$$G_+(x^\lambda, y^\lambda) - G_+(x, y)$$

$$= C_n^m \int_{t(x,y)}^{t(x^\lambda, y^\lambda)} \frac{\partial H(s(x, y), t)}{\partial t} \, dt$$

$$> C_n^m \int_{t(x,y)}^{t(x^\lambda, y^\lambda)} \frac{\partial H(s(x^\lambda, y), t)}{\partial t} \, dt$$

$$\geq C_n^m \int_{t(x,y^\lambda)}^{t(x^\lambda, y)} \frac{\partial H(s(x^\lambda, y), t)}{\partial t} \, dt$$

$$= C_n^m \left| H(s(x^\lambda, y), t(x^\lambda, y)) - H(s(x, y^\lambda), t(x, y^\lambda)) \right|$$

$$= |G_+(x^\lambda, y) - G_+(x, y^\lambda)|.$$

Here we have used the fact that $s(x^\lambda, y) = s(x, y^\lambda)$.
(ii) Noticing that for $x \in \Sigma_\lambda$ and $y \in \Sigma_\lambda^C$, we have

$$|x^\lambda - y| < |x - y| \quad \text{and} \quad t(x, y) < t(x^\lambda, y).$$

Then (7.39) follows immediately from (7.42) and (7.43).
This completes the proof of Lemma 7.3.1.

The following lemma is also a key ingredient in our integral estimates, which was obtained in [FZ] and [LZ].

Lemma 7.3.2 *For any $x \in \Sigma_\lambda$, it holds*

$$u(x) - u_\lambda(x) \leq \int_{\Sigma_\lambda} \left[G_+(x^\lambda, y^\lambda) - G_+(x, y^\lambda) \right] \left[u^p(y) - u_\lambda^p(y) \right] dy.$$

Proof.

$$u(x) = \int_{\Sigma_\lambda} G_+(x, y) u^p(y) dy + \int_{\Sigma_\lambda} G_+(x, y^\lambda) u_\lambda^p(y) dy$$
$$+ \int_{\Sigma_\lambda^C \setminus \widetilde{\Sigma}_\lambda} G_+(x, y) u^p(y) dy,$$

$$u(x^\lambda) = \int_{\Sigma_\lambda} G_+(x^\lambda, y) u^p(y) dy + \int_{\Sigma_\lambda} G_+(x^\lambda, y^\lambda) u_\lambda^p(y) dy$$
$$+ \int_{\Sigma_\lambda^C \setminus \widetilde{\Sigma}_\lambda} G_+(x^\lambda, y) u^p(y) dy,$$

where $\widetilde{\Sigma}_\lambda = \{ x^\lambda \mid x \in \Sigma_\lambda \}$. By Lemma 7.3.1, we arrive at

$$u(x) - u(x^\lambda) = \int_{\Sigma_\lambda} \left[G_+(x, y) - G_+(x^\lambda, y) \right] u^p(y) dy$$
$$+ \int_{\Sigma_\lambda} \left[G_+(x, y^\lambda) - G_+(x^\lambda, y^\lambda) \right] u_\lambda^p(y) dy$$
$$+ \int_{\Sigma_\lambda^C \setminus \widetilde{\Sigma}_\lambda} \left[G_+(x, y) - G_+(x^\lambda, y) \right] u^p(y) dy$$
$$\leq \int_{\Sigma_\lambda} \left[G_+(x, y) - G_+(x^\lambda, y) \right] u^p(y) dy$$
$$- \int_{\Sigma_\lambda} \left[G_+(x^\lambda, y^\lambda) - G_+(x, y^\lambda) \right] u_\lambda^p(y) dy$$
$$\leq \int_{\Sigma_\lambda} \left[G_+(x^\lambda, y^\lambda) - G_+(x, y^\lambda) \right] u^p(y) dy$$
$$- \int_{\Sigma_\lambda} \left[G_+(x^\lambda, y^\lambda) - G_+(x, y^\lambda) \right] u_\lambda^p(y) dy$$
$$= \int_{\Sigma_\lambda} \left[G_+(x^\lambda, y^\lambda) - G_+(x, y^\lambda) \right] \left[u^p(y) - u_\lambda^p(y) \right] dy.$$

7.3.2 Method of Moving Planes

In this subsection, we use the *method of moving planes in integral forms* to show that the positive solutions of integral equation (7.35) must depend on x_n only.

Because there is no global integrability assumption on the solutions u, one is not able to carry out the *method of moving planes* directly on u. To circumvent this difficulty, we resort to Kelvin type transforms.

For $z^0 \in \partial \mathbb{R}_+^n$, let

$$\bar{u}(x) = \frac{1}{|x - z^0|^{n-2m}} u\left(\frac{x - z^0}{|x - z^0|^2} + z^0\right) \tag{7.45}$$

be the Kelvin type transform of u centered at z^0.

The Critical Case

In the critical case $p = \frac{n+2m}{n-2m}$, we consider two possibilities.

Possibility 1. If there is a $z^0 = (z_1^0, \cdots, z_{n-1}^0, 0) \in \partial \mathbb{R}_+^n$ such that $\bar{u}(x)$ is not singular at z^0, then by (7.45), we obtain

$$u(y) = \frac{1}{|y - z^0|^{n-2m}} \bar{u}\left(\frac{y - z^0}{|y - z^0|^2} + z^0\right).$$

And we further deduce

$$u(y) = O\left(\frac{1}{|y|^{n-2m}}\right), \quad \text{as } |y| \to \infty. \tag{7.46}$$

Since $u \in L_{loc}^{\frac{n(p-1)}{2m}}(\mathbb{R}_+^n)$, by (7.46), we have

$$\int_{\mathbb{R}_+^n} u^{\frac{n(p-1)}{2m}}(y) dy < \infty. \tag{7.47}$$

In this situation, we apply the moving planes on the original solution u, and move the planes perpendicular to x_n axis all way up to show that u is monotone increasing in x_n and thus derive a contradiction with (7.46). Hence what's left is

Possibility 2. For all $z^0 = (z_1^0, \cdots, z_{n-1}^0, 0) \in \partial \mathbb{R}_+^n$, $\bar{u}(x)$ is singular at z^0. In this situation, we work on $\bar{u}$ and move the planes perpendicular to $x_1, \cdots, x_{n-1}$ axes to show that $\bar{u}$ is axially symmetric about the line parallel to x_n axis and passing through z^0. Due to the arbitrariness of the point z^0, we conclude that $\bar{u}$ must depend on x_n only, so does the original solution u.

Detailed arguments in Possibility 1.

The proof consists of two steps. In the first step, we start from the very low end of our region $\mathbb{R}_+^n$, i.e. near $x_n = 0$. We show that for λ sufficiently small,

$$w_\lambda(x) = u_\lambda(x) - u(x) \geq 0, \quad a.e. \ \forall x \in \Sigma_\lambda. \tag{7.48}$$

In the second step, we move our plane $T_\lambda = \{x \in \mathbb{R}_+^n \mid x_n = \lambda\}$ along the positive x_n-direction as long as inequality (7.48) holds.

Step 1. Define

$$\Sigma_\lambda^- = \{x \in \Sigma_\lambda \mid w_\lambda(x) < 0\}.$$

We show that for λ sufficiently small, Σ_λ^- must be measure zero. In fact, for any $x \in \Sigma_\lambda^-$, by the *mean value theorem*, Lemma 7.3.1, and Lemma 7.3.2, we obtain

$$
\begin{aligned}
0 \;<\; & u(x) - u_\lambda(x) \\
\leq\; & \int_{\Sigma_\lambda} \left[G_+(x^\lambda, y^\lambda) - G_+(x, y^\lambda) \right] \left[u^p(y) - u_\lambda^p(y) \right] dy \\
=\; & \int_{\Sigma_\lambda^-} \left[G_+(x^\lambda, y^\lambda) - G_+(x, y^\lambda) \right] \left[u^p(y) - u_\lambda^p(y) \right] dy \\
& + \int_{\Sigma_\lambda \setminus \Sigma_\lambda^-} \left[G_+(x^\lambda, y^\lambda) - G_+(x, y^\lambda) \right] \left[u^p(y) - u_\lambda^p(y) \right] dy \\
\leq\; & \int_{\Sigma_\lambda^-} \left[G_+(x^\lambda, y^\lambda) - G_+(x, y^\lambda) \right] \left[u^p(y) - u_\lambda^p(y) \right] dy \\
\leq\; & \int_{\Sigma_\lambda^-} G_+(x^\lambda, y^\lambda) \left[u^p(y) - u_\lambda^p(y) \right] dy \\
=\; & p \int_{\Sigma_\lambda^-} G_+(x^\lambda, y^\lambda) \psi_\lambda^{p-1}(y) [u(y) - u_\lambda(y)] dy \\
\leq\; & p \int_{\Sigma_\lambda^-} G_+(x^\lambda, y^\lambda) u^{p-1}(y) [u(y) - u_\lambda(y)] dy, \qquad (7.49)
\end{aligned}
$$

where $\psi_\lambda(y)$ is valued between $u(y)$ and $u_\lambda(y)$. Thus on Σ_λ^-, we have

$$0 \leq u_\lambda(y) \leq \psi_\lambda(y) \leq u(y).$$

Notice that $0 < 2m < n$,

$$
\begin{aligned}
G_+(x^\lambda, y^\lambda) &= \frac{C_{n,m}}{|x^\lambda - y^\lambda|^{n-2m}} \int_0^{\frac{4(2\lambda - x_n)(2\lambda - y_n)}{|x^\lambda - y^\lambda|^2}} \frac{z^{m-1}}{(z+1)^{\frac{n}{2}}} dz \\
&\leq \frac{C}{|x^\lambda - y^\lambda|^{n-2m}} \\
&= \frac{C}{|x - y|^{n-2m}}.
\end{aligned}
$$

By (7.49), we get

$$0 < u(x) - u_\lambda(x) \leq \int_{\Sigma_\lambda^-} \frac{C}{|x - y|^{n-2m}} |u^{p-1}(y)| |u(y) - u_\lambda(y)| dy. \qquad (7.50)$$

We apply the Hardy-Littlewood-Sobolev inequality and Hölder inequality to (7.50) to obtain, for any $q > \frac{n}{n-2m}$,

$$
\begin{aligned}
\|w_\lambda\|_{L^q(\Sigma_\lambda^-)} &\leq C \|u^{p-1} w_\lambda\|_{L^{\frac{nq}{n+2mq}}(\Sigma_\lambda^-)} \\
&\leq C \|u^{p-1}\|_{L^{\frac{n}{2m}}(\Sigma_\lambda^-)} \|w_\lambda\|_{L^q(\Sigma_\lambda^-)}. \qquad (7.51)
\end{aligned}
$$

By (7.47), we can choose sufficiently small positive λ such that

$$C\|u^{p-1}\|_{L^{\frac{n}{2m}}(\Sigma_\lambda^-)} = C\left[\int_{\Sigma_\lambda^-} u^{\frac{n(p-1)}{2m}}(y)dy\right]^{\frac{2m}{n}} \leq \frac{1}{2}.$$

Now inequality (7.51) implies

$$\|w_\lambda\|_{L^q(\Sigma_\lambda^-)} = 0,$$

and therefore Σ_λ^- must be measure zero.

Step 2. Inequality (7.48) provides a starting point to move the plane $T_\lambda = \{x \in \mathbb{R}_+^n \mid x_n = \lambda\}$. Now we start from the neighborhood of $x_n = 0$ and move the plane up as long as (7.48) holds.

Define

$$\lambda_0 = \sup\{\lambda \mid w_\rho(x) \geq 0, \ \rho \leq \lambda, \ \forall x \in \Sigma_\rho\}.$$

We prove

$$\lambda_0 = +\infty. \tag{7.52}$$

Suppose in the contrary that $\lambda_0 < \infty$, we will show that $u(x)$ is symmetric about the plane T_{λ_0}, i.e.,

$$w_{\lambda_0}(x) = 0, \quad a.e. \ \text{in } \Sigma_{\lambda_0}. \tag{7.53}$$

Otherwise, for such a λ_0, we have $w_{\lambda_0} \geq 0$, but $w_{\lambda_0} \not\equiv 0$ a.e. on Σ_{λ_0}. We show that the plane can be moved upward further. More precisely, there exists an $\epsilon > 0$ such that for all λ in $[\lambda_0, \lambda_0 + \epsilon)$, it holds

$$u(x) \leq u_\lambda(x) \ a.e. \ \text{on } \Sigma_\lambda.$$

By inequality (7.51), we have

$$\|w_\lambda\|_{L^q(\Sigma_\lambda^-)} \leq C\left[\int_{\Sigma_\lambda^-} u^{\frac{n(p-1)}{2m}}(y)dy\right]^{\frac{2m}{n}}\|w_\lambda\|_{L^q(\Sigma_\lambda^-)}. \tag{7.54}$$

By condition (7.47), we can choose ϵ sufficiently small so that for all λ in $[\lambda_0, \lambda_0 + \epsilon)$,

$$C\left[\int_{\Sigma_\lambda^-} u^{\frac{n(p-1)}{2m}}(y)dy\right]^{\frac{2m}{n}} \leq \frac{1}{2}. \tag{7.55}$$

We postpone the proof of (7.55) for a moment. Now by (7.54) and (7.55), we have $\|w_\lambda\|_{L^q(\Sigma_\lambda^-)} = 0$, and therefore Σ_λ^- must be measure zero. Hence, for these values of $\lambda > \lambda_0$, we have

$$w_\lambda(x) \geq 0, \quad a.e. \ \text{in } \Sigma_\lambda.$$

This contradicts with the definition of λ_0. Therefore (7.53) must hold.

By (7.53), we derive that the plane $x_n = 2\lambda_0$ is the symmetric image of the boundary $\partial\mathbb{R}_+^n$ with respect to the plane T_{λ_0}, and hence $u(x) = 0$ when x is on the plane $x_n = 2\lambda_0$. This contradicts with our assumption $u(x) > 0$. Therefore, (7.52) must be valid.

Now we prove inequality (7.55). For any small $\eta > 0$, we can choose R sufficiently large so that

$$\left(\int_{\mathbb{R}^n_+ \backslash B_R(0)} u^{\frac{n(p-1)}{2m}}(y) dy \right)^{\frac{2m}{n}} < \eta. \tag{7.56}$$

We fix this R and then show that the measure of $\Sigma_\lambda^- \cap B_R(0)$ is sufficiently small for λ close to λ_0. First, we have

$$w_{\lambda_0}(x) > 0 \tag{7.57}$$

in the interior of Σ_{λ_0}.

Indeed, by the first two expressions in the proof of Lemma 7.3.2 and Lemma 7.3.1, we have

$$u_{\lambda_0}(x) - u(x) \geq \int_{\Sigma_{\lambda_0}} \left[G_+(x^{\lambda_0}, y^{\lambda_0}) - G_+(x, y^{\lambda_0}) \right] \left[u_{\lambda_0}^p(y) - u^p(y) \right] dy$$

$$+ \int_{\Sigma_{\lambda_0}^C \backslash \widetilde{\Sigma}_{\lambda_0}} \left[G_+(x^{\lambda_0}, y) - G_+(x, y) \right] u^p(y) dy$$

$$\geq \int_{\Sigma_{\lambda_0}^C \backslash \widetilde{\Sigma}_{\lambda_0}} \left[G_+(x^{\lambda_0}, y) - G_+(x, y) \right] u^p(y) dy. \tag{7.58}$$

If (7.57) is violated, there exists some point $x_0 \in \Sigma_{\lambda_0}$ such that $u(x_0) = u_{\lambda_0}(x_0)$. And then by Lemma 7.3.1 and (7.58), we obtain

$$u(y) \equiv 0, \quad \forall y \in \Sigma_{\lambda_0}^C \backslash \widetilde{\Sigma}_{\lambda_0}. \tag{7.59}$$

This is a contradiction with our assumption that $u > 0$. Therefore (7.57) must be true.

For any $\gamma > 0$, let

$$E_\gamma = \{ x \in \Sigma_{\lambda_0} \cap B_R(0) \mid w_{\lambda_0}(x) > \gamma \}, \quad F_\gamma = (\Sigma_{\lambda_0} \cap B_R(0)) \backslash E_\gamma. \tag{7.60}$$

It is obvious that

$$\lim_{\gamma \to 0} \mu(F_\gamma) = 0.$$

For $\lambda > \lambda_0$, let

$$D_\lambda = (\Sigma_\lambda \backslash \Sigma_{\lambda_0}) \cap B_R(0).$$

Then it is easy to see that

$$(\Sigma_\lambda^- \cap B_R(0)) \subset (\Sigma_\lambda^- \cap E_\gamma) \cup F_\gamma \cup D_\lambda. \tag{7.61}$$

Apparently, the measure of D_λ is small for λ close to λ_0. We show that the measure of $\Sigma_\lambda^- \cap E_\gamma$ can be sufficiently small as λ goes to λ_0. In fact, for any $x \in \Sigma_\lambda^- \cap E_\gamma$, we have

$$w_\lambda(x) = u_\lambda(x) - u(x) = u_\lambda(x) - u_{\lambda_0}(x) + u_{\lambda_0}(x) - u(x) < 0.$$

Hence

$$u_{\lambda_0}(x) - u_\lambda(x) > w_{\lambda_0}(x) > \gamma.$$

It follows that

$$(\Sigma_\lambda^- \cap E_\gamma) \subset G_\gamma \equiv \{x \in B_R(0) \mid u_{\lambda_0}(x) - u_\lambda(x) > \gamma\}. \tag{7.62}$$

By the Chebyshev inequality, we have

$$\mu(G_\gamma) \le \frac{1}{\gamma^{p+1}} \int_{G_\gamma} |u_{\lambda_0}(x) - u_\lambda(x)|^{p+1} dx$$

$$\le \frac{1}{\gamma^{p+1}} \int_{B_R(0)} |u_{\lambda_0}(x) - u_\lambda(x)|^{p+1} dx.$$

For each fixed γ, as λ approaches λ_0, the right hand side of the above inequality can be made as small as we wish. Therefore, by (7.61) and (7.62), the measure of $\Sigma_\lambda^- \cap B_R(0)$ can also be made sufficiently small. Combining this with (7.56), we obtain (7.55).

Now from (7.52), u is monotone increasing with respect to x_n. This contradicts with (7.46). Hence **Possibility 1** will not happen, and what remains is the following.

Possibility 2. For all $z^0 = (z_1^0, \cdots, z_{n-1}^0, 0) \in \partial \mathbb{R}_+^n$, $\bar{u}(x)$ is singular at z^0. Here we prove that $\bar{u}(x)$ is rotationally symmetric about the line passing through z^0 and parallel to the x_n-axis. For $x \in \mathbb{R}_+^n \setminus B_\epsilon(z^0)$ with $\epsilon > 0$, we have

$$\bar{u}(x) = \frac{1}{|x-z^0|^{n-2m}} u\left(\frac{x-z^0}{|x-z^0|^2} + z^0\right)$$

$$= \frac{1}{|x-z^0|^{n-2m}} \int_{\mathbb{R}_+^n} G_+\left(\frac{x-z^0}{|x-z^0|^2} + z^0, y\right) u^p(y) dy$$

$$= \frac{1}{|x-z^0|^{n-2m}} \int_{\mathbb{R}_+^n} \frac{G_+\left(\frac{x-z^0}{|x-z^0|^2} + z^0, \frac{\tilde{y}-z^0}{|\tilde{y}-z^0|^2} + z^0\right)}{|\tilde{y}-z^0|^{2n}}$$

$$\cdot u^p\left(\frac{\tilde{y}-z^0}{|\tilde{y}-z^0|^2} + z^0\right) d\tilde{y}$$

$$= \int_{\mathbb{R}_+^n} \frac{G_+\left(\frac{x-z^0}{|x-z^0|^2} + z^0, \frac{\tilde{y}-z^0}{|\tilde{y}-z^0|^2} + z^0\right)}{|x-z^0|^{n-2m}|\tilde{y}-z^0|^{n-2m}}$$

$$\cdot \left[\frac{1}{|\tilde{y}-z^0|^{n-2m}} u\left(\frac{\tilde{y}-z^0}{|\tilde{y}-z^0|^2} + z^0\right)\right]^p \frac{1}{|\tilde{y}-z^0|^\beta} d\tilde{y}$$

$$= \int_{\mathbb{R}_+^n} G_+(x,y) \frac{\bar{u}^p(y)}{|y-z^0|^\beta} dy, \tag{7.63}$$

where $\frac{n}{n-2m} < p \le \tau$, $\beta = (n-2m)(\tau-p) \ge 0$, $\tau = \frac{n+2m}{n-2m}$.

(i) For $p = \tau = \frac{n+2m}{n-2m}$, if $u(x)$ is a solution of

$$u(x) = \int_{\mathbb{R}_+^n} G_+(x,y) u^\tau(y) dy, \tag{7.64}$$

then $\bar{u}$ is also a solution of (7.64). Since $u \in L_{loc}^{\frac{2n}{n-2m}}(\mathbb{R}_+^n)$, for any domain Ω that is a positive distance away from z^0, we have

$$\int_{\Omega} \bar{u}^{\frac{2n}{n-2m}}(y)dy < \infty. \tag{7.65}$$

From now on, we only need to deal with $\bar{u}$. For simplicity, we still denote it by u.

In this case, we will move the planes in $x_1, x_2, \cdots, x_{n-1}$ directions, hence we need to redefine Σ_λ. For a given real number λ, define

$$\hat{\Sigma}_\lambda = \{x = (x_1, \cdots, x_n) \in \mathbb{R}_+^n \mid x_1 < \lambda\}$$

and let

$$x^\lambda = (2\lambda - x_1, x_2, \cdots, x_n).$$

For $x, y \in \hat{\Sigma}_\lambda$, $x \neq y$, we have

$$G_+(x, y) = G_+(x^\lambda, y^\lambda), \quad G_+(x^\lambda, y) = G_+(x, y^\lambda), \tag{7.66}$$

and

$$G_+(x^\lambda, y^\lambda) > G_+(x, y^\lambda). \tag{7.67}$$

Obviously, we have

$$u(x) = \int_{\hat{\Sigma}_\lambda} G_+(x, y)u^\tau(y)dy + \int_{\hat{\Sigma}_\lambda} G_+(x, y^\lambda)u_\lambda^\tau(y)dy$$

$$u(x^\lambda) = \int_{\hat{\Sigma}_\lambda} G_+(x^\lambda, y)u^\tau(y)dy + \int_{\hat{\Sigma}_\lambda} G_+(x^\lambda, y^\lambda)u_\lambda^\tau(y)dy.$$

By (7.67), it is easy to see

$$u(x) - u(x^\lambda)$$
$$= \int_{\hat{\Sigma}_\lambda} \left[G_+(x, y) - G_+(x^\lambda, y)\right] u^\tau(y)dy$$
$$+ \int_{\hat{\Sigma}_\lambda} \left[G_+(x, y^\lambda) - G_+(x^\lambda, y^\lambda)\right] u_\lambda^\tau(y)dy$$
$$= \int_{\hat{\Sigma}_\lambda} \left[G_+(x, y) - G_+(x^\lambda, y)\right] \left[u^\tau(y) - u_\lambda^\tau(y)\right] dy. \tag{7.68}$$

The proof consists of two steps. In the first step, we will show that for λ sufficiently negative,

$$w_\lambda(x) = u_\lambda(x) - u(x) \geq 0, \quad a.e. \text{ in } \hat{\Sigma}_\lambda. \tag{7.69}$$

In the second step, we deduce that $\hat{T}$ can be moved to the right all the way to z^0. And furthermore, we derive $w_{z_1^0} \equiv 0$, $\forall x \in \hat{\Sigma}_{z_1^0}$.

Step 1. Define

$$\hat{\Sigma}_\lambda^- = \{x \in \hat{\Sigma}_\lambda \backslash B_\epsilon((z^0)^\lambda) \mid w_\lambda(x) < 0\},$$

where $(z^0)^\lambda$ is the reflection of z^0 about the plane $\hat{T}_\lambda = \{x \in \mathbb{R}_+^n \mid x_1 = \lambda\}$. We show that for λ sufficiently negative, $\hat{\Sigma}_\lambda^-$ must be measure zero. In fact, by the *mean value theorem*, we obtain, for $x \in \hat{\Sigma}_\lambda^-$,

$$
\begin{aligned}
0 \; < \; & u(x) - u_\lambda(x) \\
= \; & \int_{\hat{\Sigma}_\lambda^-} [G_+(x,y) - G_+(x,y^\lambda)][u^\tau(y) - u_\lambda^\tau(y)]dy \\
& + \int_{\hat{\Sigma}_\lambda \backslash \hat{\Sigma}_\lambda^-} [G_+(x,y) - G_+(x,y^\lambda)][u^\tau(y) - u_\lambda^\tau(y)]dy \\
\leq \; & \int_{\hat{\Sigma}_\lambda^-} [G_+(x,y) - G_+(x,y^\lambda)][u^\tau(y) - u_\lambda^\tau(y)]dy \\
\leq \; & \int_{\hat{\Sigma}_\lambda^-} G_+(x,y)[u^\tau(y) - u_\lambda^\tau(y)]dy \\
= \; & \tau \int_{\hat{\Sigma}_\lambda^-} G_+(x,y)\psi_\lambda^{\tau-1}(y)[u(y) - u_\lambda(y)]dy \\
\leq \; & \tau \int_{\hat{\Sigma}_\lambda^-} G_+(x,y)u^{\tau-1}(y)[u(y) - u_\lambda(y)]dy \\
\leq \; & \int_{\hat{\Sigma}_\lambda^-} \frac{C}{|x-y|^{n-2m}} |u^{\tau-1}(y)||u(y) - u_\lambda(y)|dy. \quad (7.70)
\end{aligned}
$$

We apply the Hardy-Littlewood-Sobolev inequality and Hölder inequality to (7.70) to obtain, for any $q > \frac{n}{n-2m}$,

$$
\begin{aligned}
\|w_\lambda\|_{L^q(\hat{\Sigma}_\lambda^-)} & \leq C\|u^{\tau-1}w_\lambda\|_{L^{\frac{nq}{n+2mq}}(\hat{\Sigma}_\lambda^-)} \\
& \leq C\|u^{\tau-1}\|_{L^{\frac{n}{2m}}(\hat{\Sigma}_\lambda^-)}\|w_\lambda\|_{L^q(\hat{\Sigma}_\lambda^-)}. \quad (7.71)
\end{aligned}
$$

By (7.65), we can choose N sufficiently large such that for $\lambda \leq -N$,

$$C\|u^{\tau-1}\|_{L^{\frac{n}{2m}}(\hat{\Sigma}_\lambda^-)} = C\left\{\int_{\hat{\Sigma}_\lambda^-} u^{\frac{2n}{n-2m}}(y)dy\right\}^{\frac{2m}{n}} \leq \frac{1}{2}.$$

Now inequality (7.71) implies

$$\|w_\lambda\|_{L^q(\hat{\Sigma}_\lambda^-)} = 0,$$

and therefore $\hat{\Sigma}_\lambda^-$ must be measure zero.

Step 2. Move the plane to the limiting position to derive symmetry.

Inequality (7.69) provides a starting point to move the plane $\hat{T}_\lambda = \{x \in \mathbb{R}_+^n \mid x_1 = \lambda\}$. Now we start from the neighborhood of $x_1 = -\infty$ and move the plane to the right as long as (7.69) holds to the limiting position. Define

$$\lambda_0 = \sup\{\lambda \leq z_1^0 \mid w_\rho(x) \geq 0, \; \rho \leq \lambda, \; \forall x \in \hat{\Sigma}_\rho\}.$$

We prove that $\lambda_0 = z_1^0$. Suppose on the contrary that $\lambda_0 < z_1^0$. We show that $u(x)$ is symmetric about the plane T_{λ_0}, i.e.

$$w_{\lambda_0} \equiv 0, \quad a.e. \ \forall x \in \hat{\Sigma}_{\lambda_0} \backslash B_\epsilon((z^0)^{\lambda_0}). \tag{7.72}$$

Suppose that for such a λ_0, we have $w_{\lambda_0} \geq 0$, and

$$w_{\lambda_0} \not\equiv 0 \ a.e. \ \text{on} \ \hat{\Sigma}_{\lambda_0} \backslash B_\epsilon((z^0)^{\lambda_0}).$$

We show that the plane can be moved further to the right. More precisely, there exists a $\zeta > 0$ such that for all $\lambda \in [\lambda_0, \lambda_0 + \zeta)$

$$u(x) \leq u_\lambda(x) \ a.e. \ \text{on} \ \hat{\Sigma}_\lambda \backslash B_\epsilon((z^0)^\lambda).$$

By inequality (7.71), we have

$$\|w_\lambda\|_{L^q(\hat{\Sigma}_\lambda^-)} \leq C \left\{ \int_{\hat{\Sigma}_\lambda^-} u^{\frac{2n}{n-2m}}(y)dy \right\}^{\frac{2m}{n}} \|w_\lambda\|_{L^q(\hat{\Sigma}_\lambda^-)}. \tag{7.73}$$

With condition (7.65), similar to the proof of (7.55), we can choose ζ sufficiently small so that for all $\lambda \in [\lambda_0, \lambda_0 + \zeta)$,

$$C\{ \int_{\hat{\Sigma}_\lambda^-} u^{\frac{2n}{n-2m}}(y)dy \}^{\frac{2m}{n}} \leq \frac{1}{2}. \tag{7.74}$$

We postpone the proof for a moment. Now by (7.73) and (7.74), we have $\|w_\lambda\|_{L^q(\hat{\Sigma}_\lambda^-)} = 0$, and therefore $\hat{\Sigma}_\lambda^-$ must be measure zero. Hence, for these values of $\lambda > \lambda_0$, we have

$$w_\lambda(x) \geq 0, \quad a.e. \ \text{on} \ \hat{\Sigma}_\lambda \backslash B_\epsilon((z^0)^\lambda), \ \forall \epsilon > 0.$$

This contradicts with the definition of λ_0. Therefore (7.72) must hold. That is, if $\lambda_0 < z_1^0$, for any $\epsilon > 0$,

$$\bar{u}(x) \equiv \bar{u}_{\lambda_0}(x), \quad a.e. \ \text{on} \ \hat{\Sigma}_{\lambda_0} \backslash B_\epsilon((z^0)^{\lambda_0}).$$

Since $\bar{u}$ is singular at z^0, $\bar{u}$ must also be singular at $(z^0)^\lambda$. This is impossible. So we deduce

$$\lambda_0 = z_1^0, \quad w_{\lambda_0}(x) \geq 0, \quad a.e. \ \text{in} \ \hat{\Sigma}_{\lambda_0}.$$

Similarly, we can move the plane from near $x_1 = \infty$ to the left and derive that $w_{\lambda_0}(x) \leq 0$. As a result, we have

$$w_{\lambda_0}(x) \equiv 0, \quad a.e. \ \text{in} \ \hat{\Sigma}_{\lambda_0}, \ \lambda_0 = z_1^0.$$

Now we prove inequality (7.74). For any small $\eta > 0$ and $\epsilon > 0$, we can choose R sufficiently large so that

$$\left(\int_{(\mathbb{R}_+^n \backslash B_\epsilon(z^0)) \backslash B_R(0)} u^{\frac{2n}{n-2m}}(y)dy \right)^{\frac{2m}{n}} < \eta. \tag{7.75}$$

We fix this R and then show that the measure of $\hat{\Sigma}_\lambda^- \cap B_R(0)$ can be sufficiently small for λ close to λ_0. By (7.68), we have

$$w_{\lambda_0}(x) > 0 \tag{7.76}$$

in the interior of $\hat{\Sigma}_{\lambda_0} \backslash B_\epsilon((z^0)^{\lambda_0})$.

The rest is similar to the proof of (7.55) in **Possibility 1**. We only need to use $\hat{\Sigma}_\lambda \backslash B_\epsilon((z^0)^\lambda)$ instead of Σ_λ and $\hat{\Sigma}_{\lambda_0} \backslash B_\epsilon((z^0)^{\lambda_0})$ instead of Σ_{λ_0}.

The Subcritical Case

For $\frac{n}{n-2m} < p < \frac{n+2m}{n-2m}$, since $u \in L_{loc}^{\frac{n(p-1)}{2m}}(\mathbb{R}_+^n)$, for any domain Ω that is a positive distance away from z^0, we have

$$\int_\Omega \left[\frac{\bar{u}^{p-1}(y)}{|y - z^0|^\beta} \right]^{\frac{n}{2m}} dy < \infty, \tag{7.77}$$

where $\beta = (n - 2m)(\tau - p) > 0$, $\tau = \frac{n+2m}{n-2m}$.

By (7.63), we have

$$\bar{u}(x) = \int_{\hat{\Sigma}_\lambda} G_+(x, y) \frac{\bar{u}^p(y)}{|y - z^0|^\beta} dy + \int_{\hat{\Sigma}_\lambda} G_+(x, y^\lambda) \frac{\bar{u}_\lambda^p(y)}{|y^\lambda - z^0|^\beta} dy,$$

$$\bar{u}(x^\lambda) = \int_{\hat{\Sigma}_\lambda} G_+(x^\lambda, y) \frac{\bar{u}^p(y)}{|y - z^0|^\beta} dy + \int_{\hat{\Sigma}_\lambda} G_+(x^\lambda, y^\lambda) \frac{\bar{u}_\lambda^p(y)}{|y^\lambda - z^0|^\beta} dy.$$

By (7.66), we calculate

$$\bar{u}(x) - \bar{u}_\lambda(x)$$

$$= \int_{\hat{\Sigma}_\lambda} \left[G_+(x, y) - G_+(x^\lambda, y) \right] \frac{\bar{u}^p(y)}{|y - z^0|^\beta} dy$$

$$+ \int_{\hat{\Sigma}_\lambda} \left[G_+(x, y^\lambda) - G_+(x^\lambda, y^\lambda) \right] \frac{\bar{u}_\lambda^p(y)}{|y^\lambda - z^0|^\beta} dy$$

$$= \int_{\hat{\Sigma}_\lambda} \left[G_+(x, y) - G_+(x^\lambda, y) \right] \left[\frac{\bar{u}^p(y)}{|y - z^0|^\beta} - \frac{\bar{u}_\lambda^p(y)}{|y^\lambda - z^0|^\beta} \right] dy. \tag{7.78}$$

The proof also consists of two steps.

Step 1. For any $\epsilon > 0$, define

$$\hat{\Sigma}_\lambda^- = \{x \in \hat{\Sigma}_\lambda \backslash B_\epsilon((z^0)^\lambda) \mid w_\lambda(x) = u_\lambda(x) - u(x) < 0\}.$$

We show that for λ sufficiently negative, $\hat{\Sigma}_\lambda^-$ must be measure zero.

By the *mean value theorem*, we obtain, for sufficiently negative values of λ and $x \in \hat{\Sigma}_\lambda^-$,

$$0 < \bar{u}(x) - \bar{u}_\lambda(x)$$

$$= \int_{\hat{\Sigma}_\lambda} \left[G_+(x,y) - G_+(x^\lambda, y) \right] \left[\frac{\bar{u}^p(y)}{|y - z^0|^\beta} - \frac{\bar{u}_\lambda^p(y)}{|y^\lambda - z^0|^\beta} \right] dy$$

$$= \int_{\hat{\Sigma}_\lambda^-} \left[G_+(x,y) - G_+(x^\lambda, y) \right] \left[\frac{\bar{u}^p(y)}{|y - z^0|^\beta} - \frac{\bar{u}_\lambda^p(y)}{|y^\lambda - z^0|^\beta} \right] dy$$

$$+ \int_{\hat{\Sigma}_\lambda \backslash \hat{\Sigma}_\lambda^-} \left[G_+(x,y) - G_+(x^\lambda, y) \right] \left[\frac{\bar{u}^p(y)}{|y - z^0|^\beta} - \frac{\bar{u}_\lambda^p(y)}{|y^\lambda - z^0|^\beta} \right] dy$$

$$\le \int_{\hat{\Sigma}_\lambda^-} \left[G_+(x,y) - G_+(x^\lambda, y) \right] \left[\frac{\bar{u}^p(y)}{|y - z^0|^\beta} - \frac{\bar{u}_\lambda^p(y)}{|y^\lambda - z^0|^\beta} \right] dy$$

$$= \int_{\hat{\Sigma}_\lambda^-} \left[G_+(x,y) - G_+(x^\lambda, y) \right]$$

$$\cdot \left[\frac{\bar{u}^p(y)}{|y - z^0|^\beta} - \frac{\bar{u}_\lambda^p(y)}{|y - z^0|^\beta} + \frac{\bar{u}_\lambda^p(y)}{|y - z^0|^\beta} - \frac{\bar{u}_\lambda^p(y)}{|y^\lambda - z^0|^\beta} \right] dy$$

$$= \int_{\hat{\Sigma}_\lambda^-} \left[G_+(x,y) - G_+(x^\lambda, y) \right]$$

$$\cdot \left[\frac{\bar{u}^p(y) - \bar{u}_\lambda^p(y)}{|y - z^0|^\beta} + \bar{u}_\lambda^p(y)\left[\frac{1}{|y - z^0|^\beta} - \frac{1}{|y^\lambda - z^0|^\beta} \right] \right] dy$$

$$\le \int_{\hat{\Sigma}_\lambda^-} \left[G_+(x,y) - G_+(x^\lambda, y) \right] \frac{\bar{u}^p(y) - \bar{u}_\lambda^p(y)}{|y - z^0|^\beta} dy$$

$$\le p \int_{\hat{\Sigma}_\lambda^-} G_+(x,y) \frac{\bar{u}^{p-1}(y)}{|y - z^0|^\beta} [\bar{u}(y) - \bar{u}_\lambda(y)] dy$$

$$\le \int_{\hat{\Sigma}_\lambda^-} \frac{C}{|x - y|^{n-2m}} |\frac{\bar{u}^{p-1}(y)}{|y - z^0|^\beta}| |\bar{u}(y) - \bar{u}_\lambda(y)| dy. \tag{7.79}$$

We apply the Hardy-Littlewood-Sobolev inequality and Hölder inequality to (7.79) to obtain, for any $q > \frac{n}{n-2m}$,

$$\|w_\lambda\|_{L^q(\hat{\Sigma}_\lambda^-)} \le C \|\frac{\bar{u}^{p-1}}{|y - z^0|^\beta} w_\lambda \|_{L^{\frac{nq}{n+2mq}}(\hat{\Sigma}_\lambda^-)}$$

$$\le C \|\frac{\bar{u}^{p-1}}{|y - z^0|^\beta}\|_{L^{\frac{n}{2m}}(\hat{\Sigma}_\lambda^-)} \|w_\lambda\|_{L^q(\hat{\Sigma}_\lambda^-)}. \tag{7.80}$$

By (7.77), we can choose N sufficiently large, such that for $\lambda \le -N$,

$$C \left\{ \int_{\hat{\Sigma}_\lambda^-} [\frac{\bar{u}^{p-1}}{|y - z^0|^\beta}]^{\frac{n}{2m}} dy \right\}^{\frac{2m}{n}} \le \frac{1}{2}.$$

Now inequality (7.80) implies

$$\|w_\lambda\|_{L^q(\hat{\Sigma}_\lambda^-)} = 0,$$

and therefore $\hat{\Sigma}_\lambda^-$ must be measure zero. This proves

$$w_\lambda(x) \ge 0, \quad a.e. \text{ in } \hat{\Sigma}_\lambda. \tag{7.81}$$

Step 2. Move the plane to the limiting position to derive symmetry.

Inequality (7.81) provides a starting point to move the plane $\hat{T}_\lambda = \{x \in \mathbb{R}^n_+ \mid x_1 = \lambda\}$. Now we start from the neighborhood of $x_1 = -\infty$ and move the plane to the right as long as (7.81) holds to the limiting position. Define

$$\lambda_0 = \sup\{\lambda \leq z_1^0 \mid w_\rho(x) \geq 0, \ \rho \leq \lambda, \ \forall x \in \hat{\Sigma}_\rho\}.$$

The rest is entirely similar to the case $p = \frac{n+2m}{n-2m}$. We only need to use $\int [\frac{\bar{u}^{p-1}(y)}{|y-z^0|^\beta}]^{\frac{n}{2m}} dy$ instead of $\int u^{\frac{2n}{n-2m}}(y)dy$. We also conclude

$$w_{\lambda_0}(x) \equiv 0, \quad a.e. \ \text{in} \ \hat{\Sigma}_{\lambda_0}, \quad \lambda_0 = z_1^0.$$

In **Case 2**, for $\frac{n}{n-2m} < p \leq \frac{n+2m}{n-2m}$, since we can choose any direction that is perpendicular to the x_n-axis as the x_1 direction, we have actually shown that the solution $\bar{u}(x)$ is rotationally symmetric about the line parallel to x_n-axis and passing through z^0. Now, for any two points X^1 and X^2, with

$$X^i = (x^i, x_n) \in R^{n-1} \times [0, \infty), \quad i = 1, 2.$$

Let z^0 be the projection of $\bar{X} = \frac{X^1+X^2}{2}$ on $\partial \mathbb{R}^n_+$. Set

$$Y^i = \frac{X^i - z^0}{|X^i - z^0|^2} + z^0, \quad i = 1, 2.$$

From the above arguments, it is easy to see $\bar{u}(X^1) = \bar{u}(X^2)$, hence $u(Y^1) = u(Y^2)$. This implies that u is independent of $(x_1, \cdots, x_{n-1})$.

7.3.3 Deriving a Contradiction if $u = u(x_n)$

So far, in both critical and subcritical cases, we have shown that the solution u depends on x_n only. Next, we derive a contradiction with the finiteness of the integral. The approaches are different depending on whether m is an integer or not.

The case m is an integer.

We prove that $u(x)$ is monotone and then $u \equiv 0$. For $x = (x', x_n)$, $y = (y', y_n) \in R^{n-1} \times [0, +\infty)$, we assume that $u(x) = u(x_n)$ is a solution of

$$u(x) = \int_{\mathbb{R}^n_+} G_+(x,y)u^p(y)dy, \tag{7.82}$$

where

$$G_+(x,y) = \frac{C_{n,m}}{|x-y|^{n-2m}} \int_0^{\frac{4x_n y_n}{|x-y|^2}} \frac{z^{m-1}}{(z+1)^{n/2}}dz,$$

on $\mathbb{R}^n_+$.

For each fixed $x \in \mathbb{R}^n_+$, set $|x_n - y_n|^2 = a^2$, $|x' - y'|^2 = r^2$. By elementary calculations, we have

$$+\infty > u(x) = u(x_n)$$

$$= \int_{\mathbb{R}^n_+} \frac{c_{n,m}}{|x-y|^{n-2m}} \int_0^{\frac{4x_n y_n}{|x-y|^2}} \frac{z^{m-1}}{(z+1)^{n/2}} dz u^p(y) dy$$

$$\sim C \int_{\mathbb{R}^n_+} \frac{1}{|x-y|^{n-2m}} \int_0^{\frac{4x_n y_n}{|x-y|^2}} z^{m-1} dz u^p(y) dy$$

$$\geq C \int_{x_n+1}^{\infty} u^p(y_n) y_n^m \int_{R^{n-1}} \frac{1}{\left[|x'-y'|^2 + |x_n-y_n|^2\right]^{\frac{n}{2}}} dy' dy_n$$

$$= C \int_{x_n+1}^{\infty} u^p(y_n) y_n^m \int_0^{\infty} \frac{r^{n-2}}{(r^2+a^2)^{\frac{n}{2}}} dr dy_n$$

$$= C \int_{x_n+1}^{\infty} u^p(y_n) y_n^m \int_0^{\infty} \frac{a^{n-2}\tau^{n-2}}{a^n(\tau^2+1)^{\frac{n}{2}}} a d\tau dy_n$$

$$= C \int_{x_n+1}^{\infty} \frac{u^p(y_n) y_n^m}{|x_n-y_n|} dy_n. \tag{7.83}$$

It follows that there exists a sequence $\{y_n^i\}$ such that

$$u^p(y_n^i)(y_n^i)^m \to 0, \text{ as } y_n^i \to \infty.$$

Hence we have

$$u(y_n^i) \to 0, \text{ as } y_n^i \to \infty. \tag{7.84}$$

For simplicity, we set $u(x) = u(x_n) = u(t)$. Suppose otherwise that $u \not\equiv 0$. Then there is a $t_0 > 0$ such that $u(t_0) > 0$. Due to the continuity of $u(t)$ and (7.82), we have $u(t) > 0$ in $(0, \infty)$.

For $m = 2k$, $k \in N$ we have

$$u^{(2m)}(t) = (-1)^m u^{(2m)}(x_n)$$
$$= (-\triangle)^m u(x)$$
$$= \int_{\mathbb{R}^n_+} (-\triangle)^m G_+(x,y) u^p(y) dy$$
$$= u^p(x) > 0. \tag{7.85}$$

It implies that

$$u^{(2m-1)}(t) \text{ is monotone increasing.} \tag{7.86}$$

We can prove

$$u^{(2m-1)}(t) \leq 0. \tag{7.87}$$

If not, there is a $t_0 > 0$ such that $u^{(2m-1)}(t_0) > 0$. By (7.86), we have

$$u^{(2m-1)}(t) \geq u^{(2m-1)}(t_0) > 0, \text{ for } t \geq t_0 > 0.$$

Integrating several times, and let $t \to \infty$, we have $u(t) \to \infty$. This is a contradiction with (7.84). Inequality (7.87) implies

$$u^{(2m-2)}(t) \text{ is nonincreasing.} \tag{7.88}$$

We can prove
$$u^{(2m-2)}(t) \geq 0.$$

If not, there is a $t_0 > 0$ such that $u^{(2m-2)}(t_0) < 0$. By (7.88), we have

$$u^{(2m-2)}(t) \leq u^{(2m-2)}(t_0) < 0, \quad \text{for } t \geq t_0 > 0.$$

Integrating several times, and let $t \to \infty$, we have $u(t) \to -\infty$. This is a contradiction with $u(x) > 0$. Repeat the process, we actually conclude that

$$u(t) \text{ is nonincreasing.} \tag{7.89}$$

Since $u(x)$ is a nonnegative solution, $u(0) = 0$, by (7.89), it is easy to see $u(x) \equiv 0$.

Similarly, for $m = 2k + 1$, $k \in N$, we obtain

$$u(t) \text{ is nondecreasing.}$$

By (7.83), we have

$$+\infty > u(x) = u(x_n)$$
$$\sim C \int_0^\infty \frac{u^p(y_n)y_n^m}{|x_n - y_n|} dy_n$$
$$\geq C \int_1^\infty \frac{u^p(y_n)y_n^m}{|x_n - y_n|} dy_n$$
$$\geq C u^p(1) \int_1^\infty \frac{y_n^m}{|x_n - y_n|} dy_n = +\infty.$$

This is a contradiction.

The case m is any real number between 0 and 1.

In the sequel, we denote $m = \alpha/2$. Similar to the previous case, we have

$$u^p(y_n^i)(y_n^i)^{\alpha/2} \to 0, \tag{7.90}$$

and consequently,

$$+\infty > u(x_n) \geq C_0 \int_0^\infty u^p(y_n)y_n^{\alpha/2}\frac{1}{|x_n - y_n|}dy_n x_n^{\alpha/2}. \tag{7.91}$$

Let $x_n = 2R$ be sufficiently large. By (7.91), we deduce that

$$+\infty > u(x_n) \geq C_0 \int_0^1 u^p(y_n)y_n^{\alpha/2}\frac{1}{|x_n - y_n|}dy_n x_n^{\alpha/2}$$
$$\geq \frac{C_0}{2R}(2R)^{\alpha/2} \int_0^1 u^p(y_n)y_n^{\alpha/2}dy_n$$
$$\geq C_1(2R)^{\alpha/2-1} = C_1 x_n^{\alpha/2-1}. \tag{7.92}$$

Then by (7.91) and (7.92), for $x_n = 2R$ sufficiently large, we also obtain

$$u(x_n) \geq C_0 \int_{\frac{R}{2}}^{R} (C_1 y_n)^{p(\alpha/2-1)} y_n^{\alpha/2} \frac{1}{|x_n - y_n|} dy_n x_n^{\alpha/2}$$

$$\geq C_0 (C_1)^{p(\frac{\alpha}{2}-1)} R^{p(\frac{\alpha}{2}-1)} \frac{2}{3R} (2R)^{\alpha/2} \int_{\frac{R}{2}}^{R} y_n^{\alpha/2} dy_n$$

$$\geq C_0 (C_1)^{p(\frac{\alpha}{2}-1)} \frac{2^{\alpha/2+2}}{3(\alpha+2)} \left(1 - \frac{1}{2^{\alpha/2+1}}\right) R^{p(\frac{\alpha}{2}-1)+\alpha}$$

$$:= A R^{p(\frac{\alpha}{2}-1)+\alpha}$$

$$= \frac{A}{2^{p(\frac{\alpha}{2}-1)+\alpha}} x_n^{p(\frac{\alpha}{2}-1)+\alpha}$$

$$:= A_1 x_n^{p(\frac{\alpha}{2}-1)+\alpha}.$$

Repeating the argument above for m times, for $x_n = 2R$, we have

$$u(x_n) \geq A(m, p, \alpha) x_n^{p^m(\frac{\alpha}{2}-1)+\frac{p^m-1}{p-1}\alpha}. \tag{7.93}$$

For any fixed $0 < \alpha < 2$, we choose m to be an integer greater than $\frac{3-\alpha^2}{\alpha}$. That is

$$m \geq \left\lceil \frac{3-\alpha^2}{\alpha} \right\rceil + 1, \tag{7.94}$$

where $\lceil a \rfloor$ is the integer part of a.

We claim that for such choice of m, it holds

$$\tau(p) := \left[p^m (\frac{\alpha}{2}-1) + \frac{p^m-1}{p-1}\alpha \right] p + \frac{\alpha}{2} \geq 0. \tag{7.95}$$

We postpone the proof of (7.95) for a moment. Now by (7.93) and (7.95), we derive that

$$u^p(x_n) x_n^{\alpha/2} \geq A(m, p, \alpha) x_n^{\tau(p)} \geq A(m, p, \alpha) > 0, \tag{7.96}$$

for all x_n sufficiently large. This contradicts (7.90).

Now what's left is to verify (7.95). In fact, if we let

$$f(p) := \tau(p)(p-1) = p^{m+2}(\frac{\alpha}{2}-1) + (\frac{\alpha}{2}+1)p^{m+1} - \frac{\alpha}{2}p - \frac{\alpha}{2},$$

then

$$f'(p) = p^m[(m+2)(\frac{\alpha}{2}-1)p + (m+1)(\frac{\alpha}{2}+1)] - \frac{\alpha}{2}.$$

We show that

$$f'(p) > 0, \text{ for } 1 < p \leq \frac{n+\alpha}{n-\alpha}.$$

Since $p > 1$, it suffices to show

$$(m+2)(\frac{\alpha}{2}-1)p + (m+1)(\frac{\alpha}{2}+1) \geq \frac{\alpha}{2}.$$

Due to the fact $\frac{\alpha}{2} - 1 < 0$, $n \geq 3$, and $p \leq \frac{n+\alpha}{n-\alpha}$, we only need to verify that

$$(m+2)(\frac{\alpha}{2}-1)\frac{3+\alpha}{3-\alpha} + (m+1)(\frac{\alpha}{2}+1) \geq \frac{\alpha}{2}$$

which can be derived directly from (7.94). $\quad \square$

7.4 Equivalence Between PDE Systems and Integral Systems in $\mathbb{R}^n$

In this section, we show that, besides deriving symmetry of solutions, the *method of moving planes in integral forms* can also be utilized to establish equivalence between systems of PDEs and of integral equations.

For $0 < \alpha < 2$, we consider the following PDE system for the fractional Laplacian:

$$\begin{cases} (-\triangle)^{\alpha/2} u_i(x) = f_i(u_1(x), \cdots, u_m(x)), \\ u_i \geq 0, \qquad\qquad\qquad\qquad\qquad\qquad x \in \mathbb{R}^n, \end{cases} \tag{7.97}$$

where $i = 1, \cdots, m$.

In order to guarantee that the system contains no independent subsystem, for example, to avoid a situation like below:

$$(-\triangle)^{\alpha/2} u_1 = u_1^p$$
$$(-\triangle)^{\alpha/2} u_2 = u_2^q,$$

we introduce the following assumption. We say system (7.97) is *interrelated* if

$$(f_{i_1}(u), f_{i_2}(u), \ldots, f_{i_l}(u)) \neq (f_{i_1}(v), f_{i_2}(v), \ldots, f_{i_l}(v))$$

whenever

$$(u_{i_1}, u_{i_2}, \ldots, u_{i_l}) = (v_{i_1}, v_{i_2}, \ldots, v_{i_l})$$

and

$$u_{i_{l+1}} > v_{i_{l+1}}, \ u_{i_{l+2}} > v_{i_{l+2}}, \ldots, u_{i_m} > v_{i_m},$$

where $i_1, i_2, \ldots, i_m$ is a permutation of $1, 2, \ldots, m$.

We also assume

(f) $f_1, \ldots, f_m$ are real-valued, non-negative, continuous, homogeneous functions of degree $1 < \gamma \leq \frac{n+\alpha}{n-\alpha}$, and nondecreasing with respect to the variables $u_1, \ldots, u_m$.

Let $u = (u_1, \ldots, u_m)$. We will establish the equivalence between (7.97) and the following integral system:

$$\begin{cases} u_i(x) = \int_{\mathbb{R}^n} G(x, y) f_i(u(y)) dy, \\ u_i(x) \geq 0, \qquad\qquad\qquad\qquad\qquad x \in \mathbb{R}^n. \end{cases} \tag{7.98}$$

where $G(x, y) = \frac{C_{n,\alpha}}{|x-y|^{n-\alpha}}$ is the Green's function associated with $(-\triangle)^{\frac{\alpha}{2}}$ in $\mathbb{R}^n$.

Theorem 7.4.1 *Assume that system (7.97) is interrelated and condition* **(f)** *holds. Let* $u = (u_1, \ldots, u_m) \in L_\alpha$ *be a locally bounded and lower semi-continuous positive solution of (7.97). Then, for* $1 < \gamma \leq \frac{n+\alpha}{n-\alpha}$, $u = (u_1, \ldots, u_m)$ *also satisfies integral system (7.98), and vice versa.*

As a byproduct of the proof of Theorem 7.4.1, we have:

Corollary 7.4.1   *1. In the critical case* $\gamma = \frac{n+\alpha}{n-\alpha}$, *u is radially symmetric and monotone decreasing about some point in $\mathbb{R}^n$.*
2. *In the subcritical case* $1 < \gamma < \frac{n+\alpha}{n-\alpha}$, $u \equiv 0$.

The proof of the theorem is very long, and we divide it into three subsections. The first one is a general preparation, the second and the third ones deal with critical and subcritical cases separately.

7.4.1 The General Preparation

We first show that for $1 \leq i \leq m$,

$$u_i(x) = c_i + \int_{\mathbb{R}^n} \frac{c_n}{|x-y|^{n-\alpha}} f_i(u(y))dy. \tag{7.99}$$

Let

$$v_i^R(x) = \int_{B_R} G_R(x,y) f_i(u(y))dy,$$

where $G_R(x,y)$ is the Green's function on the ball $B_R(0)$. It is easy to see that

$$\begin{cases} (-\triangle)^{\alpha/2} v_i^R(x) = f_i(u(x)), & x \in B_R(0), \\ v_i^R = 0, & x \in B_R^c(0). \end{cases} \tag{7.100}$$

Let $w_i^R(x) = u_i(x) - v_i^R(x)$, by (7.97) and (7.100), we have

$$\begin{cases} (-\triangle)^{\alpha/2} w_i^R(x) = 0, & x \in B_R(0), \\ w_i^R \geq 0, & x \in B_R^c(0). \end{cases}$$

By the *maximum principle*, we have

$$w_i^R(x) \geq 0, \quad x \in \mathbb{R}^n.$$

It's easy to see that as $R \to \infty$,

$$v_i^R(x) \to v_i(x) = \int_{\mathbb{R}^n} \frac{c_n}{|x-y|^{n-\alpha}} f_i(u(y))dy.$$

Obviously,

$$(-\triangle)^{\alpha/2} v_i(x) = f_i(u(x)), \quad x \in \mathbb{R}^n.$$

Let

$$w_i(x) = u_i(x) - v_i(x).$$

Then for $x \in \mathbb{R}^n$,

$$\begin{cases} (-\triangle)^{\alpha/2} w_i(x) = 0, \\ w_i \geq 0. \end{cases}$$

By the *Liouville theorem* (see Proposition 5.1.1 in Chapter 5), we derive that $w_i(x) = c_i$. This proves (7.99).

Next we show that
$$c_i \equiv 0, \quad i = 1, \ldots, m.$$

It requires much more effort for a system than for a single equation as illustrated below.

Recall that in Chapter 5, for the single equation

$$(-\triangle)^{\alpha/2} u = u^p, \quad x \in \mathbb{R}^n, \tag{7.101}$$

we obtained

$$u(x) = c + \int_{\mathbb{R}^n} G(x,y) u^p(y) dy, \tag{7.102}$$

with $c \geq 0$. Then it is quite easy to derive that c must be 0. Otherwise, if $c > 0$, then since $u \geq c$, and the integral on the right hand side of (7.102) would become infinity, because

$$\int_{\mathbb{R}^n} G(x,y) u^p(y) dy \geq \int_{\mathbb{R}^n} G(x,y) c^p dy = \infty.$$

A contradiction with the finiteness of $u(x)$ on the left hand side.

Now the situation is much more complicated in the case of systems. Take the following simple system in $\mathbb{R}^n$ as an example:

$$\begin{cases} (-\triangle)^{\alpha/2} u = u^{p_1} v^{q_1} \\ (-\triangle)^{\alpha/2} v = u^{p_2} v^{q_2}. \end{cases} \tag{7.103}$$

Using the above *maximum principle and Liouville theorem* arguments, we immediately arrive at

$$\begin{cases} u(x) = c_1 + \int_{\mathbb{R}^n} G(x,y) u^{p_1}(y) v^{q_1}(y) dy \\ v(x) = c_2 + \int_{\mathbb{R}^n} G(x,y) u^{p_2}(y) v^{q_2}(y) dy, \end{cases} \tag{7.104}$$

with $c_1, c_2 \geq 0$.

If both c_1 and c_2 are positive, then it is trivial to derive a contradiction as we did for the single equation. However, there are other two possibilities:

$$c_1 > 0 \text{ and } c_2 = 0, \text{ or } c_1 = 0 \text{ and } c_2 > 0.$$

In these two cases, the above argument for single equation (7.101) does not give us any contradiction. It is even more complicated for a system of m equations.

To circumvent this complexity, we apply the *method of moving planes in integral forms* on each equation to show that c_i must be zero for each $i = 1, 2, \cdots, m$.

Because we do not assume any decay on the solutions near infinity, we employ the Kelvin transform.

For $\lambda < 0$, let

$$\Sigma_\lambda = \{x = (x_1, x_2, \cdots, x_n) \in \mathbb{R}^n \mid x_1 \leq \lambda\},$$

$$T_\lambda = \{x \in \mathbb{R}^n \mid x_1 = \lambda\},$$

and

$$x^\lambda = (2\lambda - x_1, x_2, \cdots, x_n)$$

be the reflection of the point $x = (x_1, x_2, \cdots, x_n)$ about the plane T_λ. Let

$$u_i^\lambda(x) = u_i(x^\lambda),$$

and

$$u^\lambda(x) = u(x^\lambda).$$

In the proof, we'll use an equivalent form of the classical Hardy-Littlewood-Sobolev inequality in Lemma 7.2.1.

For any $x^0 \in \mathbb{R}^n$, consider the Kelvin transform centered at x^0:

$$\bar{u}_i(x) = \frac{1}{|x - x^0|^{n-\alpha}} u_i\left(\frac{x - x^0}{|x - x^0|^2} + x^0\right).$$

Without loss of generality, we choose x^0 to be the origin. Then

$$\bar{u}_i(x) = \frac{1}{|x|^{n-\alpha}} u_i\left(\frac{x}{|x|^2}\right).$$

Since $u_i \in L^\infty_{loc}(\mathbb{R}^n)$, for any domain Ω that is a positive distance away from the origin, we have

$$\int_\Omega \bar{u}_i^{\frac{2n}{n-\alpha}}(y)dy < \infty, \quad i = 1, \ldots, m. \tag{7.105}$$

From (7.99), we derive

$$
\begin{aligned}
&\bar{u}_i(x) \\
&= \frac{1}{|x|^{n-\alpha}} u_i\left(\frac{x}{|x|^2}\right) \\
&= \frac{c_i}{|x|^{n-\alpha}} + \frac{1}{|x|^{n-\alpha}} \int_{\mathbb{R}^n} G\left(\frac{x}{|x|^2}, y\right) f_i(u(y))dy, \\
&= \frac{c_i}{|x|^{n-\alpha}} + \frac{1}{|x|^{n-\alpha}} \int_{\mathbb{R}^n} G\left(\frac{x}{|x|^2}, \frac{z}{|z|^2}\right) f_i\left(u\left(\frac{z}{|z|^2}\right)\right) \frac{1}{|z|^{2n}} dz, \quad y = \frac{z}{|z|^2} \\
&= \frac{c_i}{|x|^{n-\alpha}} + \int_{\mathbb{R}^n} \frac{G(x,z)|x|^{n-\alpha}|z|^{n-\alpha}}{|x|^{n-\alpha}|z|^{2n}} |z|^{\gamma(n-\alpha)} f_i(\bar{u}(z))dz \\
&= \frac{c_i}{|x|^{n-\alpha}} + \int_{\mathbb{R}^n} \frac{G(x,z)}{|z|^{2n-(\gamma+1)(n-\alpha)}} f_i(\bar{u}(z))dz, \tag{7.106}
\end{aligned}
$$

where γ is the homogeneous degree of f_i.

7.4.2 The Critical Case

For $\gamma = \frac{n+\alpha}{n-\alpha}$, equation (7.106) becomes

$$\bar{u}_i(x) = \frac{c_i}{|x|^{n-\alpha}} + \int_{\mathbb{R}^n} G(x,y)f_i(\bar{u}(y))dy. \tag{7.107}$$

To show that

$$c_i = 0, \quad i = 1,\ldots,m,$$

we argue by contradiction. Suppose that

$$c_{i_o} > 0, \text{ for some } i_o.$$

By (7.107), we have

$$\bar{u}_i(x) = \frac{c_i}{|x|^{n-\alpha}} + \int_{\Sigma_\lambda} G(x,y)f_i(\bar{u}(y))dy + \int_{\Sigma_\lambda} G(x,y^\lambda)f_i(\bar{u}^\lambda(y))dy$$

$$= \frac{c_i}{|x|^{n-\alpha}} + \int_{\Sigma_\lambda} G(x,y)f_i(\bar{u}(y))dy + \int_{\Sigma_\lambda} G(x^\lambda,y)f_i(\bar{u}^\lambda(y))dy, \tag{7.108}$$

and

$$\bar{u}_i^\lambda(x) = \frac{c_i}{|x^\lambda|^{n-\alpha}} + \int_{\Sigma_\lambda} G(x^\lambda,y)f_i(\bar{u}(y))dy + \int_{\Sigma_\lambda} G(x^\lambda,y^\lambda)f_i(\bar{u}^\lambda(y))dy$$

$$= \frac{c_i}{|x^\lambda|^{n-\alpha}} + \int_{\Sigma_\lambda} G(x^\lambda,y)f_i(\bar{u}(y))dy + \int_{\Sigma_\lambda} G(x,y)f_i(\bar{u}^\lambda(y))dy. \tag{7.109}$$

Hence,

$$\bar{u}_i(x) - \bar{u}_i^\lambda(x)$$
$$= \frac{c_i}{|x|^{n-\alpha}} - \frac{c_i}{|x^\lambda|^{n-\alpha}} + \int_{\Sigma_\lambda} [G(x,y) - G(x^\lambda,y)][f_i(\bar{u}) - f_i(\bar{u}^\lambda)]dy. \tag{7.110}$$

Let

$$\Gamma_i^\lambda = \{x \in \Sigma_\lambda \backslash B_\epsilon(0^\lambda) \mid \bar{u}_i^\lambda(x) < \bar{u}_i(x)\},$$

and

$$\Sigma_i^\lambda = \{x \in \Sigma_\lambda \backslash B_\epsilon(0^\lambda) \mid f_i(\bar{u}^\lambda(x)) < f_i(\bar{u}(x))\},$$

for ϵ sufficiently small.

Let

$$w_i^\lambda(y) = \begin{cases} 0, & \text{for } \bar{u}_i(y) < \bar{u}_i^\lambda(y), \\ \bar{u}_i(y) - \bar{u}_i^\lambda(y), & \text{for } \bar{u}_i(y) > \bar{u}_i^\lambda(y), \end{cases}$$

and

$$w^\lambda(y) = (w_1^\lambda(y), \ldots, w_m^\lambda(y)).$$

By (7.110), we have

$$\bar{u}_i(x) - \bar{u}_i^\lambda(x) \leq \int_{\Sigma_i^\lambda} [G(x, y) - G(x^\lambda, y)][f_i(\bar{u}(y)) - f_i(\bar{u}^\lambda(y))] dy.$$

$$(7.111)$$

Using the *method of moving planes in integral forms*, we show that the solution is radially symmetric about the origin. The proof consists of two steps.

Step 1. Prepare to move the plane from near $x_1 = -\infty$.

For λ sufficiently negative, and $\epsilon > 0$ sufficiently small, we show that

$$\bar{u}_i^\lambda(x) \geq \bar{u}_i(x), \quad a.e. \text{ in } \Sigma_\lambda \backslash B_\epsilon(0^\lambda). \tag{7.112}$$

In other words, Γ_i^λ is measure zero.

Without loss of generality, we consider $\bar{u}_1$.

For any $y \in \Sigma_1^\lambda$, since f_1 is nondecreasing, there exists some $i_0 \in [1, m]$ such that

$$\bar{u}_{i_0}(y) > \bar{u}_{i_0}^\lambda(y).$$

For convenience's sake, we may assume that

$$\bar{u}_i(y) > \bar{u}_i^\lambda(y), \quad i = 1, \ldots, k,$$

and

$$\bar{u}_i(y) \leq \bar{u}_i^\lambda(y), \quad i = k+1, \ldots, m.$$

For $y \in \Sigma_1^\lambda$ and $j = 1, \ldots, k$, there are two possible cases:

i. For some $j_0 \in [1, k]$, $\bar{u}_{j_0}^\lambda(y) \leq \frac{1}{2}\bar{u}_{j_0}(y)$.

ii. For all $j \in [1, k]$, $\bar{u}_j(y) > \bar{u}_j^\lambda(y) > \frac{1}{2}\bar{u}_j(y)$.

In case *i*,

$$f_1(\bar{u}(y)) - f_1(\bar{u}^\lambda(y)) \leq f_1(\bar{u}(y)) \leq C|\bar{u}(y)|^{\frac{n+\alpha}{n-\alpha}}. \tag{7.113}$$

Since $u_i \in L_{loc}^\infty(\mathbb{R}^n)$, by (7.107) it's easy to see that

$$\frac{a_1}{|x|^{n-\alpha}} \leq \bar{u}_i(x) \leq \frac{a_2}{|x|^{n-\alpha}}, \quad i = 1, \ldots, m, \tag{7.114}$$

where a_1 and a_2 are constants. Thus

$$|w_{j_0}^\lambda(y)| \geq \frac{1}{2}\bar{u}_{j_0}(y) \geq \frac{a_1}{|y|^{n-\alpha}} \cong |\bar{u}(y)|. \tag{7.115}$$

Combining (7.113) with (7.115), it yields

$$f_1(\bar{u}(y)) - f_1(\bar{u}^\lambda(y)) \le f_1(\bar{u}(y))$$
$$\le C|\bar{u}(y)|^{\frac{2\alpha}{n-\alpha}}|w_{j_o}^\lambda(y)|$$
$$\le C|\bar{u}(y)|^{\frac{2\alpha}{n-\alpha}}|w^\lambda(y)|. \tag{7.116}$$

In case *ii*, from the monotonicity of f_1, using the *mean value theorem*, we derive that

$$f_1(\bar{u}(y)) - f_1(\bar{u}^\lambda(y))$$
$$= f_1(\bar{u}_1, \bar{u}_2, \ldots, \bar{u}_m) - f_1(\bar{u}_1^\lambda, \ldots, \bar{u}_k^\lambda, \bar{u}_{k+1}^\lambda, \ldots, \bar{u}_m^\lambda)$$
$$= f_1(\bar{u}_1, \bar{u}_2, \ldots, \bar{u}_m) - f_1(\bar{u}_1^\lambda, \bar{u}_2, \ldots, \bar{u}_m)$$
$$+ f_1(\bar{u}_1^\lambda, \bar{u}_2, \ldots, \bar{u}_m) - f_1(\bar{u}_1^\lambda, \bar{u}_2^\lambda, \ldots, \bar{u}_m)$$
$$+ \cdots - f_1(\bar{u}_1^\lambda, \ldots, \bar{u}_{k-1}^\lambda, \bar{u}_k, \ldots, \bar{u}_m)$$
$$+ f_1(\bar{u}_1^\lambda, \ldots, \bar{u}_{k-1}^\lambda, \bar{u}_k, \ldots, \bar{u}_m) - f_1(\bar{u}_1^\lambda, \ldots, \bar{u}_k^\lambda, \bar{u}_{k+1}, \ldots, \bar{u}_m)$$
$$+ f_1(\bar{u}_1^\lambda, \ldots, \bar{u}_k^\lambda, \bar{u}_{k+1}, \ldots, \bar{u}_m) - f_1(\bar{u}_1^\lambda, \ldots, \bar{u}_k^\lambda, \bar{u}_{k+1}^\lambda, \ldots, \bar{u}_m^\lambda)$$
$$\le f_1(\bar{u}_1, \bar{u}_2, \ldots, \bar{u}_m) - f_1(\bar{u}_1^\lambda, \bar{u}_2, \ldots, \bar{u}_m)$$
$$+ f_1(\bar{u}_1^\lambda, \bar{u}_2, \ldots, \bar{u}_m) - f_1(\bar{u}_1^\lambda, \bar{u}_2^\lambda, \ldots, \bar{u}_m)$$
$$+ \cdots - f_1(\bar{u}_1^\lambda, \ldots, \bar{u}_{k-1}^\lambda, \bar{u}_k, \ldots, \bar{u}_m)$$
$$+ f_1(\bar{u}_1^\lambda, \ldots, \bar{u}_{k-1}^\lambda, \bar{u}_k, \ldots, \bar{u}_m) - f_1(\bar{u}_1^\lambda, \ldots, \bar{u}_k^\lambda, \bar{u}_{k+1}, \ldots, \bar{u}_m)$$
$$= \frac{\partial f_1}{\partial \bar{u}_1}(\xi_1, \bar{u}_2, \ldots, \bar{u}_m)w_1^\lambda(y) + \cdots$$
$$+ \frac{\partial f_1}{\partial \bar{u}_k}(\bar{u}_1^\lambda, \ldots, \bar{u}_{k-1}^\lambda, \xi_k, \bar{u}_{k+1}, \ldots, \bar{u}_m)w_k^\lambda(y) \tag{7.117}$$

where ξ_j is valued between $\bar{u}_j$ and $\bar{u}_j^\lambda$, $j = 1, \ldots, k$. Since

$$\bar{u}_j(y) > \bar{u}_j^\lambda(y) > \frac{1}{2}\bar{u}_j(y), \ \ j = 1, \ldots, k, \tag{7.118}$$

we have

$$\bar{u}_j(y) > \xi_j > \bar{u}_j^\lambda(y) > \frac{1}{2}\bar{u}_j(y), \ \ j = 1, \ldots, k. \tag{7.119}$$

Since f_1 is homogeneous and non-decreasing, by (7.119), we have

$$\frac{\partial f_1}{\partial \bar{u}_j}(\bar{u}_1, \bar{u}_2, \ldots, \xi_j, \ldots, \bar{u}_m)w_j^\lambda(y)$$

$$\le C\frac{f_1}{\xi_j}(\bar{u}_1, \bar{u}_2, \ldots, \xi_j, \ldots, \bar{u}_m)w_j^\lambda(y)$$

$$\le C\frac{f_1}{\bar{u}_j}(\bar{u}_1, \bar{u}_2, \ldots, \xi_j, \ldots, \bar{u}_m)w_j^\lambda(y)$$

$$\le C|\bar{u}(y)|^{\frac{2\alpha}{n-\alpha}}|w^\lambda(y)|.$$

Together with (7.117), it's easy to see that

$$f_1(\bar{u}(y)) - f_1(\bar{u}^\lambda(y)) \leq C|\bar{u}(y)|^{\frac{2\alpha}{n-\alpha}}|w^\lambda(y)|. \qquad (7.120)$$

In both cases, we have proved that for any $y \in \Sigma_1^\lambda$,

$$0 < f_1(\bar{u}(y)) - f_1(\bar{u}^\lambda(y)) \leq C|\bar{u}(y)|^{\frac{2\alpha}{n-\alpha}}|w^\lambda(y)|.$$

Then it follows from (7.111) that for any $x \in \Gamma_1^\lambda$,

$$\begin{aligned}
0 < \bar{u}_1(x) - &\bar{u}_1^\lambda(x) \\
&\leq \int_{\Sigma_1^\lambda} [G(x,y) - G(x^\lambda,y)][f_1(\bar{u}(y)) - f_1(\bar{u}^\lambda(y))]dy \\
&\leq C\int_{\Sigma_1^\lambda} [G(x,y) - G(x^\lambda,y)]|\bar{u}(y)|^{\frac{2\alpha}{n-\alpha}}|w^\lambda(y)|dy \\
&\leq C\int_{\Sigma_1^\lambda} \frac{1}{|x-y|^{n-\alpha}}|\bar{u}(y)|^{\frac{2\alpha}{n-\alpha}}|w^\lambda(y)|dy. \qquad (7.121)
\end{aligned}$$

Noticing that for any $i \in [1,m]$, there exists at least one $k \in [1,m]$ such that $\Sigma_i^\lambda \subseteq \Gamma_k^\lambda$. Hence

$$\Sigma_i^\lambda \subset \bigcup_{j=1}^m \Gamma_j^\lambda := \Gamma_\lambda.$$

Since $supp\, w_1^\lambda \subseteq \Gamma_1^\lambda$, we can write (7.121) as

$$0 \leq -w_1^\lambda(x) \leq C\int_{\Gamma_\lambda} \frac{1}{|x-y|^{n-\alpha}}|\bar{u}(y)|^{\frac{2\alpha}{n-\alpha}}|w^\lambda(y)|dy. \qquad (7.122)$$

Through similar argument, for all $i = 1, 2, \cdots, m$, one can obtain (7.122) for $w_i^\lambda(x)$ in Γ^λ. Therefore, for $x \in \Gamma_\lambda$, we have

$$|w^\lambda(x)| \leq C\int_{\Gamma_\lambda} \frac{1}{|x-y|^{n-\alpha}}|\bar{u}(y)|^{\frac{2\alpha}{n-\alpha}}|w^\lambda(y)|dy. \qquad (7.123)$$

Since $supp\, w^\lambda \subset \Gamma_\lambda$, applying Hardy-Littlewood-Sobolev inequality (see Lemma 7.2.1) and Hölder inequality to (7.123), we have

$$\|w^\lambda\|_{L^{\frac{2n}{n-\alpha}}(\Gamma_\lambda)} \leq C\|\bar{u}\|_{L^{\frac{2n}{n-\alpha}}(\Gamma_\lambda)}^{\frac{2\alpha}{n-\alpha}} \|w^\lambda\|_{L^{\frac{2n}{n-\alpha}}(\Gamma_\lambda)}. \qquad (7.124)$$

By (7.105), for λ sufficiently negative and for $\epsilon > 0$ sufficiently small, it holds that

$$C\|\bar{u}\|_{L^{\frac{2n}{n-\alpha}}(\Gamma_\lambda)}^{\frac{2\alpha}{n-\alpha}} < \frac{1}{2}.$$

Combining this with (7.124), we conclude that

$$\|w^\lambda\|_{L^{\frac{2n}{n-\alpha}}(\Sigma_\lambda)} = 0.$$

This proves (7.112).

Step 2. Move the plane to the limiting position to derive symmetry.

Step 1 provides a starting point to move the plane T_λ. Now we continue to move the plane T_λ to the right as long as (7.112) holds.

Define

$$\lambda_0 = \sup\{\lambda \le 0 \mid \bar{u}_i^{\lambda'}(x) \ge \bar{u}_i(x),\ a.e.\ \text{in}\ \Sigma_{\lambda'}\backslash B_\epsilon(0^{\lambda'}),\ \lambda' \le \lambda, i = 1, \cdots, m\}.$$

We will show that

$$\lambda_0 = 0.$$

Suppose on the contrary, $\lambda_0 < 0$, then we will be able to prove that $\bar{u}(x)$ is symmetric about T_{λ_0}, that is

$$\bar{u}(x) \equiv \bar{u}^{\lambda_0}(x),\quad a.e.\ \text{in}\ \Sigma_{\lambda_0}\backslash B_\epsilon(0^{\lambda_0}). \tag{7.125}$$

Indeed, if (7.125) does not hold, then for all $i = 1, \cdots, m$, we have

$$\bar{u}_i^{\lambda_0}(x) > \bar{u}_i(x),\quad a.e.\ \text{in}\ \Sigma_{\lambda_0}\backslash B_\epsilon(0^{\lambda_0}). \tag{7.126}$$

In fact, when (7.125) does not hold, there exists some $j_0 \in [1, m]$ such that

$$\bar{u}_{j_0}^{\lambda_0}(y) > \bar{u}_{j_0}(y),\ \text{on a set}\ D\ \text{of positive measure.} \tag{7.127}$$

By the definition of λ_0, we have

$$\bar{u}_i^{\lambda_0}(y) > \bar{u}_i(y),\quad y \in \Sigma_{\lambda_0}\backslash B_\epsilon(0^{\lambda_0}).$$

Since the system (7.97) is interrelated and f_i is non-decreasing, there must exist at least one $l \ne j_0$, such that

$$f_l(\bar{u}^{\lambda_0}(y)) > f_l(\bar{u}(y)),\quad y \in D.$$

And thus for any $x \in \Sigma_{\lambda_0}\backslash B_\epsilon(0^{\lambda_0})$,

$$\bar{u}_l^{\lambda_0}(x) - \bar{u}_l(x) \ge \int_{\Sigma_{\lambda_0}} [G(x, y) - G(x_\lambda, y)][f_l(\bar{u}^{\lambda_0}(y)) - f_l(\bar{u}(y))]dy$$
$$> 0.$$

Hence for $i \in [1, m]$,

$$\bar{u}_i^{\lambda_0}(x) - \bar{u}_i(x) \ge \int_{\Sigma_{\lambda_0}} [G(x, y) - G(x_\lambda, y)][f_i(\bar{u}^{\lambda_0}(y)) - f_i(\bar{u}(y))]dy$$
$$> 0.$$

With (7.126), we can move the plane further to the right, i.e., for $\lambda > \lambda_0$ and sufficiently close to λ_0,

$$\bar{u}_i^\lambda(x) \ge \bar{u}_i(x),\quad a.e.\ \text{in}\ \Sigma_\lambda\backslash B_\epsilon(0^\lambda),\quad i \in [1, m]. \tag{7.128}$$

In fact, by inequality (7.124), we have

$$\|w^\lambda\|_{L^{\frac{2n}{n-\alpha}}(\Gamma_\lambda)} \le C\left(\int_{\Gamma_\lambda} |\bar{u}(y)|^{\frac{2n}{n-\alpha}} dy\right)^{\frac{\alpha}{n}} \|w^\lambda\|_{L^{\frac{2n}{n-\alpha}}(\Gamma_\lambda)}. \tag{7.129}$$

By (7.105), for any small $\eta > 0$, we can choose R sufficiently large, such that

$$\int_{(\mathbb{R}^n \setminus B_\varepsilon(0^\lambda)) \setminus B_R(0)} |\bar{u}(y)|^{\frac{2n}{n-\alpha}} \, dy < \eta. \qquad (7.130)$$

For any $\tau > 0$, define

$$E_i^\tau = \left\{ x \in (\Sigma_{\lambda_0} \setminus B_\epsilon(0^{\lambda_0})) \cap B_R(0) \mid \bar{u}_i^{\lambda_0}(x) - \bar{u}_i(x) > \tau \right\},$$

and

$$F_i^\tau = \left\{ (\Sigma_{\lambda_0} \setminus B_\epsilon(0^{\lambda_0})) \cap B_R(0) \right\} \setminus E_i^\tau.$$

Obviously,

$$\lim_{\tau \to 0} \mu(F_i^\tau) = 0.$$

For $\lambda > \lambda_0$, let

$$D_\tau = \left\{ (\Sigma_\lambda \setminus B_\epsilon(0^\lambda)) \setminus (\Sigma_{\lambda_0} \setminus B_\epsilon(0^{\lambda_0})) \right\} \cap B_R(0).$$

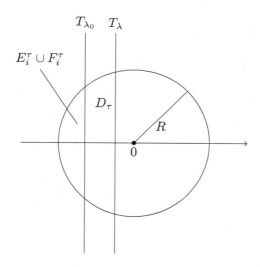

It's easy to see that

$$\left\{ \Gamma_i^\lambda \cap B_R(0) \right\} \subset (\Gamma_i^\lambda \cap E_i^\tau) \cup F_i^\tau \cup D_\tau. \qquad (7.131)$$

For λ sufficiently close to λ_0, $\mu(D_\tau)$ is very small. We show that $\mu(\Gamma_i^\lambda \cap E_i^\tau)$ can be arbitrarily small when λ is sufficiently close to λ_0.

In fact, for $x \in \Gamma_i^\lambda \cap E_i^\tau$,

$$0 > \bar{u}_i^\lambda(x) - \bar{u}_i(x)$$
$$= \bar{u}_i^\lambda(x) - \bar{u}_i^{\lambda_0}(x) + \bar{u}_i^{\lambda_0}(x) - \bar{u}_i(x).$$

Therefore,

$$\bar{u}_i^{\lambda_0}(x) - \bar{u}_i^{\lambda}(x) > \bar{u}_i^{\lambda_0}(x) - \bar{u}_i(x) > \tau, \ \forall x \in \Gamma_i^\lambda \cap E_i^\tau.$$

It follows that

$$(\Gamma_i^\lambda \cap E_i^\tau) \subset H_i^\tau = \left\{ x \in B_R(0) \mid \bar{u}_i^{\lambda_0}(x) - \bar{u}_i^{\lambda}(x) > \tau \right\}. \qquad (7.132)$$

By Chebyshev inequality, we arrive at

$$\mu(H_i^\tau) \le \frac{1}{\tau^{p+1}} \int_{H_i^\tau} |\bar{u}_i^{\lambda_0}(y) - \bar{u}_i^{\lambda}(y)|^{p+1} dy$$

$$\le \frac{1}{\tau^{p+1}} \int_{B_R(0)} |\bar{u}_i^{\lambda_0}(y) - \bar{u}_i^{\lambda}(y)|^{p+1} dy. \qquad (7.133)$$

For fixed τ, the right hand side of inequality (7.133) can be as small as we want when λ is sufficiently close to λ_0. Same is true for $\mu(\Gamma_i^\lambda \cap B_R(0))$. Combining this with (7.130), we deduce that

$$C\left(\int_{\Gamma_\lambda} |\bar{u}(y)|^{\frac{2n}{n-\alpha}} dy \right)^{\frac{\alpha}{n}} \le \frac{1}{2}. \qquad (7.134)$$

Together with (7.129) and (7.134), we obtain

$$\|w^\lambda\|_{L^{\frac{2n}{n-\alpha}}(\Gamma_\lambda)} = 0. \qquad (7.135)$$

It thus proves (7.128), which contradicts the definition of λ_0. This completes the proof of (7.125).

Recall that, by our assumption, $c_{i_o} > 0$ and

$$\bar{u}_{i_o}(x) = \frac{c_{i_o}}{|x|^{n-\alpha}} + \int_{\mathbb{R}^n} G(x, z) f_{i_o}(\bar{u}(z)) dz.$$

This implies that $\bar{u}_{i_o}$ is singular at 0. With (7.125), we know that $\bar{u}_{i_o}$ is also singular at 0^{λ_0}. This is impossible. Therefore it must be true that

$$\lambda_0 = 0.$$

Similarly, we can move the plane from $x_1 = +\infty$ to the left, and prove that

$$\lambda_0 = 0, \quad \bar{u}_{i_o}^{\lambda_0}(x) \le \bar{u}_{i_o}(x).$$

Now we can conclude that

$$\lambda_0 = 0, \quad \bar{u}_{i_o}^{\lambda_0}(x) = \bar{u}_{i_o}(x), \quad a.e. \ in \ \Sigma_0. \qquad (7.136)$$

Since the direction of x_1-axis is arbitrary, we derive that the solution $\bar{u}$ of (7.107) is radially symmetric about the origin.

For any $x^0 \in \mathbb{R}^n$, employing the Kelvin transform centered at x^0

$$\bar{u}_{i_0}(x) = \frac{1}{|x - x^0|^{n-\alpha}} u_{i_0}\left(\frac{x - x^0}{|x - x^0|^2} + x^0 \right),$$

and through a similar argument, one can show that $\bar{u}(x)$ is radially symmetric about x^0. We show that the solution u of (7.99) must be constant.

For any x^1, x^2 in $\mathbb{R}^n$, we choose their midpoint $\frac{x^1+x^2}{2}$ as the center of the Kelvin transform. From the radial symmetry of $\bar{u}$ about x^0, we know that

$$\bar{u}_{i_0}\left(\frac{x^1 - x^0}{|x^1 - x^0|^2} + x^0\right) = \bar{u}_{i_0}\left(\frac{x^2 - x^0}{|x^2 - x^0|^2} + x^0\right),$$

and consequently $u_{i_0}(x^1) = u_{i_0}(x^2)$. This shows that $u_{i_0}(x)$ is constant. Together with the positiveness assumption on u, we have $u_{i_0}(x) = b_0 > 0$. By (7.97), we have

$$0 = (-\triangle)^{\frac{\alpha}{2}} u_{i_0}(x) = f_{i_0}(u_1, \ldots, b_0, \ldots, u_m) > 0. \tag{7.137}$$

This is impossible. Hence, in (7.99), c_i must be zero. This shows that the positive solutions of pseudo-differential system (7.97) also satisfy integral system (7.98). The converse can be derived easily by applying the operator $(-\triangle)^{\frac{\alpha}{2}}$ to both sides of the integral system (7.98).

This completes the proof.

7.4.3 The Subcritical Case

Previously, we proved that if $u = (u_1, \cdots, u_m)$ is a solution of pseudo-differential system (7.97), then there exists $c_i \geq 0$, such that

$$u_i(x) = c_i + \int_{\mathbb{R}^n} \frac{C_{n,\alpha}}{|x - y|^{n-\alpha}} f_i(u(y)) dy, \quad i = 1, \ldots, m. \tag{7.138}$$

Let the Kelvin transform of u_i be

$$\bar{u}_i(x) = \frac{1}{|x|^{n-\alpha}} u_i\left(\frac{x}{|x|^2}\right).$$

Since $u_i \in L^\infty_{loc}(\mathbb{R}^n)$, for any domain $\tilde{\Omega}$ that is a positive distance away from the origin, for $i \in [1, m]$, we have

$$\int_{\tilde{\Omega}} \left(\frac{\bar{u}_i^{\gamma-1}(y)}{|y|^\beta}\right)^{\frac{n}{\alpha}} dy = \int_\Omega u_i^{\frac{n(\gamma-1)}{\alpha}}(y) dy < \infty, \tag{7.139}$$

where $\tilde{\Omega}$ is the image of Ω under the inversion.

In the subcritical case where $1 < \gamma < \frac{n+\alpha}{n-\alpha}$, $\bar{u}_i$ satisfies

$$\bar{u}_i(x) = \frac{c_i}{|x|^{n-\alpha}} + \int_{\mathbb{R}^n} \frac{C_{n,\alpha}}{|x - y|^{n-\alpha}|y|^\beta} f_i(\bar{u}(y)) dy \tag{7.140}$$

where $\beta = 2n - (\gamma + 1)(n - \alpha)$, and γ is the homogeneous degree of f_i.

In the critical case $\gamma = \frac{n+\alpha}{n-\alpha}$, $\beta = 0$; while in the subcritical case $1 < \gamma < \frac{n+\alpha}{n-\alpha}$, $\beta > 0$. We will use the presence of the singular term $\frac{1}{|y|^\beta}$ to show that no matter whether c_i are zero or not, $\bar{u}$ is always symmetric about the origin.

The argument is quite similar to, but not entirely the same as that in the critical case, hence we still present some details.

It's easy to see that

$$\bar{u}_i(x) - \bar{u}_i^\lambda(x) = c_i\left(\frac{1}{|x|^{n-\alpha}} - \frac{1}{|x^\lambda|^{n-\alpha}}\right)$$
$$+ \int_{\Sigma_\lambda} [G(x,y) - G(x^\lambda,y)] \left[\frac{f_i(\bar{u}(y))}{|y|^\beta} - \frac{f_i(\bar{u}^\lambda(y))}{|y^\lambda|^\beta}\right] dy.$$
$$(7.141)$$

The proof also consists of two steps.

Step 1. Prepare to move the plane from near $x_1 = -\infty$.

For any $\epsilon > 0$, define

$$\Gamma_i^\lambda = \{x \in \Sigma_\lambda \backslash B_\epsilon(0^\lambda) \mid \bar{u}_i^\lambda(x) < \bar{u}_i(x)\},$$

and

$$\Sigma_i^\lambda = \{x \in \Sigma_\lambda \backslash B_\epsilon(0^\lambda) \mid f_i(\bar{u}^\lambda(x)) < f_i(\bar{u}(x))\}.$$

By (7.141), we have

$$\bar{u}_i(x) - \bar{u}_i^\lambda(x)$$
$$\leq \int_{\Sigma_i^\lambda} [G(x,y) - G(x^\lambda,y)] \left[\frac{f_i(\bar{u}(y))}{|y|^\beta} - \frac{f_i(\bar{u}^\lambda(y))}{|y^\lambda|^\beta}\right] dy. \quad (7.142)$$

By an elementary calculation, for $x \in \Sigma_i^\lambda$,

$$\bar{u}_i(x) - \bar{u}_i^\lambda(x)$$
$$\leq \int_{\Sigma_i^\lambda} [G(x,y) - G(x^\lambda,y)]$$
$$\cdot \left[\frac{f_i(\bar{u}(y))}{|y|^\beta} - \frac{f_i(\bar{u}^\lambda(y))}{|y|^\beta} + \frac{f_i(\bar{u}^\lambda(y))}{|y|^\beta} - \frac{f_i(\bar{u}^\lambda(y))}{|y^\lambda|^\beta}\right] dy$$
$$= \int_{\Sigma_i^\lambda} [G(x,y) - G(x^\lambda,y)]$$
$$\cdot \left[\frac{f_i(\bar{u}(y)) - f_i(\bar{u}^\lambda(y))}{|y|^\beta} + f_i(\bar{u}^\lambda(y))\left(\frac{1}{|y|^\beta} - \frac{1}{|y^\lambda|^\beta}\right)\right] dy$$
$$\leq \int_{\Sigma_i^\lambda} [G(x,y) - G(x^\lambda,y)]\frac{f_i(\bar{u}(y)) - f_i(\bar{u}^\lambda(y))}{|y|^\beta} dy. \quad (7.143)$$

Similar to (7.121), we have

$$\bar{u}_1(x) - \bar{u}_1^\lambda(x) \leq C \int_{\Sigma_1^\lambda} \frac{1}{|x-y|^{n-\alpha}} \frac{|\bar{u}(y)|^{\gamma-1}}{|y|^\beta} |w^\lambda(y)| dy. \quad (7.144)$$

Combining Lemma 7.2.1 and the Hölder inequality, we obtain

$$\|w^\lambda\|_{L^{\frac{2n}{n-\alpha}}(\Gamma_\lambda)} \le C\left\{\int_{\Gamma_\lambda}\left(\frac{\bar{u}_i^{\gamma-1}(y)}{|y|^\beta}\right)^{\frac{n}{\alpha}}dy\right\}^{\frac{\alpha}{n}}\|w^\lambda\|_{L^{\frac{2n}{n-\alpha}}(\Gamma_\lambda)}. \qquad (7.145)$$

By (7.139), we can choose N sufficiently large, such that for $\lambda \le -N$ and $\epsilon > 0$ sufficiently small,

$$C\left\{\int_{\Gamma_\lambda}\left(\frac{\bar{u}_i^{\gamma-1}(y)}{|y|^\beta}\right)^{\frac{n}{\alpha}}dy\right\}^{\frac{\alpha}{n}} \le \frac{1}{2}. \qquad (7.146)$$

Combining (7.145) with (7.146), we deduce that

$$\|w^\lambda\|_{L^{\frac{2n}{n-\alpha}}(\Gamma_\lambda)} = 0.$$

This implies that

$$\bar{u}_i^\lambda(x) \ge \bar{u}_i(x), \quad a.e. \text{ in } \Sigma_\lambda \backslash B_\epsilon(0^\lambda). \qquad (7.147)$$

Step 2. Move the plane to the limiting position to derive symmetry.

Step 1 provides a starting point to move the plane T_λ. We will keep moving the plane T_λ to the right as long as (7.147) holds till the limiting position. Define

$$\lambda_0 = \sup\{\lambda < 0 \mid \bar{u}_i^{\lambda'}(x) \ge \bar{u}_i(x), \ a.e. \text{ in } \Sigma_{\lambda'} \backslash B_\epsilon(0^{\lambda'}), \ \lambda' \le \lambda\}.$$

The rest of the proof is identical to that of the critical case, except that we need to use $\int_{\Gamma_\lambda}\left(\frac{\bar{u}_i^{\gamma-1}(y)}{|y|^\beta}\right)^{\frac{n}{\alpha}}dy$ instead of $\int_{\Gamma_\lambda}|\bar{u}|^{\frac{2n}{n-\alpha}}(y)dy$. Then we conclude that

$$\lambda_0 = 0, \ \bar{u}(x) = \bar{u}^{\lambda_0}(x), \quad a.e. \text{ in } \Sigma_0.$$

Now it is easy to see that each positive solution $\bar{u}$ of (7.140) is radially symmetric about the origin in the subcritical case. Through a similar argument to that in the critical case, one can derive that $\bar{u}$ is radially symmetric about any point $x^0 \in \mathbb{R}^n$ and hence $u(x) \equiv 0$. This shows that there is no positive solution of (7.97) in the subcritical case, and hence completes the proofs of Theorem 7.4.1 and Corollary 7.4.1.

7.5 Various Applications of MMP in Integral Forms

Since the introduction of the *method of moving planes in integral forms*, it has been applied to many integral equations and systems as well as differential equations and systems to obtain symmetry and a priori estimates of solutions, including systems dealing with fully nonlinear Wolff potentials, Bessel potentials, boundary values problems on half spaces and unit balls, as well as problems on complete Riemannian manifolds.

7.5.1 System of Integral Equations and Fully Nonlinear Wolff Potentials

In [CL2], the authors considered a system of integral equations for $0 < \alpha < n$,

$$\begin{cases} u_i(x) = \int_{\mathbb{R}^n} \dfrac{1}{|x-y|^{n-\alpha}} f_i(u(y)) dy, \quad x \in \mathbb{R}^n, \quad i = 1, \ldots, m, \\ u(x) = (u_1(x), u_2(x), \cdots, u_m(x)). \end{cases} \tag{7.148}$$

Here $f_1(u)$, $f_2(u)$, $\cdots$, $f_m(u)$ are real-valued functions, non-negative, and continuous.

Using the *method of moving planes in integral forms*, the authors were able to derive the radial symmetry of the positive solutions and prove

Theorem 7.5.1 *Assume that system (7.148) is interrelated (see [CL2]). Let $f_i(u)$ be real-valued homogeneous functions of degree $\frac{n+\alpha}{n-\alpha}$ and*

$$\frac{\partial f_i}{\partial u_j} \geq 0 \quad for \ 1 \leq i,j \leq m \quad and \ for \ u_i > 0.$$

Then any positive solution u of the system (7.148), with $u_i(x) \in L^\infty_{loc}(\mathbb{R}^n)$ for $i = 1, \cdots, m$, is composed of radially symmetric functions with the same center, and assumes the form

$$u_i(x) = c_i \phi_{x_o, t_i}(x) := \left(\frac{t_i}{t_i^2 + |x - x_o|^2} \right)^{(n-\alpha)/2}$$

for some $c_i > 0$, $t_i > 0$, and $x_o \in \mathbb{R}^n$.

An example of (7.148) is the following system:

$$\begin{cases} -\triangle u(x) = u(x)^\alpha v(x)^\beta \\ -\triangle v(x) = u(x)^\beta v(x)^\alpha. \end{cases} \tag{7.149}$$

It was studied in [LM] by Li and Ma. In particular it includes the quintic Schrödinger equation considered by Bourgain [B] as a special case. In general, system (7.148) with $\alpha = 2$ is related to the stationary Schrödinger system with critical exponents for Bose-Einstein condensate. The results in [LM] were later improved, in [MZ1] by Ma and Zhao, to include the cases where the right hand sides can be positive linear combinations of the nonlinear terms appeared in (7.149).

In [CL5], the authors considered fully nonlinear integral systems involving Wolff potentials:

$$\begin{cases} u(x) = W_{\beta,\gamma}(v^q)(x), \\ v(x) = W_{\beta,\gamma}(u^p)(x), \quad x \in \mathbb{R}^n, \end{cases} \tag{7.150}$$

where

$$W_{\beta,\gamma}(f)(x) = \int_0^\infty \left[\frac{\int_{B_t(x)} f(y) dy}{t^{n-\beta\gamma}} \right]^{\frac{1}{\gamma-1}} \frac{dt}{t}.$$

After modifying and refining the techniques on the *method of moving planes in integral forms*, the authors obtained radial symmetry and monotonicity for the positive solutions to systems (7.150).

This system includes many known systems as special cases, in particular, when $\beta = \frac{\alpha}{2}$ and $\gamma = 2$, system (7.150) reduces to the previously considered system

$$\begin{cases} u(x) = \int_{\mathbb{R}^n} \frac{1}{|x-y|^{n-\alpha}} v(y)^q dy, \\ v(x) = \int_{\mathbb{R}^n} \frac{1}{|x-y|^{n-\alpha}} u(y)^p dy, \quad x \in \mathbb{R}^n. \end{cases} \tag{7.151}$$

It is also known that when $\beta = 1$, $\gamma = p$,

$$u(x) = W_{1,p}(u^q)(x) \tag{7.152}$$

corresponds to the p-Laplacian equation

$$-div(|\nabla u|^{p-2}\nabla u) = u^q(x); \tag{7.153}$$

and when $\beta = \frac{2k}{k+1}$ and $\gamma = k+1$,

$$u(x) = W_{\frac{2k}{k+1}, k+1}(u^q)(x) \tag{7.154}$$

corresponds to the well-known k-Hessian equation

$$F_k[-u] = u^q(x), \quad k = 1, 2, \cdots, n. \tag{7.155}$$

Here

$$F_k[u] = S_k(\lambda(D^2 u)) \quad \text{and} \quad \lambda(D^2 u) = (\lambda_1, \cdots, \lambda_n)$$

with λ_i being eigenvalues of the Hessian matrix $(D^2 u)$, and $S_k(\cdot)$ is the k-th symmetric function:

$$S_k(\lambda) = \sum_{1 \le i_1 < \cdots < i_k \le n} \lambda_{i_1} \lambda_{i_2} \cdots \lambda_{i_k}.$$

In two special cases when $k = 1$ and $k = n$, we have

$$F_1[u] = \triangle u \quad \text{and} \quad F_n[u] = det(D^2 u).$$

7.5.2 Bessel Potentials

In [MC1], Ma and Chen considered equation

$$(I - \triangle)^{\frac{\alpha}{2}} u = f(u), \quad x \in \mathbb{R}^n, \tag{7.156}$$

where I is the identity operator. It is equivalent to the integral equation

$$u(x) = g_\alpha * f(u), \quad x \in \mathbb{R}^n, \tag{7.157}$$

where $*$ is the convolution and

$$g_\alpha(x) = \frac{1}{(4\pi)^\alpha \Gamma(\frac{\alpha}{2})} \int_0^\infty \exp\left\{ -\frac{\pi|x|^2}{t} - \frac{t}{4\pi} \right\} \frac{dt}{t^{(n-\alpha)/2+1}}$$

is the kernel of Bessel potential. Using the *method of moving planes in integral forms*, they were able to obtained the radial symmetry for positive solutions of (7.157).

It is known that the nonlinear Choquard equation arises in various domains of mathematical physics such as physics of many-particle systems, quantum mechanics, physics of laser beams and so on.

In [Lie1], Lieb classified the minimizers of the corresponding functional of the following stationary Choquard equation

$$\triangle u - u + 2u \cdot \left(\frac{1}{|x|} * |u|^2 \right) = 0, \ \ u \in H^1(R^3) \tag{7.158}$$

then he posed the classification of all the positive solutions of equation (7.158) as an open question.

To answer this question, Ma and Zhao [MZ] proved

Theorem 7.5.2 ([MZ]) *The positive solutions of the stationary Choquard equation (7.158) are uniquely determined up to translations.*

Their idea was to use Bessel potential to write equation (7.158) as an equivalent integral system

$$\begin{cases} u(x) = \int_{\mathbb{R}^3} g_2(x-y)u(y)v(y)dy \\ v(x) = \int_{\mathbb{R}^3} \frac{2}{|x-y|}|u^2(y)|dy. \end{cases} \tag{7.159}$$

Then they were able to apply the *method of moving planes in integral forms* to derive the radial symmetry of the positive solutions of (7.159). Actually, the system they considered was more general than (7.159).

7.5.3 Boundary Value Problems

Consider the poly-harmonic equation

$$(-\triangle)^m u = f(u), \ \text{in} \ \Omega, \tag{7.160}$$

with boundary conditions

$$u = \frac{\partial u}{\partial \nu} = \cdots = \frac{\partial^{m-1} u}{\partial \nu^{m-1}} = 0, \ \text{on} \ \partial\Omega, \tag{7.161}$$

where $\frac{\partial u}{\partial \nu}$ is the outward normal derivative.

When Ω is a ball or the upper half Euclidean space $\mathbb{R}^n_+$, it is known that the corresponding Green's function $G_+(x, y)$ exists (see [Bo]), and the boundary value problem described by (7.160) and (7.161) is equivalent to the integral equation

$$u(x) = \int_\Omega G_+(x, y) f(u(y)) dy. \tag{7.162}$$

On the upper half space, one may also consider

$$u(x) = \int_{\mathbb{R}_+^n} \left(\frac{1}{|x-y|^{n-\alpha}} - \frac{1}{|x^*-y|^{n-\alpha}} \right) f(u(y)) dy \qquad (7.163)$$

where x^* is the reflection of the point x about the boundary plane $x_n = 0$, and $0 < \alpha < n$. When $\alpha = 2m$ is an even integer, it corresponds to the boundary value problem

$$\begin{cases} (-\triangle)^m u = f(u), & x \in \mathbb{R}_+^n \\ u = \triangle u = \cdots = \triangle^{m-1} u = 0, & x \in \partial \mathbb{R}_+^n. \end{cases}$$

For integral equations (7.162) and (7.163), one can apply the *method of moving planes in integral forms* to obtain symmetry of positive solutions (see [CC] [CFY] [CZ] [FC] [LZ]).

7.5.4 Problems on Riemannian Manifolds

The *method of moving planes in integral forms* can also be used to obtain the a priori estimate for solutions of certain partial differential or integral equations on Riemannian manifolds. In [QR], Qing and Raske considered a family of smooth equations

$$P_\alpha[g]u = u^{\frac{n+\alpha}{n-\alpha}}, \qquad (7.164)$$

where

$$P_\alpha[g] = 2^\alpha \frac{\Gamma(\frac{\alpha}{2})}{\Gamma(-\frac{\alpha}{2})} S(\frac{n+\alpha}{2})[g],$$

$\alpha \in [2, n)$, and the scattering operator $S(z)$ is a family of meromorphic peusdo-differential operators on M. In particular, when k is a positive integer, P_{2k} is a conformally invariant power of Laplacian. For instance

$$P_2[g] = -\triangle_g + \frac{n-2}{4(n-1)} R[g]$$

is the conformal Laplacian, and the corresponding equation (7.164) is the well-known Yamabe equation.

$$P_4[g] = (-\triangle)^2 - div_g \left(\left(\frac{(n-2)^2+4}{2(n-1)(n-2)} R[g] - \frac{4}{n-2} Ric[g] \right) d \right)$$
$$+ \frac{n-4}{2} Q[g]$$

is the Paneitz operator, where $R[g]$ is the scalar curvature, $Ric[g]$ the Ricci curvature, and $Q[g]$ the so-called Q-curvature. In this case (7.164) becomes the Paneitz-Branson equation.

If the Poincaré exponent of the holonomy representation of the fundamental group $\pi_1(M)$ is less than $\frac{n-\alpha}{2}$, then there is a conformally covariant integral operator I_α, which is a right inverse to P_α. The existence of such integral operators allows one to consider the conformally invariant integral equation

$$u = I_\alpha[g](u^{\frac{n+\alpha}{n-\alpha}}). \tag{7.165}$$

Qing and Raske applied the *method of moving planes in integral forms* to establish a priori estimates for solutions of equation (7.165) and obtained

Proposition 7.5.1 *(Qing-Raske) Suppose that (M^n, g) is a locally conformally flat manifold with positive Yamabe constant and Poincaré exponent less than $\frac{n-\alpha}{2}$ for some $\alpha \in [2, n)$. And suppose that (M^n, g) is not conformally equivalent to the standard round sphere. Then there exists a constant $C = C(n, \alpha, k)$, such that, for any smooth positive solution u to integral equation (7.165), we have*

$$\|u\|_{C^k(M)} + \|\frac{1}{u}\|_{C^k(M)} \leq C.$$

Based on this a priori estimate and the degree theory described in [Sc], Qing and Raske can prove the existence of positive solutions to both PDE (7.164) and integral equation (7.165).

8

A Method of Moving Spheres for the Fractional Laplacian

8.1 Introduction — Outline of the Method

In this chapter, we introduce another direct method – the *method of moving spheres* on the fractional Laplacian, which is more convenient to apply than the method of moving planes in some contexts, for instance, in the prescribing scalar curvature problem.

Outline of the Method of Moving Spheres

In the *method of moving planes*, we move parallel planes T_λ along a chosen direction to the limiting position to obtain symmetry of the solutions about the limiting plane. While in the *method of moving spheres*, we fix a center and increase or decrease the radius of the spheres to show some kind of monotonicity or symmetry of the solutions along the radial directions of the spheres. Here, the monotonicity or symmetry is in the sense that we compare the solution with its Kelvin transform. As we will see in the examples in Section 3, if such a monotonicity holds for arbitrary centers, then we will be able to deduce non-existence of solutions; or if such symmetry holds for arbitrary centers, then all the solutions can be classified. However, from this process, one will not able to derive the usual symmetry of the solutions directly.

To illustrate the general idea, consider a simple example in $\mathbb{R}^n$:

$$(-\triangle)^{\alpha/2} u(x) = u^p(x)$$

in the subcritical case $1 < p < \frac{n+\alpha}{n-\alpha}$.

Given $\lambda > 0$ and $x_0 \in \mathbb{R}^n$, we define the Kelvin transform of u centered at x_0 with respect to a sphere of radius λ as follows,

$$u_\lambda(x) \equiv (\frac{\lambda}{|x - x_0|})^{n-\alpha} u(x^\lambda), \tag{8.1}$$

where

$$x^\lambda \equiv \frac{\lambda^2(x - x_0)}{|x - x_0|^2} + x_0$$

is the inversion of x with respect to the sphere

$$S_\lambda(x_0) \equiv \{x \mid |x - x_0| = \lambda\}.$$

For simplicity of writing, here we take $x_0 = 0$.

We compare the value of $u(x)$ with its Kelvin transform $u_\lambda(x)$ with respect to the sphere $S_\lambda(0)$ centered at the origin. Notice that for $x \in B_\lambda(0)$, $u_\lambda(x)$ actually is the value of u at x^λ outside the ball multiplied by $\left(\frac{\lambda}{|x|}\right)^{n-\alpha}$.

Let $w(x) = u_\lambda(x) - u(x)$, then one can easily verify that w is anti-symmetric in the sense

$$w(x) = -w_\lambda(x).$$

Also with Lemma A.2.1, we can show that w satisfies the equation

$$(-\triangle)^{\alpha/2}w(x) + c(x)w(x) = 0,$$

for some $c(x)$ depending on $u(x)$.

In step 1, we show that for λ sufficiently small, it holds

$$w(x) \geq 0, \quad \forall\, x \in B_\lambda(0). \tag{8.2}$$

This will be accomplished by the *narrow region principle*:

Theorem 8.1.1 *Let $w \in L_\alpha \cap C^{1,1}_{loc}(\Omega)$ be lower semi-continuous on $\bar{\Omega}$. If $c(x)$ is bounded from below in Ω and*

$$\begin{cases} (-\triangle)^{\alpha/2}w(x) + c(x)w(x) \geq 0, & x \in \Omega \subset B_\lambda(0), \\ w(x) \geq 0, & x \in B_\lambda(0)\backslash\Omega, \\ w(x) = -w_\lambda(x), & x \in B_\lambda(0). \end{cases}$$

Then for some sufficiently small $\delta > 0$, if Ω is a spherically narrow region, i.e.,

$$\Omega \subset A_{\lambda-\delta,\lambda}(0) \equiv \{x \in \mathbb{R}^n \mid \lambda - \delta < |x| < \lambda\},$$

then we have

$$\inf_{x \in \Omega} w(x) \geq 0.$$

Furthermore, if $w(x) = 0$ for some $x \in \Omega$, then $w(x) = 0$ in $\mathbb{R}^n$.

Here we take Ω to be the whole $B_\lambda(0)$.

Step 1 provides a starting point to move the sphere. Now we increase λ, the radius of the sphere, as long as inequality (8.2) holds. Define

$$\lambda_o = \sup\{\lambda \mid w_\mu(x) \geq 0, x \in B_\mu(0); \forall\, \mu \leq \lambda\}.$$

In Step 2, we show that $\lambda_o = \infty$, by employing the *narrow region principle* again. Because each time we only increase the radius λ a little bit, the region in between the two spheres is a spherically narrow region.

Steps 1 and 2 can be carried through for any center x_0, hence we have actually proved that

$$\left(\frac{\lambda}{|x - x_0|}\right)^{n-\alpha} u(x^\lambda) \geq u(x), \quad \forall \, \lambda > 0.$$

This is the kind of the monotonicity we mentioned earlier.

Due to the arbitrariness of x_0, one will be able to show that

$$\nabla u(x) = 0, \quad \forall x \in \mathbb{R}^n.$$

Consequently, u must be constant, which contradicts the equation if u is a positive solution. This proves non-existence of positive solutions. We will present detailed argument in Section 8.3.

8.2 Various Maximum Principles

Similar to the *method of moving planes*, the *method of moving spheres* is a continuous application of the *maximum principles*.

We start with a simple maximum principle for anti-symmetric functions.

8.2.1 Maximum Principle for Anti-Symmetric Functions

Theorem 8.2.1 *Let Ω be an open subset of $B_\lambda(0)$. Assume that $w \in L_\alpha \cap C_{loc}^{1,1}(\Omega)$ is lower semi-continuous on $\bar{\Omega}$. If*

$$\begin{cases} (-\triangle)^{\alpha/2} w(x) \geq 0, \, x \in \Omega, \\ w(x) \geq 0, & x \in B_\lambda(0) \backslash \Omega, \\ w(x) = -w_\lambda(x), & x \in B_\lambda(0), \end{cases}$$

then $w(x) \geq 0$ for every $x \in \Omega$. Moreover, if $w(x) = 0$ for some $x \in \Omega$, then $w(x) \equiv 0$ for every $x \in \mathbb{R}^n$.

Proof. We argue by contradiction.

Since w is lower semi-continuous on $\bar{\Omega}$, we may assume that there is some $x_0 \in \Omega$ such that

$$w(x_0) \equiv \min_{x \in \Omega} w(x) < 0.$$

Let $\tilde{w}(x) \equiv w(x) - w(x_0)$, then

$$\begin{cases} (-\triangle)^{\alpha/2} \tilde{w}(x) = (-\triangle)^{\alpha/2} w(x) \geq 0, & x \in \Omega, \\ \tilde{w}(x) \geq 0, & x \in B_\lambda(0), \\ \tilde{w}(x_0) = 0. \end{cases}$$

By the anti-symmetry assumption $w(x) = -w_\lambda(x)$, it holds that

$$\left(\frac{\lambda}{|x|}\right)^{n-\alpha} \tilde{w}(x^\lambda) = \left(\frac{\lambda}{|x|}\right)^{n-\alpha} w(x^\lambda) - \left(\frac{\lambda}{|x|}\right)^{n-\alpha} w(x_0)$$

$$= (-w(x)) + w(x_0) - \left[1 + \left(\frac{\lambda}{|x|}\right)^{n-\alpha}\right] w(x_0)$$

$$> -\tilde{w}(x). \tag{8.3}$$

By straightforward calculations,

$$
\begin{aligned}
(-\triangle)^{\alpha/2}\tilde{w}(x_0) &= \int_{\mathbb{R}^n} \frac{\tilde{w}(x_0) - \tilde{w}(y)}{|x_0 - y|^{n+\alpha}} dy \\
&= \int_{B_\lambda(0)} \frac{-\tilde{w}(y)}{|x_0 - y|^{n+\alpha}} dy + \int_{\mathbb{R}^n \setminus B_\lambda(0)} \frac{-\tilde{w}(y)}{|x_0 - y|^{n+\alpha}} dy \\
&\equiv I_1 + I_2.
\end{aligned}
\tag{8.4}
$$

Let $y \equiv \frac{\lambda^2 z}{|z|^2}$, then by (8.3), it holds that

$$
\begin{aligned}
I_2 &= \int_{B_\lambda(0)} \frac{-\tilde{w}(\frac{\lambda^2 z}{|z|^2})}{\left|x_0 - \frac{\lambda^2 z}{|z|^2}\right|^{n+\alpha}} \cdot \frac{\lambda^{2n}}{|z|^{2n}} dz \\
&< \int_{B_\lambda(0)} \frac{\tilde{w}(z)}{\left|x_0 - \frac{\lambda^2 z}{|z|^2}\right|^{n+\alpha}} \cdot \left(\frac{\lambda}{|z|}\right)^{n+\alpha} dz.
\end{aligned}
\tag{8.5}
$$

By (8.4), we have

$$
(-\triangle)^{\alpha/2}\tilde{w}(x_0) < \int_{B_\lambda(0)} \left(\frac{1}{\left|\frac{|z|x_0}{\lambda} - \frac{\lambda z}{|z|}\right|^{n+\alpha}} - \frac{1}{|x_0 - z|^{n+\alpha}}\right) \tilde{w}(z) dz.
$$

Notice that, for $z \in B_\lambda(0)$,

$$
\left|\frac{|z|x_0}{\lambda} - \frac{\lambda z}{|z|}\right|^2 - |x_0 - z|^2 = \frac{(|x_0|^2 - \lambda^2)(|z|^2 - \lambda^2)}{\lambda^2} > 0,
$$

which implies that

$$
(-\triangle)^{\alpha/2}\tilde{w}(x_0) < 0.
$$

A contradiction.

Now we assume that $w(x_0) = 0$. Then by the same calculations as above, we have

$$
0 \leq (-\triangle)^{\alpha/2}w(x_0) = \int_{B_\lambda(0)} \left(\frac{1}{\left|\frac{|z|x_0}{\lambda} - \frac{\lambda z}{|z|}\right|^{n+\alpha}} - \frac{1}{|x_0 - z|^{n+\alpha}}\right) w(z) dz \leq 0.
$$

Therefore, $w(x) = 0$ for almost every $x \in B_\lambda(0)$. By the anti-symmetry assumption $w(x) = -w_\lambda(x)$, it holds that $w(x) = 0$, for almost every $x \in \mathbb{R}^n$.

Remark 8.2.1 We introduce $\tilde{w}$ at the beginning of the proof for the sake of a simpler calculation when changing variables for (8.5).

8.2.2 Narrow Region Principle

Theorem 8.2.2 *Let $w \in L_\alpha \cap C_{loc}^{1,1}(\Omega)$ be lower semi-continuous on $\bar{\Omega}$. If $c(x)$ is bounded from below in Ω and*

$$\begin{cases} (-\triangle)^{\alpha/2}w(x) + c(x)w(x) \geq 0, \ x \in \Omega \subset B_\lambda(0), \\ w(x) \geq 0, & x \in B_\lambda(0)\backslash\Omega, \\ w(x) = -w_\lambda(x), & x \in B_\lambda(0). \end{cases}$$

Then for some sufficiently small $\delta > 0$, if Ω is a spherically narrow region, i.e.,

$$\Omega \subset A_{\lambda-\delta,\lambda}(0) \equiv \{x \in \mathbb{R}^n \mid \lambda - \delta < |x| < \lambda\},$$

then we have

$$\inf_{x\in\Omega} w(x) \geq 0.$$

Furthermore, if $w(x) = 0$ for some $x \in \Omega$, then $w(x) = 0$ in $\mathbb{R}^n$.

Proof. We argue by contradiction. Suppose there exists some $x_0 \in \Omega$ such that

$$w(x_0) = \min_{x\in\Omega} w(x) < 0.$$

Let $\tilde{w}(x) = w(x) - w(x_0)$, then $\tilde{w}(x_0) = 0$ and

$$(-\triangle)^{\alpha/2}\tilde{w}(x) = (-\triangle)^{\alpha/2}w(x).$$

It follows from the anti-symmetry condition of w that

$$\left(\frac{\lambda}{|x|}\right)^{n-\alpha}\tilde{w}(x^\lambda) = -\tilde{w}(x) - \left[1 + \left(\frac{\lambda}{|x|}\right)^{n-\alpha}\right]w(x_0). \qquad (8.6)$$

By straightforward calculations,

$$\begin{aligned} (-\triangle)^{\alpha/2}\tilde{w}(x_0) &= \int_{\mathbb{R}^n} \frac{\tilde{w}(x_0) - \tilde{w}(y)}{|x_0 - y|^{n+\alpha}}dy \\ &= \int_{B_\lambda(0)} \frac{-\tilde{w}(y)}{|x_0 - y|^{n+\alpha}}dy + \int_{\mathbb{R}^n\backslash B_\lambda(0)} \frac{-\tilde{w}(y)}{|x_0 - y|^{n+\alpha}}dy \\ &\equiv J_1 + J_2. \end{aligned} \qquad (8.7)$$

By (8.6), we have

$$\begin{aligned} J_2 &= \int_{\mathbb{R}^n\backslash B_\lambda(0)} \frac{-\tilde{w}(y)}{|x_0 - y|^{n+\alpha}}dy \\ &= \int_{\mathbb{R}^n\backslash B_\lambda(0)} \frac{\left(\frac{\lambda}{|y|}\right)^{n-\alpha}\tilde{w}(y^\lambda)}{|x_0 - y|^{n+\alpha}}dy + \int_{\mathbb{R}^n\backslash B_\lambda(0)} \frac{\left[1 + \left(\frac{\lambda}{|y|}\right)^{n-\alpha}\right]w(x_0)}{|x_0 - y|^{n+\alpha}}dy \\ &\equiv J_{21} + J_{22}. \end{aligned}$$

Let $y = \frac{\lambda^2 z}{|z|^2}$, then

$$J_1 + J_{21} = \int_{B_\lambda(0)} \left(\frac{1}{\left|\frac{|z|x_0}{\lambda} - \frac{\lambda z}{|z|}\right|^{n+\alpha}} - \frac{1}{|x_0 - z|^{n+\alpha}}\right)\tilde{w}(z)dz.$$

Since $\tilde{w}(z) \geq 0$ for every $z \in B_\lambda(0)$, so $J_1 + J_{21} \leq 0$. Notice that $w(x_0) < 0$, then it holds that

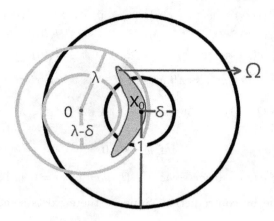

Fig. 8.1

$$(-\triangle)^{\alpha/2}\tilde{w}(x_0) = (J_1 + J_{21}) + J_{22} \le J_{22} \le w(x_0) \int_{\mathbb{R}^n \setminus B_\lambda(0)} \frac{1}{|x_0 - y|^{n+\alpha}} dy.$$

Now we estimate the right hand side of the above inequality,

$$\int_{\mathbb{R}^n \setminus B_\lambda(0)} \frac{1}{|x_0 - y|^{n+\alpha}} dy \ge \int_{(\mathbb{R}^n \setminus B_\lambda(0)) \cap (B_1(x_0) \setminus B_\delta(x_0))} \frac{1}{|x_0 - y|^{n+\alpha}} dy$$

$$\ge \frac{1}{4} \int_{B_1(x_0) \setminus B_\delta(x_0)} \frac{1}{|x_0 - y|^{n+\alpha}} dy$$

$$\ge \frac{C}{\delta^\alpha}.$$

The contradiction arises when $\delta > 0$ is sufficiently small.

8.3 Applications

Using the *maximum principles* developed in the previous section, we are able to classify non-negative solutions to semi-linear elliptic equations and prove a non-existence result for the prescribing fractional order Q_α curvature equation on $\mathbb{S}^n$. We also give an alternative proof of a Liouville-type theorem in $\mathbb{R}^n_+$.

8.3.1 Classification of Solutions to Semi-Linear Fractional Equations in $\mathbb{R}^n$

Theorem 8.3.1 *Let* $g : \mathbb{R}^1_+ \to \mathbb{R}^1_+$ *be a locally bounded function such that* $\frac{g(r)}{r^p}$ *is non-increasing with* $p \equiv \frac{n+\alpha}{n-\alpha}$.

If $u \in L_\alpha(\mathbb{R}^n) \cap C^{1,1}_{loc}(\mathbb{R}^n)$ *is a nonnegative solution to*

$$(-\triangle)^{\alpha/2}u(x) = g(u(x)), \ x \in \mathbb{R}^n, \tag{8.8}$$

then one of the following holds:

(1) *For some* $C_0 \geq 0$, $u(x) \equiv C_0$ *for every* $x \in \mathbb{R}^n$ *and* $g(C_0) = 0$.

(2) *There exists* $\beta_1 > 0$, $\beta_2 > 0$, $x_0 \in \mathbb{R}^n$ *such that*

$$u(x) = \frac{\beta_1}{(|x - x_0|^2 + \beta_2^2)^{\frac{n-\alpha}{2}}}, \ \forall x \in \mathbb{R}^n, \tag{8.9}$$

and $g(r)$ *is a multiple of* r^p *for every* $r \in (0, \max\limits_{x \in \mathbb{R}^n} u(x)]$.

The main ingredient in the proof of Theorem 8.3.1 is to establish the following key lemma by applying the *method of moving spheres*. Precisely,

Lemma 8.3.1 *Let* u *be a positive solution to equation (8.8) and let* u_λ *be the Kelvin transform of* u *in the sense of (8.1). Then exactly one of the following holds.*

(A) *For every* $x_0 \in \mathbb{R}^n$, *for all* $\lambda \in (0, +\infty)$, *it holds that*

$$u_\lambda(x) \geq u(x), \quad \forall x \in B_\lambda(x_0) \backslash \{x_0\}.$$

(B) *For every* $x_0 \in \mathbb{R}^n$, *there exists* $\lambda_0 \in (0, +\infty)$ *which depends on* x_0 *such that*

$$u_{\lambda_0}(x) \equiv u(x), \quad \forall x \in B_{\lambda_0}(x_0) \backslash \{x_0\}.$$

Proof of Lemma 8.3.1   For $\lambda > 0$, let

$$B_\lambda^-(x_0) = \{x \in B_\lambda(x_0) \backslash \{x_0\} \mid w_\lambda(x) < 0\},$$

and

$$w_\lambda(x) \equiv u_\lambda(x) - u(x).$$

Note that the definition of $w_\lambda(x)$ here is different from the one in the previous two sections. By this definition of w_λ, for $x \in B_\lambda(x_0) \backslash \{x_0\}$ we have

$$\left(\frac{\lambda}{|x - x_0|}\right)^{n-\alpha} w_\lambda(x^\lambda) = \left(\frac{\lambda}{|x - x_0|}\right)^{n-\alpha} [u_\lambda(x^\lambda) - u(x^\lambda)]$$

$$= \left(\frac{\lambda}{|x - x_0|}\right)^{n-\alpha} \left[\left(\frac{|x - x_0|}{\lambda}\right)^{n-\alpha} u(x) - u(x^\lambda)\right]$$

$$= u(x) - \left(\frac{\lambda}{|x - x_0|}\right)^{n-\alpha} u(x^\lambda)$$

$$= -w_\lambda(x). \tag{8.10}$$

From Lemma A.2.1, we have

$$(-\triangle)^{\alpha/2} w_\lambda(x) = \frac{g\left(\left(\frac{|x-x_0|}{\lambda}\right)^{n-\alpha} u_\lambda(x)\right)}{\left(\left(\frac{|x-x_0|}{\lambda}\right)^{n-\alpha} u_\lambda(x)\right)^p} u_\lambda^p(x) - \frac{g(u(x))}{u^p(x)} u^p(x).$$

For $x \in B_\lambda^-(x_0)$, $(\frac{|x-x_0|}{\lambda})^{n-\alpha} u_\lambda(x) < u(x)$. By assumption, $g(r)/r^p$ is non-increasing, so

$$\frac{g\left((\frac{|x-x_0|}{\lambda})^{n-\alpha} u_\lambda(x)\right)}{\left((\frac{|x-x_0|}{\lambda})^{n-\alpha} u_\lambda(x)\right)^p} \geq \frac{g(u(x))}{u^p(x)},$$

and thus

$$(-\triangle)^{\alpha/2} w_\lambda(x) \geq \frac{g(u(x))}{u^p(x)}(u_\lambda^p(x) - u^p(x))$$
$$= \varphi(x) \cdot w_\lambda(x), \tag{8.11}$$

where $\varphi(x) \equiv p \cdot \frac{g(u(x))}{u^p(x)} \cdot \psi(x)$, for some $u_\lambda^{p-1}(x) \leq \psi(x) \leq u^{p-1}(x)$. Notice that $\varphi \geq 0$ and hence

$$|\varphi(x)| \leq p \cdot \frac{g(u(x))}{u^p(x)} \cdot u^{p-1}(x) = p \cdot \frac{g(u(x))}{u(x)}. \tag{8.12}$$

We prove the lemma using the *method of moving spheres* in two steps.

Step 1. For $\lambda > 0$ sufficiently small,

$$w_\lambda(x) \geq 0, \quad x \in B_\lambda(x_0)\backslash\{x_0\}.$$

It suffices to prove that $B_\lambda^-(x_0) = \emptyset$ for $\lambda > 0$ small.

First, we claim that for every λ sufficiently small, there exists $\delta \ll \lambda$, such that

$$w_\lambda(x) \geq C > 0, \quad \forall x \in B_\delta(x_0)\backslash\{x_0\}. \tag{8.13}$$

We prove (8.13) in Lemma A.2.2 in the appendix.

Let $M_\lambda \equiv \sup_{x \in B_\lambda(x_0)} u(x) < +\infty$, $N_\lambda \equiv \sup_{x \in B_\lambda(x_0)} u^{-1}(x) < +\infty$ and $G_\lambda \equiv \sup_{t \in (0,M_\lambda)} g(t)$, then inequality (8.12) implies that

$$\|\varphi\|_{L^\infty(B_\lambda(x_0))} \leq p \cdot G_\lambda \cdot N_\lambda < +\infty. \tag{8.14}$$

It then follows from Theorem 8.2.2 and the local continuity of w_λ that for λ sufficiently small,

$$B_\lambda^-(x_0) = \emptyset.$$

This completes the proof of *Step 1*.

In fact, *Step 1* provides a starting point to carry out the *method of moving spheres* for any given center $x_0 \in \mathbb{R}^n$. Then we continuously increase the radius λ of the sphere $S_\lambda(x_0) \equiv \partial B_\lambda(x_0)$ such that

$$w_\lambda(x) \geq 0, \quad x \in B_\lambda(x_0)\backslash\{x_0\}.$$

For a given center $x_0 \in \mathbb{R}^n$, the *critical scale* $\lambda_0 \in (0, +\infty]$ is defined as follows,

$$\lambda_0 \equiv \sup\{\lambda > 0 \mid w_\mu(x) \geq 0, x \in B_\mu(x_0)\backslash\{x_0\}, \forall 0 < \mu \leq \lambda\}. \tag{8.15}$$

Immediately, by definition, it holds that

$$w_{\lambda_0}(x) \geq 0, x \in B_{\lambda_0}(x_0)\backslash\{x_0\}. \tag{8.16}$$

The lemma follows from the analysis of the *global finiteness* of the critical scale. More precisely,

Step 2. Given $x_0 \in \mathbb{R}^n$, let the critical scale $\lambda_0 > 0$ be defined by (8.15), then exactly one of the following holds:

(a) for every $x_0 \in \mathbb{R}^n$, the corresponding critical scale $\lambda_0 > 0$ is finite, or

(b) for every $x_0 \in \mathbb{R}^n$, the corresponding critical scale $\lambda_0 = +\infty$, that is, for every $\lambda > 0$

$$w_\lambda(x) \geq 0, x \in B_\lambda(x_0)\backslash\{x_0\}. \tag{8.17}$$

The statement of *Step 2* can be reduced to the following: if there exists $z_0 \in \mathbb{R}^n$ such that the corresponding critical scale $\mu_0 < \infty$, then for every $x_0 \in \mathbb{R}^n$ the corresponding critical scale $\lambda_0 < \infty$.

First, we need the following proposition.

Proposition 8.3.1 *If* $\mu_0 < \infty$, *then* $u_{\mu_0}(x) = u(x)$ *for every* $x \in B_{\mu_0}(z_0)\backslash\{z_0\}$.

Proof of Proposition 8.3.1 We prove by ruling out the case that there exists $x' \in B_{\mu_0}(z_0)\backslash\{z_0\}$ such that $w_{\mu_0}(x') \neq 0$.

Later in Lemma A.2.3 in the appendix, we show that if $w_{\mu_0} \not\equiv 0$, then for ε sufficiently small, it holds that

$$w_{\mu_0}(x) \geq c > 0, \quad x \in B_\varepsilon(x_0)\backslash\{x_0\}.$$

By the definition of the critical scale,

$$w_{\mu_0}(x) \geq 0, \ x \in B_{\mu_0}(z_0)\backslash\{z_0\}.$$

If there exists some $x' \in B_{\mu_0}(z_0)\backslash\{z_0\}$, such that

$$w_{\mu_0}(x') > 0, \tag{8.18}$$

then we have

$$w_{\mu_0}(x) > 0, \quad x \in B_{\mu_0}(z_0)\backslash\{z_0\}. \tag{8.19}$$

Indeed, if (8.19) does not hold, then there exists some $x'' \in B_{\mu_0}(z_0)\backslash\{z_0\}$ such that

$$0 = w_{\mu_0}(x'') = \min_{x \in B_{\mu_0}(z_0)\backslash\{z_0\}} w_{\mu_0}(x).$$

That is, $u_{\mu_0}(x'') = u(x'')$. By the calculations in the proof of *Step 1*,

$$(-\triangle)^{\alpha/2} w_{\mu_0}(x'') = \frac{g\left(\left(\frac{|x''-z_0|}{\mu_0}\right)^{n-\alpha} u_{\mu_0}(x'')\right)}{\left(\left(\frac{|x''-z_0|}{\mu_0}\right)^{n-\alpha} u_{\mu_0}(x'')\right)^p} u_{\mu_0}^p(x'') - \frac{g(u(x''))}{u^p(x'')} u^p(x'')$$

$$= \left(\frac{g\left(\left(\frac{|x''-z_0|}{\mu_0}\right)^{n-\alpha} u(x'')\right)}{\left(\left(\frac{|x''-z_0|}{\mu_0}\right)^{n-\alpha} u(x'')\right)^p} - \frac{g(u(x''))}{u^p(x'')}\right) u^p(x'')$$

$$\geq 0,$$

where the last inequality is due to the monotonicity of $g(t)/t^p$. Hence, by Theorem 8.2.1,

$$w_{\mu_0}(x) \equiv 0, \ x \in B_{\mu_0}(z_0)\backslash\{z_0\},$$

which contradicts (8.18). This proves

$$w_{\mu_0}(x) > 0, \ x \in B_{\mu_0}(z_0)\backslash\{z_0\}. \tag{8.20}$$

With (8.20), we can show that for $\delta_1 > 0$ sufficiently small, it holds that for any $\lambda \in [\mu_0, \mu_0 + \delta_1)$,

$$w_\lambda(x) > 0, \ x \in B_\lambda(z_0)\backslash\{z_0\}. \tag{8.21}$$

It contradicts the definition of μ_0.

To verify (8.21), observe from (8.20) that, for $\varepsilon > 0$ small,

$$\min_{x \in B_{\mu_0 - \varepsilon}(z_0)\backslash\{z_0\}} w_{\mu_0}(x) := m_0 > 0.$$

From the continuity of w_λ in λ, we know that

$$w_\lambda(x) \geq m_0/2 > 0, \ x \in B_{\mu_0 - \varepsilon}(z_0)\backslash\{z_0\}. \tag{8.22}$$

Through computations similar to that in *Step 1*, with Theorem 8.2.2 and the local boundedness of g, we arrive at (8.21), and hence complete the proof.

To finish the proof of the reduced statement of *Step 2*, we argue by contradiction.

Suppose there exists some $x_0 \in \mathbb{R}^n$ such that $\lambda_{x_0} = +\infty$, and some $z_0 \in \mathbb{R}^n$ such that $\lambda_{z_0} < +\infty$. Choose a sequence $\{\lambda_j\}_{j \in \mathbb{Z}_+}$ such that $\lim_{j \to +\infty} \lambda_j = +\infty$. Then for any given $\bar{x} \in B_{\lambda_{z_0}}(z_0)\backslash\{z_0\}$, for j large enough, $\bar{x} \in B_{\lambda_j}\backslash\{x_0\}$. Therefore,

$$u_{\lambda_j}(\bar{x}) \geq u(\bar{x}). \tag{8.23}$$

By Proposition 8.3.1,

$$\left(\frac{\lambda_{z_0}}{|x - z_0|}\right)^{n-\alpha} \cdot u\left(\frac{\lambda_{z_0}^2(x - z_0)}{|x - z_0|^2} + z_0\right) = u(x), \ x \in B_{\lambda_{z_0}}(z_0)\backslash\{z_0\}.$$

Then

$$\lim_{|x| \to \infty} |x|^{n-\alpha} \cdot u(x) = \lambda_{z_0}^{n-\alpha} \cdot \lim_{|x| \to \infty} u\left(\frac{\lambda_{z_0}^2(x - z_0)}{|x - z_0|^2} + z_0\right)$$
$$= \lambda_{z_0}^{n-\alpha} u(z_0) \equiv C > 0. \tag{8.24}$$

On the other hand, (8.23) implies that

$$u(\bar{x}) \leq \left(\frac{\lambda_j}{|\bar{x} - x_0|}\right)^{n-\alpha} \cdot u\left(\frac{\lambda_j^2(\bar{x} - x_0)}{|\bar{x} - x_0|^2} + x_0\right). \tag{8.25}$$

Let $j \to \infty$ in (8.25), by (8.24) we have

$$u(\bar{x}) \leq \lim_{j \to +\infty} \Big(\frac{\lambda_j}{|\bar{x} - x_0|} \Big)^{n-\alpha} \cdot u\Big(\frac{\lambda_j^2(\bar{x} - x_0)}{|\bar{x} - x_0|^2} + x_0 \Big)$$

$$= \lim_{j \to +\infty} \frac{1}{\lambda_j^{n-\alpha}} \Big(\frac{\lambda_j^2(\bar{x} - x_0)}{|\bar{x} - x_0|^2} + x_0 \Big)^{n-\alpha} \cdot u\Big(\frac{\lambda_j^2(\bar{x} - x_0)}{|\bar{x} - x_0|^2} + x_0 \Big)$$

$$= \lim_{j \to +\infty} \frac{1}{\lambda_j^{n-\alpha}} \cdot C$$

$$= 0.$$

This contradicts $u > 0$ everywhere in $\mathbb{R}^n$.

We thus finish the proofs of *Step 2* and Lemma 8.3.1.

Now we prove Theorem 8.3.1 through a combination of Lemma 8.3.1 and the following fact.

Proposition 8.3.2 ([LiZ]) *Let $u \in C^{1,1}(\mathbb{R}^n)$ and let u_λ be its Kelvin transform. In the notation of Lemma 8.3.1, we have the following:*

(C) *If (A) holds, then u is constant.*

(D) *If (B) holds, then there exists $\beta_1 > 0$, $\beta_2 > 0$ such that*

$$u(x) = \frac{\beta_1}{(|x - x_0|^2 + \beta^2)^{\frac{n-2}{2}}}, \quad x \in \mathbb{R}^n. \tag{8.26}$$

Proof of (C). It suffices to show that for every $z \in \mathbb{R}^n$, $\nabla u(z) = 0$. This actually is a result of [LZ], however, for the reader's convenience, we outline it here.

For $y \in \mathbb{R}^n$, let

$$\lambda \equiv |z - y|.$$

Choose a point in the line segment joining y and z:

$$x \equiv \frac{|x - y|(z - y)}{\lambda} + y.$$

Let

$$t \equiv \frac{|x - y|}{\lambda}.$$

Consider the Kelvin transform about the center $y \in \mathbb{R}^n$,

$$w_\lambda(x) = u_\lambda(x) - u(x)$$
$$= (\frac{1}{t})^{n-\alpha} u(\frac{z - y}{t} + y) - u(t(z - y) + y) \equiv h(t).$$

It is clear that $h(1) = 0$. By (A), we know that $h(t) > 0$ for every $t < 1$, and thus

$$0 \geq \frac{d}{dt}\Big|_{t=1^-} h(t) = -(n - \alpha)u(z) - 2\langle \nabla u(z), z - y \rangle.$$

Hence, for every $y \in \mathbb{R}^n$ with $z \neq y$, we have

$$\frac{-(n - \alpha)u(z) - 2\langle \nabla u(z), z - y \rangle}{|z - y|} \leq 0. \tag{8.27}$$

Let $\nu \equiv \frac{z-y}{|z-y|}$, as $|y| \to +\infty$, (8.27) becomes

$$\langle \nabla u(z), \nu \rangle \geq 0.$$

Since y is arbitrary, one can also show that

$$\langle \nabla u(z), -\nu \rangle \geq 0.$$

Hence

$$\nabla u(z) = 0.$$

This completes the proof.

Proof of (D). If (B) holds, then for any $z \in \mathbb{R}^n$, there exists $\lambda_z \in R$ such that

$$u(x) = \left(\frac{\lambda_z}{|x-z|}\right)^{n-\alpha} u\left(\frac{\lambda_z^2(x-z)}{|x-z|^2} + z\right), \quad x \in \mathbb{R}^n \backslash \{z\}. \tag{8.28}$$

It follows from (8.28) that for any fixed z, there exists a constant A independent of z such that

$$A := \lim_{|x| \to \infty} u(x)|x|^{n-\alpha} = \lambda_z^{n-\alpha} u(z). \tag{8.29}$$

If $A = 0$, then (8.29) implies that $u \equiv 0$. This conflicts with our assumption that u is positive. If $A > 0$, then without any loss of generality, we assume that $A = 1$.

For $|x|$ sufficiently large, doing Taylor expansion to u in (8.28) at z gives

$$
\begin{aligned}
u(x) &= \left(\frac{\lambda_z}{|x-z|}\right)^{n-\alpha}\left[u(z) + \sum_i \frac{\partial u}{\partial z_i}(z) \cdot \frac{\lambda_z^2(x_i - z_i)}{|x-z|^2} + O\left(\frac{1}{|x-z|^2}\right)\right] \\
&= \frac{1}{|x-z|^{n-\alpha}} + \sum_i \frac{\partial u}{\partial z_i}(z) \cdot (x_i - z_i)\left(\frac{\lambda_z}{|x-z|}\right)^{n-\alpha+2} + O\left(\frac{1}{|x-z|^{n-\alpha+2}}\right).
\end{aligned}
\tag{8.30}
$$

Substituting z by 0 in (8.30) yields

$$u(x) = \frac{1}{|x|^{n-\alpha}} + \sum_i \frac{\partial u}{\partial z_i}(0) \cdot x_i \left(\frac{\lambda_0}{|x|}\right)^{n-\alpha+2} + O\left(\frac{1}{|x|^{n-\alpha+2}}\right). \tag{8.31}$$

Again, from the Taylor expansion at $z = 0$, we have

$$\frac{1}{|x-z|^{n-\alpha}} = \frac{1}{|x|^{n-\alpha}} + (n-\alpha)\sum_i \frac{x_i \cdot z_i}{|x|^{n-\alpha+2}} + O\left(\frac{1}{|x|^{n-\alpha+2}}\right). \tag{8.32}$$

Let $x = (0, \ldots, 0, x_i, 0, \ldots, 0)$, $i = 1, 2, \ldots, n$ in (8.30), (8.31) and (8.32). For each i and $|x|$ sufficiently large, after combining terms from those identities, we arrive at

$$(n-\alpha)\frac{x_i \cdot z_i}{|x|} + \lambda_z^{n-\alpha+2}\frac{\partial u}{\partial z_i}(z) \cdot \frac{x_i - z_i}{|x-z|}(\frac{|x|}{|x-z|})^{n-\alpha+1} + O(\frac{1}{|x|})$$
$$= \lambda_0^{n-\alpha+2}\frac{\partial u}{\partial z_i}(0) \cdot \frac{x_i}{|x|}. \tag{8.33}$$

As $|x| \to \infty$, equation (8.33) becomes

$$(n-\alpha)z_i + \lambda_z^{n-\alpha+2}\frac{\partial u}{\partial z_i}(z) = \lambda_0^{n-\alpha+2}\frac{\partial u}{\partial z_i}(0).$$

Together with (8.29), we have

$$(n-\alpha)z_i + u(z)^{-\frac{n-\alpha+2}{n-\alpha}}\frac{\partial u}{\partial z_i}(z) = u(0)^{-\frac{n-\alpha+2}{n-\alpha}}\frac{\partial u}{\partial z_i}(0).$$

Through an elementary calculation, one can show that, for some $\beta_2 > 0$ and $x_0 \in \mathbb{R}^n$,

$$u(z) = (\frac{1}{|z-x_0|^2 + \beta_2})^{(n-\alpha)/2}.$$

For $A \neq 1$, it's easy to see that (8.26) is true.

With the above results, we can finish the proof of Theorem 8.3.1.

Proof of Theorem 8.3.1.   By Lemma 8.3.1 and Proposition 8.3.2, if $u \in L_\alpha(\mathbb{R}^n) \cap C_{loc}^{1,1}(\mathbb{R}^n)$ is a non-negative solution to (8.8), then either (C) or (D) holds. If (C) holds, there exists $C_0 \geq 0$ such that $u \equiv C_0$ and thus $g(C_0) = (-\triangle)^{\alpha/2}C_0 = 0$. Now assume that (D) holds. Notice that (8.26) is a solution to

$$(-\triangle)^{\alpha/2}u = g(u).$$

It's easy to see that, for some $\gamma_0 > 0$, the function g satisfies

$$g(r) = \gamma_0 \cdot r^p, \quad r \in (0, \max_{x \in \mathbb{R}^n} u(x)].$$

The completes the proof of the theorem.

8.3.2 A Non-Existence Result for a Prescribing Q_α Curvature Equation on $\mathbb{S}^n$

A natural question in conformal geometry is the following:

Let $(\mathbb{S}^n, g_1)$ be the round sphere of dimension $n \geq 3$ such that $\sec_{g_1} \equiv 1$. Given $0 < \alpha < 2$ and a function $Q \in C^\infty(\mathbb{S}^n)$, does a conformal metric $\tilde{g} = u^{\frac{4}{n-\alpha}}g_1$ exist such that $Q_\alpha(x) \equiv Q_\alpha[g](x) = Q(x)$ for every $x \in \mathbb{S}^n$?

When $\alpha = 2$, Q_α is the scalar curvature up to a dimensional constant, and the corresponding prescribing scalar curvature problem is called the Nirenberg problem. The study of such a problem leads to a very active research area in the last decade. The pioneering work in the direction of prescribing Q_α curvature problem can be found in [JLX1] and [JLX2]. Here we extend a non-existence result in [CL6] to the fractional setting.

We focus on the fractional GJMS operator P_α on a round sphere $(\mathbb{S}^n, g_1)$ with $\sec_{g_1} \equiv 1$ in the sense of [CG]. For the convenience of readers, let us briefly introduce the preliminary materials.

Fractional GJMS Operators and a Prescribing Fractional Curvature Problem

We start with the definition of fractional GJMS operator. Let $(\mathbb{H}^{n+1}, g_H)$ be the $(n+1)$-dimensional hyperbolic space with $\sec_{g_H} \equiv -1$. It is standard that the n-sphere $\mathbb{S}^n$ with the conformal structure $[g_1]$ can be viewed as the conformal infinity of $(\mathbb{H}^{n+1}, g_H)$. Specifically, the conformal compactification is given by the following coordinates,

$$g_H = \frac{dt^2 + (1 - \frac{t^2}{4})^2 g_1}{t^2}, \ t \in [0, 2].$$

Clearly, $(t^2 g_H)|_{t=0} = g_1$. By [CG], for any representative $h \in [g_1]$, the fractional GJMS operator $P_\alpha[h] \equiv P_\alpha[h, g_H]$ can be defined in terms of the scattering operator. Precisely, consider the Poisson equation with $s \in (0, n)$,

$$-\triangle_{g_H} v - s(n - s)v = 0,$$

and the solution of the form

$$v = y^{n-s} F + y^s H,$$

with $F, G \in C^\infty(\mathbb{H}^{n+1})$ and y is a geodesic defining function of $(\mathbb{S}^n, h)$ such that

1. there is some $\epsilon > 0$ such that $|\nabla y|_{y^2 g_+} \equiv 1$ holds on $\mathbb{S}^n \times [0, \epsilon)$,
2. $y^2 g_H|_{y=0} = h$.

By standard calculations (see [CG] or [GZ]), we have the following expansion of F and G in terms of y,

$$F = f_0 + f_2 \cdot y^2 + \cdots,$$

and

$$H = h_0 + h_2 \cdot y^2 + \cdots.$$

Then the scattering operator $S(s)$ can be defined as follows, if $F|_{y=0} = f$, then $S(s)(f) \equiv h$ with $h = H|_{y=0}$. Now for any $\alpha \in (0, n)$, the fractional GJMS operator $P_\alpha[h]$ is defined by

$$P_\alpha[h](f) \equiv 2^\alpha \cdot \frac{\Gamma(\frac{\alpha}{2})}{\Gamma(-\frac{\alpha}{2})} \cdot S\left(\frac{n+\alpha}{2}\right)(f).$$

In the above definition, the order α can be any real number in $(0, n)$. Moreover, the above procedure can be generalized to any conformally compact Einstein manifold (X^{n+1}, g_+) with a conformal infinity $(M^n, [h])$ (for more details, see

[CG]). In general, the fractional GJMS operator has the following conformal covariance property: if $\hat{h} = u^{\frac{4}{n-\alpha}}h$, then for every $v \in C^\infty(M^n)$,

$$P_\alpha[\hat{h}, g_+](v) = u^{-\frac{n+\alpha}{n-\alpha}} P_\alpha[h, g_+](uv). \tag{8.34}$$

For $0 < \alpha < n$, we define the Q_α curvature

$$Q_\alpha \equiv \frac{n-\alpha}{2} \cdot P_\alpha[h, g_+](1)$$

with respect to the metric h on M^n.

In the rest of this section, we will briefly recall the prescribing Q_α curvature problem with $0 < \alpha < 2$. By (8.34), the existence of the conformal metric $h = \tilde{u}^{\frac{4}{n-\alpha}} g_1$ with $Q_\alpha[h] = Q$ is equivalent to the existence of positive solution to

$$P_\alpha[g_1](\tilde{u})(x) = \frac{2}{n-\alpha} \cdot Q(x) \cdot \tilde{u}^{\frac{n+\alpha}{n-\alpha}}(x). \tag{8.35}$$

Next, we will see that the prescribing Q_α curvature problem on $\mathbb{S}^n$ can be reduced to the equation on Euclidean space via stereographic projection. Let $\pi_S : \mathbb{S}^n \backslash \{S\} \to \mathbb{R}^n$ be the stereographic projection, then the pullback metric $(\pi^{-1})^* g_1$ is conformal to g_0, the Euclidean metric on $\mathbb{R}^n$. Indeed, for every $x \in \mathbb{R}^n$,

$$(\pi_S^{-1})^* g_1 \equiv \left(\frac{2}{1+|x|^2}\right)^2 \cdot g_0.$$

Therefore, g_1 can be viewed as a conformal metric of $\mathbb{R}^n$. Let $\eta(x) \equiv \left(\frac{2}{1+|x|^2}\right)^{\frac{n-\alpha}{2}}$, then

$$g_1 = \eta^{\frac{4}{n-\alpha}} g_0.$$

By the conformal covariance property (8.34), it holds that

$$P_\alpha[g_1](\tilde{u})(x) = \eta^{-\frac{n+\alpha}{n-\alpha}}(-\triangle)^{\alpha/2}(\tilde{u} \cdot \eta)(x), \; x \in \mathbb{R}^n.$$

Now let $\tilde{u}$ be a positive solution to (8.35) and $u(x) := \tilde{u}(x) \cdot \eta(x)$ for every $x \in \mathbb{R}^n$, then the above identity and equation (8.35) imply that

$$(-\triangle)^{\alpha/2} u(x) = \frac{2}{n-\alpha} \cdot Q(x) \cdot u^{\frac{n+\alpha}{n-\alpha}}(x), \; x \in \mathbb{R}^n.$$

For convenience, we consider the following equation,

$$(-\triangle)^{\alpha/2} u(x) = Q(x) \cdot u^p(x), \; x \in \mathbb{R}^n, \tag{8.36}$$

with $p \equiv \frac{n+\alpha}{n-\alpha}$. By our assumption, $\tilde{u}$ is smooth and thus bounded on $\mathbb{S}^n$, which implies that

$$|x|^{n-\alpha} u(x) = \tilde{u}(x) \cdot \left(\frac{2|x|^2}{1+|x|^2}\right)^{\frac{n-\alpha}{2}} \longrightarrow 2^{\frac{n-\alpha}{2}} \cdot \tilde{u}(S) \equiv M_0, \text{ as } |x| \to \infty. \tag{8.37}$$

From now on, we focus on the equation (8.36) in the Euclidean space with decay condition (8.37).

In the study of prescribing Q_α curvature problem, it is very natural to start with some smooth function Q with certain symmetry. In this section, we assume that the smooth function Q is rotationally symmetric. In [CL6], the authors proved the following non-existence theorem for prescribing scalar curvature problem.

Theorem 8.3.2 ([CL6]) *Let Q be smooth and rotationally symmetric. If Q is monotone in the region where $Q > 0$ and $Q \not\equiv C$, then the prescribing scalar curvature problem on sphere does not admit any positive solution.*

We extend Theorem 8.3.2 to the fractional setting. First we reduce the corresponding assumption on the radial function $Q(x) \equiv Q(|x|)$ to the following

$$
\begin{cases}
Q \in C^\infty(\mathbb{R}^n), \\
Q(r) > 0, \ Q'(r) \leq 0, & r < 1, \\
Q(r) \leq 0, & r \geq 1,
\end{cases}
\tag{8.38}
$$

where $r \equiv |x|$. With the above preparations, we are ready to state and prove the non-existence result of the prescribing Q_α curvature problem in the following.

A Non-Existence Result

We apply the method of moving spheres to prove the non-existence result of the prescribing Q_α curvature equation on the sphere.

Theorem 8.3.3 *Let Q be continuous and rotationally symmetric on the round sphere $(\mathbb{S}^n, g_1)$ of curvature 1. Assume that Q is monotone in the region where $Q > 0$ and $Q \not\equiv C$. Then for every $0 < \alpha \leq 2$, the prescribing Q-curvature equation on $(\mathbb{S}^n, g_1)$*

$$
P_\alpha(\tilde{u}) = Q \cdot \tilde{u}^p, \ p \equiv \frac{n + \alpha}{n - \alpha},
\tag{8.39}
$$

does not admit any positive solution $\tilde{u} \in \mathcal{D}(P_\alpha) \cap C^{1,1}(\mathbb{S}^n)$.

Proof. To prove the theorem, it suffices to show the non-existence of positive $C^{1,1}$ solution to equation (8.36) with (8.37) and (8.38).

Let

$$
x^\lambda \equiv \frac{\lambda^2 x}{|x|^2}, \ u_\lambda(x) \equiv \left(\frac{\lambda}{|x|} \right)^{n - \alpha} u(x^\lambda),
$$

and

$$
w_\lambda(x) \equiv u_\lambda(x) - u(x).
$$

By (8.10) we have

$$
\left(\frac{\lambda}{|x|} \right)^{n - \alpha} w_\lambda(x^\lambda) = -w_\lambda(x),
$$

and

$$(-\triangle)^{\alpha/2}w_\lambda(x) = Q\Big(\frac{\lambda^2}{r}\Big)u_\lambda^p(x) - Q(r)u^p(x). \tag{8.40}$$

Step 1. We begin with $\lambda = 1$.

Since $Q(r) > 0$, $Q(1/r) \le 0$ for every $0 < r < 1$, by (8.40),

$$(-\triangle)^{\alpha/2}w_1(x) < Q(1/r)u_1^p(x) \le 0.$$

By the *maximum principle* (Theorem 8.2.1), we know that

$$w_1(x) < 0, \quad x \in B_1(0)\backslash\{0\}.$$

This provides a starting point. And we can keep moving the spheres.

Step 2. Keep shrinking the sphere B_λ until the limiting scale

$$\lambda_0 \equiv \inf\{\lambda \ge 0 \mid w_\mu(x) \le 0,\ x \in B_\mu(0)\backslash\{0\},\ \ \forall \lambda \le \mu \le 1\}.$$

We show that

$$\lambda_0 = 0. \tag{8.41}$$

By definition,

$$w_\lambda(x) \le 0, \quad x \in B_\lambda(0)\backslash\{0\}.$$

It follows from the *maximum principle* (Theorem 8.2.1) that either

$$w_{\lambda_0}(x) < 0, \quad x \in B_{\lambda_0}(0)\backslash\{0\}, \tag{8.42}$$

or

$$w_{\lambda_0}(x) \equiv 0, \quad x \in B_{\lambda_0}(0)\backslash\{0\}. \tag{8.43}$$

On the other hand, by Lemma A.2.2, we know that for $0 < \delta \ll \lambda_0$ small

$$w_{\lambda_0}(x) < -1, \quad x \in B_\delta(0)\backslash\{0\}.$$

Hence (8.43) will not happen. If (8.42) holds, when $\lambda_0 > 0$, we can continue to move the sphere B_{λ_0} and show that for $\varepsilon > 0$ small,

$$w_{\lambda_0-\varepsilon}(x) \le 0, \quad x \in B_{\lambda_0-\varepsilon}(0)\backslash\{0\}. \tag{8.44}$$

This is a contradiction with the definition of λ_0. Therefore (8.42) will not happen either. It thus proves (8.41). Next we prove (8.44).

By (8.42), for $\varepsilon > 0$ small, we have

$$w_{\lambda_0}(x) \le -C < 0, \quad x \in B_{\lambda_0-\varepsilon}(0)\backslash\{0\}.$$

Since w_λ is continuous in λ, for $\delta > 0$ small it holds that

$$w_{\lambda_0-\delta}(x) \le -C < 0, \quad x \in B_{\lambda_0-\varepsilon}(0)\backslash\{0\}. \tag{8.45}$$

- If $\delta \ge \varepsilon$, then (8.45) yields that

$$w_{\lambda_0-\delta}(x) \le -C < 0, \quad x \in B_{\lambda_0-\delta}(0)\backslash\{0\}.$$

This proves (8.44).

- If $\delta < \varepsilon$, then $B_{\lambda_0 - \delta}(0) \backslash B_{\lambda_0 - \varepsilon}(0)$ forms a narrow region. By the *narrow region principle* (Theorem 8.2.2) we have

$$w_{\lambda_0 - \delta}(x) \leq 0, \quad x \in B_{\lambda_0 - \delta}(0) \backslash B_{\lambda_0 - \varepsilon}(0). \tag{8.46}$$

Combining (8.45) and (8.46), we arrive at (8.44).

Take a sequence $\{\lambda_j\} \to 0$. Since $\lambda_0 = 0$, we know that for every $j \in \mathbb{Z}_+$,

$$u(x) > \frac{1}{\lambda_j^{n-\alpha}} \Big(\frac{\lambda_j^2}{|x|} \Big)^{n-\alpha} u\Big(\frac{\lambda_j^2 \cdot x}{|x|^2} \Big), \quad x \in B_{\lambda_j}(0) \backslash \{0\}.$$

By the decay condition (8.37),

$$u(0) = \lim_{|x| \to 0} u(x) > \lim_{|x| \to 0} \frac{1}{\lambda_j^{n-\alpha}} \Big(\frac{\lambda_j^2}{|x|} \Big)^{n-\alpha} u\Big(\frac{\lambda_j^2 \cdot x}{|x|^2} \Big) > \frac{M_0}{\lambda_j^{n-\alpha}}.$$

Thus

$$u(0) > \lim_{j \to \infty} \frac{M_0}{\lambda_j^{n-\alpha}} = +\infty.$$

A contradiction.

In fact, by applying a decay estimate in [CFY], Theorem 8.3.3 can be strengthened by removing the decay condition on u. That is,

Corollary 8.3.1 *Let Q be continuous and rotationally symmetric on the Euclidean space $\mathbb{R}^n$. In addition, assume that Q is monotone in the region where $Q > 0$ and $Q \not\equiv C$. Then for every $0 < \alpha \leq 2$, the equation*

$$(-\triangle)^{\alpha/2} u(x) = Q(x) \cdot u^p(x), \quad p \equiv \frac{n+\alpha}{n-\alpha}, \quad x \in \mathbb{R}^n,$$

does not admit any positive solution $u \in L_\alpha \cap C^{1,1}(\mathbb{R}^n)$.

8.3.3 A Liouville-Type Theorem in a Half Space

In this section, we prove a Liouville-type theorem in the upper half space $\mathbb{R}_+^n = \{x \in \mathbb{R}^n \mid x_n > 0\}$, for the non-negative solutions to the problem

$$\begin{cases} (-\triangle)^{\alpha/2} u(x) = u^p(x), & x \in \mathbb{R}_+^n, \\ u(x) = 0, & x \notin \mathbb{R}_+^n, \end{cases} \tag{8.47}$$

in both subcritical and critical cases $1 < p \leq \frac{n+\alpha}{n-\alpha}$. This result was first proved in [CFY] by an integral equation method. Our proof here is quite different and much simpler.

Theorem 8.3.4 *If $u \in L_\alpha(\mathbb{R}^n) \cap C_{loc}^{1,1}(\mathbb{R}_+^n) \cap C(\mathbb{R}^n)$ is a nonnegative solution to (8.47), then $u \equiv 0$ in $\mathbb{R}^n$.*

Proof. First, by the strong *maximum principle*, either $u(x) > 0$ for every $x \in \mathbb{R}^n_+$ or $u(x) \equiv 0$ for every $x \in \mathbb{R}^n$. So without loss of generality, from now on, we assume that the solution u is strictly positive everywhere.

Given $x_0 \in \partial \mathbb{R}^n_+ = \mathbb{R}^{n-1} \times \{x_n = 0\}$, define the Kelvin transform of the function u as

$$u_\lambda(x) \equiv \left(\frac{\lambda}{|x - x_0|} \right)^{n-\alpha} u(x^\lambda), \quad x^\lambda \equiv \frac{\lambda^2 (x - x_0)}{|x - x_0|^2} + x_0.$$

By definition, if the center $x_0 \in \partial \mathbb{R}^n_+$, then for any $x \in \mathbb{R}^n_+$, its inversion point x^λ is still in $\mathbb{R}^n_+$. Let

$$w_\lambda(x) = u_\lambda(x) - u(x),$$

then it is straightforward that,

$$(-\triangle)^{\alpha/2} w_\lambda(x) = \left(\frac{\lambda}{|x - x_0|} \right)^\tau u_\lambda^p(x) - u^p(x), \quad x \in \mathbb{R}^n_+, \qquad (8.48)$$

where $\tau \equiv n + \alpha - p(n - \alpha) \geq 0$.

To apply the *method of moving spheres*, we take any $x \in B_\lambda(x_0) \backslash \{x_0\}$. Denote

$$B_\lambda^-(x_0) \equiv \{x \in B_\lambda(x_0) \backslash \{x_0\} \mid w_\lambda(x) < 0\}.$$

Notice that, for each $x_0 \in \partial \mathbb{R}^n_+$, one half of $B_\lambda(x_0)$ is in the upper half space $\mathbb{R}^n_+$, while the other half is in the lower half space $\mathbb{R}^n_- \equiv \{x \mid x_n < 0\}$. Both upper and lower half spaces are invariant under the inversion $x \mapsto x^\lambda$. Since we assume that

$$u(x) = 0, \quad \forall x \in \mathbb{R}^n_- \cup \partial \mathbb{R}^n_+,$$

w_λ is identically zero in the lower half of $B_\lambda(x_0)$, hence $B_\lambda^-(x_0)$ is contained in the upper half of $B_\lambda(x_0)$, in which equation (8.48) holds. Recall that, when proving the *narrow region principle* in the whole space $\mathbb{R}^n$, we derived a contradiction at the negative minimum of w_λ by using equation (8.48), hence via an entirely similar argument, one will be able to establish the same *narrow region principle* for such a ball $B_\lambda(x_0)$. This is true only when the center x_0 lies on the boundary.

Step 1. We show that there exists some $\delta_0 > 0$ such that for every $0 < \lambda < \delta_0$, it holds that

$$w_\lambda(x) \geq 0, \quad x \in B_\lambda(x_0) \backslash \{x_0\}.$$

The proof of *Step 1* is almost identical to that in the proof of Lemma 8.3.1 except that (8.13) in Lemma 8.3.1 is no longer true here, since $u = 0$ in $\partial \mathbb{R}^n$. Notice that our argument is always carried out at the negative minimum point of w_λ, it suffices to show there exists some $x^o \in B_\lambda(0)$ such that

$$w_\lambda(x^o) = \min_{B_\lambda(0) \backslash \{0\}} w_\lambda(x) < 0. \qquad (8.49)$$

Indeed, on one hand,

$$\inf_{B_\lambda(0)\backslash\{0\}} w_\lambda(x) > -\infty. \tag{8.50}$$

Since u is continuous on $B_\lambda(0)$, there exists some C such that

$$u(x) < C, \quad x \in B_\lambda(0).$$

Thus

$$w_\lambda(x) = u_\lambda(x) - u(x) > 0 - C > -\infty.$$

On the other hand, from (8.50) we deduce that there exists a sequence of points $\{x_k\}$ in $B_\lambda(0)$ such that

$$w_\lambda(x_k) \to \inf_{B_\lambda(0)\backslash\{0\}} w_\lambda(x),$$

and a subsequence, still denoted by $\{x_k\}$, that converges to some x^o. If $x^o = 0$, then

$$\lim_{x \to 0} w_\lambda(x) = \lim_{x \to 0} u_\lambda(x) - u(0) \geq 0.$$

This contradicts $w_\lambda(x^o) < 0$. Hence $x^o \neq 0$. This proves (8.49).

Also note that in this case $g(t)/t^p = t^{-\frac{\tau}{n-\alpha}}$ with $\tau \equiv n+\alpha-p(n-\alpha) \geq 0$, is non-increasing. The rest of the proof is exactly the same as that in Lemma 8.3.1.

Step 2. We show that the solution u depends only on x_n.

To this end, we analyze the critical scale for the moving spheres, as that in the proof of Lemma 8.3.1. For a given center $x_0 \in \partial\mathbb{R}^n_+$, the critical scale $\lambda_0 \in (0, +\infty]$ is defined as follows,

$$\lambda_0 \equiv \sup\{\lambda > 0 \mid w_\mu(x) \geq 0, \quad x \in B_\mu(x_0)\backslash\{x_0\}, \quad \forall 0 < \mu \leq \lambda\}. \tag{8.51}$$

As in the proof of *Step 2* in Lemma 8.3.1, we have similar conclusions: Given $x_0 \in \partial\mathbb{R}^n_+$, exactly one of the following holds:

(a) for every $x_0 \in \partial\mathbb{R}^n_+$, the corresponding critical scale $\lambda_0 > 0$ is finite, or

(b) for every $x_0 \in \partial\mathbb{R}^n_+$, the corresponding critical scale $\lambda_0 = +\infty$, i.e., for every $\lambda > 0$

$$w_\lambda(x) \geq 0, \quad x \in B_\lambda(x_0)\backslash\{x_0\}.$$

We first argue that case (a) is impossible. In fact, if (a) holds, then we have the decay property

$$\lim_{|x|\to\infty} |x|^{n-\alpha} \cdot u(x) = c_0 > 0, \quad x \in \mathbb{R}^n_+. \tag{8.52}$$

We also have

$$u(x) = \left(\frac{\lambda_0}{|x|}\right)^{n-\alpha} u\left(\frac{\lambda_0^2 x}{|x|^2}\right). \tag{8.53}$$

Here we take $x_0 = 0$.

Let $|x|\to 0$. Then $\frac{\lambda_0}{|x|}\to\infty$. Hence by (8.52), the right hand side of (8.53)

$$\left(\frac{\lambda_0}{|x|}\right)^{n-\alpha} u\left(\frac{\lambda_0^2 x}{|x|^2}\right) \to c_0 \lambda_0^{\alpha-n} > 0;$$

while the left hand side of (8.53) becomes

$$u(x) \to 0.$$

The above contradiction shows that case (a) does not happen. Now we deal with case (b). We look back to the proof of Proposition 8.3.2. There we showed that for all $y, z \in \partial \mathbb{R}_+^n$,

$$\frac{(n-\alpha)u(z) + 2\langle \nabla u(z), (z-y)\rangle}{|y|} \geq 0.$$

Let $|y| \to \infty$, we have

$$\langle \nabla u(z), \nu \rangle \geq 0,$$

where $\nu \equiv \frac{z-y}{|z-y|}$. Since y is arbitrary, it also holds that

$$\langle \nabla u(z), -\nu \rangle \geq 0.$$

Hence the vector $\nabla u(z)$ must be perpendicular to the hyperplane $x_n = 0$. That is, the function u depends only on x_n.

With the result in *Step 2*, we can use the arguments in Theorem 5.1 or Section 5.3 of [CFY] to show that u has to be identically zero, which gives the desired contradiction.

9

A Priori Estimates and the Existence of Solutions

It is well-known that the a priori estimates play important roles in establishing the existence of solutions. Once there is such an a priori estimate, one can use various approaches, such as continuation methods or topological degree arguments, to derive the existence of solutions.

In the first section of this chapter, we develop a direct *blowing-up and rescaling* argument for nonlinear equations involving nonlocal elliptic operators which include but are not limited to the fractional Laplacian.

The rest of the chapter will be devoted to introductions on the topological degree theories and their beautiful applications in deriving the existence of solutions.

The fundamental problems for linear equations are the existence and uniqueness of solutions; while for nonlinear equations, solutions are usually not unique. Hence besides the existence, people are also interested in how many solutions are there. Naturally, mathematicians are concerned in establishing a theory dealing with the multiplicity of solutions, and giving a rough estimate about the numbers of solutions.

It's worth pointing out that the number of solutions is unstable in the sense that a small perturbation in the equation may lead to a big change in the number of solutions. So one can hardly expect a theory that can accurately describe the total number of solutions.

The theories associated with the existence and multiplicity of solutions for nonlinear equations mainly consist of the topological degree theory and the critical point theory. In this chapter, we discuss the former.

The topological degree is an integer-valued function that depends on a mapping f and its domain D, and represents certain "algebraic number" of zeros of f in Ω. The "algebraic number" has some basic properties, for instance, the additivity of domains, homotopy invariance, and the assurance of the existence of solutions to the equation

$$f(x) = 0, \text{ for some } x \in \Omega,$$

when this "algebraic number" is not zero. From these properties, scientists are able to derive such good results as existence and multiplicity of solutions to nonlinear equations, and further apply these findings to physics, mechanics, partial differential equations and many other fields.

The topological degree theory was first introduced by L. Brouwer in 1912. Yet Brouwer's work was only on the continuous mapping defined in finite dimensional spaces, and is known today as the Brouwer degree. Later, with the establishment of the framework for linear functional analysis, J. Leray and J. Schauder generalized the Brouwer degree to a certain family of continuous mappings on Banach spaces, known as the Leray-Schauder degree.

In Section 9.2, we introduce the Brouwer degree in finite dimensional space $\mathbb{R}^n$ and prove some of its properties, such as the invariance under small perturbation via homotopy. Then in Section 9.3, we generalize this definition to infinite dimensional Banach spaces and define the Leray-Schauder degree. Finally in Section 9.4, we apply the Leray-Schauder degree to obtain the existence of positive solutions for some nonlinear equations on Banach spaces.

9.1 The A Priori Estimates

On a bounded domain $\Omega \subset \mathbb{R}^n$ with C^2 boundary, we consider Dirichlet problems involving both the fractional Laplacian $(-\triangle)^{\alpha/2}$ and a more general fractional operator

$$(-\triangle)^{\alpha/2}_a u(x) = \frac{C_{n,\alpha}}{2} \int_{\mathbb{R}^n} \frac{2u(x) - u(x+y) - u(x-y)}{|y|^{n+\alpha}} a(y) dy,$$

where

$$0 < c_o \leq a(x) \leq C_o, \quad x \in \Omega.$$

This is the uniformly elliptic nonlocal operator introduced in [CS1].

Instead of using the conventional extension method introduced by Caffarelli and Silvestre to localize the problem, we work directly on the nonlocal operator. Using the defining integral, by an elementary approach, we carry on a *blowing-up and rescaling* argument directly on the nonlocal equations and thus obtain a priori estimates on the positive solutions. The results and proofs in this chapter have mostly appeared in articles [CLL1] and [BPGQ].

We believe that these ideas can be applied to problems involving more general nonlocal operators.

9.1.1 The A Priori Estimate for a Simple Equation

Theorem 9.1.1 *For $0 < \alpha < 2$ and $1 < p < \frac{n+\alpha}{n-\alpha}$, suppose*

$$u \in L_\alpha \cap C^{1,1}_{loc}(\Omega) \text{ is upper semi-continuous on } \bar{\Omega},$$

and is a positive solution of

$$\begin{cases} (-\triangle)^{\alpha/2} u(x) = u^p(x), \ x \in \Omega, \\ u(x) \equiv 0, \qquad\qquad x \notin \Omega. \end{cases} \tag{9.1}$$

Then

$$\|u\|_{L^\infty(\Omega)} \le C, \tag{9.2}$$

for some positive constant C independent of u.

Proof. Suppose (9.2) does not hold, then there exists a sequence of solutions $\{u_k\}$ to (9.1) and a sequence of points $\{x^k\} \subset \Omega$ such that

$$u_k(x^k) = \max_\Omega u_k := m_k \to \infty. \tag{9.3}$$

Let

$$\lambda_k = m_k^{\frac{1-p}{\alpha}} \text{ and } v_k(x) = \frac{1}{m_k} u_k(\lambda_k x + x^k), \tag{9.4}$$

then we have

$$(-\triangle)^{\alpha/2} v_k(x) = v_k^p(x), \quad x \in \Omega_k := \{x \in \mathbb{R}^n \mid x = \frac{y - x^k}{\lambda_k}, y \in \Omega\}. \tag{9.5}$$

Let $d_k = dist(x^k, \partial\Omega)$. We carry out the proof using the contradiction argument while exhausting all three possibilities.

Case i. $\lim_{k\to\infty} \frac{d_k}{\lambda_k} = \infty.$

It's not difficult to see that

$$\Omega_k \to \mathbb{R}^n \text{ as } k\to\infty.$$

We prove that there exists a function v such that as $k\to\infty$,

$$v_k(x) \to v(x) \text{ and } (-\triangle)^{\alpha/2} v_k(x) \to (-\triangle)^{\alpha/2} v(x), \tag{9.6}$$

thus

$$(-\triangle)^{\alpha/2} v(x) = v^p(x), \quad x \in \mathbb{R}^n. \tag{9.7}$$

Eq. (9.4) implies that

$$v(0) = \lim_{k\to\infty} v_k(0) = 1.$$

It follows from the *maximum principle* that

$$v(x) > 0, \forall x \in \mathbb{R}^n.$$

On the other hand, by the Liouville theorem in Chapter 6, we know that (9.7) has no positive solution. This is a contradiction. Hence this case cannot happen.

In the following, for convenience's sake, define

$$C^{0,\gamma} = \begin{cases} C^\gamma, & \text{if } 0 < \gamma < 1 \\ C^{1,\gamma-1}, & \text{if } 1 < \gamma \le 2. \end{cases}$$

To verify (9.6), we need to establish a uniform $C^{0,\alpha+\sigma}$ estimate, which is independent of both k and x, for v_k in a neighborhood of any point $x \in \mathbb{R}^n$. This is done in two steps. We first obtain a C^σ estimate for some $0 < \sigma < 1$, then by using the equation satisfied by v_k, we boost the uniform regularity up to $C^{0,\alpha+\sigma}$.

We need the following proposition.

Proposition 9.1.1 ([Si]) *Let $v_k \in C^{m,\sigma}$ and suppose that $m + \sigma - \alpha$ is not an integer. Then $(-\triangle)^{\alpha/2} v_k \in C^{l,\gamma}$, where l is the integer part of $m + \sigma - \alpha$ and $\gamma = m + \sigma - \alpha - l$.*

The idea in the following arguments is similar to that in the proof of Proposition 2.8 in [Si].

Henceforth we use C to denote positive constants whose values may vary from line to line.

For any $x^o \in \mathbb{R}^n$, there exists an $N > 0$ such that for $k > N$, $B_3(x^o) \subset \Omega_k$. Let φ be a smooth cutoff function such that $0 \le \varphi(x) \le 1$ in $\mathbb{R}^n$, $supp\,\varphi \subset B_3(x^o)$ and $\varphi(x) \equiv 1$ in $B_2(x^o)$. Let $(-\triangle)^{\alpha/2} v_k = g_k$. Define

$$\bar{v}_k(x) := c_{n,-\alpha} \int_{\mathbb{R}^n} \frac{\varphi(y)g_k(y)}{|x-y|^{n-\alpha}} dy. \tag{9.8}$$

Then

$$\bar{v}_k = (-\triangle)^{-\alpha/2}(\varphi g_k) = (-\triangle)^{1-\alpha/2} \circ (-\triangle)^{-1}(\varphi g_k).$$

Let

$$h_k(x) = (-\triangle)^{-1}(\varphi g_k),$$

then

$$h_k(x) = c_{n,-2} \int_{\mathbb{R}^n} \frac{\varphi(y)g_k(y)}{|x-y|^{n-2}} dy.$$

By the $C^{1,\sigma}$–estimates for the Poisson's equation (see [GT]), we derive that there exists a constant C independent of k and x^o such that $\forall 0 < \tau < 1$,

$$\|h_k\|_{C^{1,\tau}_{B_3(x^o)}} \le C(\|\varphi g_k\|_{L^\infty(B_3(x^o))} + \|h_k\|_{L^\infty(B_3(x^o))}) < C.$$

Applying Proposition 9.1.1 to h_k, we have $\forall 0 < \sigma < \min\{1, \tau + \alpha - 1\}$,

$$\|\bar{v}_k\|_{C^{0,\sigma}(B_3(x^o))} < C, \quad C \text{ is independent of } k \text{ and } x^o.$$

Since

$$(-\triangle)^{\alpha/2}(\bar{v}_k(x) - v_k(x)) = 0, \ x \in B_2(x^o),$$

$\bar{v}_k - v_k$ is smooth in $B_1(x^o)$, hence

$$\|v_k\|_{C^{0,\sigma}(B_1(x^o))} < C, \quad \text{here } C \text{ is independent of } k \text{ and } x^o. \tag{9.9}$$

Due to the above uniform regularity estimate, it then follows from the *Arzelà-Ascoli theorem* that there exists a converging subsequence of $\{v_k\}$

in $B_1(0)$, denoted by $\{v_{1m}\}$. Then, one can find a subsequence of $\{v_{1m}\}$, denoted by $\{v_{2m}\}$, that converges in $B_2(0)$, and then a subsequence of $\{v_{2m}\}$, denoted as $\{v_{3m}\}$, converging in $B_3(0)$. By induction, we get a chain of subsequences

$$\{v_{1m}\} \supset \{v_{2m}\} \supset \{v_{3m}\}\dots$$

such that $\{v_{jm}\}$ converges in $B_j(0)$ as $m\to\infty$. Now take the diagonal sequence $\{v_{jj}\}$, then it converges at all points in any $B_R(0)$. Thus we have constructed a subsequence of solutions (still denoted by $\{v_k\}$) that converges point-wise in $\mathbb{R}^n$ to a function $v(x)$.

Next we show that, along such a subsequence, $(-\triangle)^{\alpha/2}v_k$ also converges point-wise to $(-\triangle)^{\alpha/2}v$. To do so, we need a higher uniform regularity estimate.

Using the equation

$$g_k := (-\triangle)^{\alpha/2}v_k = v_k^p,$$

from (9.9), we immediately derive, for any $x^o \in \mathbb{R}^n$ and for some $0 < \sigma < 1$,

$$\|g_k\|_{C^{0,\sigma}(B_1(x^o))} < C, \quad \text{with } C \text{ independent of } k \text{ and } x^o.$$

Now we can apply the Schauder estimate to the equation

$$h_k = (-\triangle)^{-1}(\phi g_k).$$

Similar to the argument in deriving (9.9), we show that there exists a constant C independent of k and x^o, such that

$$\|v_k\|_{C^{0,\sigma+\alpha}(B_1(x^o))} \leq C. \tag{9.10}$$

By the definition of the fractional Laplacian, we have

$$\begin{aligned}
&(-\triangle)^{\alpha/2}v_k(x) \\
&= \frac{C_{n,\alpha}}{2}\int_{\mathbb{R}^n} \frac{2v_k(x) - v_k(x+y) - v_k(x-y)}{|y|^{n+\alpha}}dy \\
&= \frac{C_{n,\alpha}}{2}\left[\int_{\mathbb{R}^n\backslash B_1(0)} \frac{2v_k(x) - v_k(x+y) - v_k(x-y)}{|y|^{n+\alpha}}dy\right. \\
&\quad \left. + \int_{B_1(0)} \frac{2v_k(x) - v_k(x+y) - v_k(x-y)}{|y|^{n+\alpha}}dy\right] \\
&= \frac{C_{n,\alpha}}{2}(I_1 + I_2).
\end{aligned}$$

It follows from (9.10) that for $y \in B_1(0)$,

$$\frac{|2v_k(x) - v_k(x+y) - v_k(x-y)|}{|y|^{n+\alpha}} \leq \frac{C|y|^{\alpha+\sigma}}{|y|^{n+\alpha}} = \frac{C}{|y|^{n-\sigma}},$$

and for $y \in B_1^c(0)$,

$$\frac{|2v_k(x) - v_k(x+y) - v_k(x-y)|}{|y|^{n+\alpha}} \leq \frac{4C}{1+|y|^{n+\alpha}}.$$

Therefore, for $y \in \mathbb{R}^n$,

$$\left| \frac{2v_k(x) - v_k(x+y) - v_k(x-y)}{|y|^{n+\alpha}} \right| \leq \frac{C}{1 + |y|^{n+\alpha}} \left(1 + \frac{1}{|y|^{n-\sigma}} \right).$$

Since

$$\int_{\mathbb{R}^n} \frac{C}{1 + |y|^{n+\alpha}} \left(1 + \frac{1}{|y|^{n-\sigma}} \right) dy < \infty,$$

from the *Lebesgue's dominated convergence theorem* it yields that

$$\lim_{k \to \infty} (-\triangle)^{\alpha/2} v_k(x)$$

$$= \lim_{k \to \infty} \frac{C_{n,\alpha}}{2} \int_{\mathbb{R}^n} \frac{2v_k(x) - v_k(x+y) - v_k(x-y)}{|y|^{n+\alpha}} dy$$

$$= \frac{C_{n,\alpha}}{2} \int_{\mathbb{R}^n} \lim_{k \to \infty} \frac{2v_k(x) - v_k(x+y) - v_k(x-y)}{|y|^{n+\alpha}} dy$$

$$= (-\triangle)^{\alpha/2} v(x).$$

This proves (9.6) and (9.7).

Case ii. $\lim\limits_{k \to \infty} \frac{d_k}{\lambda_k} = C > 0.$

In this case,

$$\Omega_k \to \mathbb{R}^n_{+C} := \{x_n \geq -C \mid x \in \mathbb{R}^n\} \text{ as } k \to \infty.$$

Similar to *Case i*, here we're able to establish the existence of a function v and a subsequence of $\{v_k\}$, such that as $k \to \infty$,

$$v_k(x) \to v(x) \text{ and } (-\triangle)^{\alpha/2} v_k(x) \to (-\triangle)^{\alpha/2} v(x), \tag{9.11}$$

and thus

$$(-\triangle)^{\alpha/2} v(x) = v^p(x), \quad x \in \mathbb{R}^n_{+C}. \tag{9.12}$$

It's known that (9.12) has no positive solution (see [CFY]). Meanwhile, it follows from (9.4) that

$$v(0) = \lim_{k \to \infty} v_k(0) = 1.$$

This is a contradiction. Next we prove (9.11) and (9.12).

Let $D_1 = B_1(0) \cap \{x_n \geq 0\}$. Then similar to the argument in *Case i*, one can show that there exists a converging subsequence $\{v_{1k}\}$ of $\{v_k\}$ such that,

$$v_{1k}(x) \to v(x) \text{ and } (-\triangle)^{\alpha/2} v_{1k}(x) \to (-\triangle)^{\alpha/2} v(x).$$

Let $D_2 = B_2(0) \cap \{x_n \geq -\frac{C}{2}\}$. We can find a subsequence $\{v_{2k}\}$ of $\{v_{1k}\}$, such that

$$v_{2k}(x) \to v(x) \text{ and } (-\triangle)^{\alpha/2} v_{2k}(x) \to (-\triangle)^{\alpha/2} v(x).$$

Repeating the above process m times yields such a point-wise converging subsequence $\{v_{mk}\}$ in $D_m = B_m(0) \cap \{x_n \geq -\frac{(m-1)}{m}C\}$. Now we take the diagonal sequence $\{v_{mm}\}$. It's easy to see that for any $x \in \mathbb{R}^n_{+C}$,

$$v_{mm}(x) \to v(x) \text{ and } (-\triangle)^{\alpha/2}v_{mm}(x) \to (-\triangle)^{\alpha/2}v(x).$$

This verifies (9.11) and (9.12).

Case iii. $\lim\limits_{k \to \infty} \frac{d_k}{\lambda_k} = 0.$

In this case, there exists a point $x^o \in \partial\Omega$ and a subsequence $\{x^k\}$, still denoted by $\{x^k\}$, such that

$$x^k \to x^o, \quad k \to \infty.$$

Let $p^k = \frac{x^o - x^k}{\lambda_k}$. Obviously,

$$v_k(p^k) = \frac{1}{m_k}u_k(x^o) = 0$$

and

$$p^k \to 0, \quad k \to \infty.$$

Next we show that v_k is uniformly Hölder continuous near p^k, i.e.

$$|v_k(x) - v_k(p^k)| \leq C|x - p^k|^{\alpha/2}. \tag{9.13}$$

We postpone the proof of (9.13) for a moment. Notice that

$$v_k(0) - v_k(p^k) = 1. \tag{9.14}$$

Meanwhile, (9.13) implies that

$$|v_k(0) - v_k(p^k)| \to 0, \text{ as } k \to \infty.$$

This is a contradiction with (9.14), and thus rules out the possibility of *Case iii.*

To prove (9.13), as a motivating example, we first consider a simple case when $\Omega_k = B_1(0)$.

It's well-known that with a proper constant $C > 0$,

$$\varphi(x) = \begin{cases} C(1 - |x|^2)^{\alpha/2}, & x \in B_1(0), \\ 0, & x \in B_1^c(0), \end{cases} \tag{9.15}$$

is a solution to

$$(-\triangle)^{\alpha/2}\varphi(x) = 1, \quad x \in B_1(0).$$

Obviously, $\varphi(x)$ is Hölder continuous near ∂B_1, since for any $y \in \partial B_1$ and $x \in B_1(y) \cap B_1(0)$,

$$\varphi(x) - \varphi(y) = \varphi(x)$$
$$= C(1 - |x|)^{\alpha/2}(1 + |x|)^{\alpha/2}$$
$$\leq C(1 - |x|)^{\alpha/2}$$
$$\leq C(|x - y|)^{\alpha/2}.$$

Intuitively, if $v_k(x)$ can be controlled by $\varphi(x)$ from above near ∂B_1, then $v_k(x)$ may have similar Hölder continuity near the boundary. To be more rigorous, let $w = \varphi - v_k$ and we have

$$\begin{cases} (-\triangle)^{\alpha/2}w(x) \geq 0, & x \in B_1(0), \\ w = 0, & x \in B_1^c(0). \end{cases}$$

From the *maximum principle* it follows that

$$w(x) > 0, \quad x \in B_1(0).$$

Or

$$\varphi(x) > v_k(x), \quad x \in B_1(0).$$

Thus for any $y \in \partial B_1$, $x \in B_1(y) \cap B_1(0)$,

$$v_k(x) - v_k(y) < \varphi(x) - \varphi(y) \leq C|x - y|^{\alpha/2}.$$

This shows that $v_k(x)$ is Hölder continuous near ∂B_1. Notice that this auxiliary function works perfectly in obtaining desired regularity near the boundary of the unit ball. When dealing with more general domains Ω_k, we will employ other auxiliary functions.

Consider $\Omega_k \subset \mathbb{R}^n$ with C^2 boundary. If we still use the same φ in (9.15) in a tangent ball in Ω_k at $y \in \partial \Omega_k$, we will not get sufficient help from φ in a full neighbourhood of y in Ω_k, however small it is, because it is only partially covered by the tangent ball. Hence we place a modified φ in a tangent ball outside of Ω_k.

Without loss of generality, let $B_1(0)$ be the tangent ball at $y \in \partial \Omega_k$ in $\mathbb{R}^n \backslash \Omega_k$. We start the construction of the auxiliary function with φ in (9.15). The Kelvin transform of $\varphi(x)$ is denoted by

$$\psi(x) = \frac{1}{|x|^{n-\alpha}}\varphi\left(\frac{x}{|x|^2}\right).$$

Then

$$(-\triangle)^{\alpha/2}\psi(x) = \frac{C}{|x|^{n+\alpha}}, \quad x \in B_1^c(0).$$

Let $D = (B_3 \backslash B_1) \cap \Omega_k$. By choosing sufficiently large $t > 0$, we have

$$(-\triangle)^{\alpha/2}(t\psi(x) - v_k(x)) \geq 0, \quad x \in D.$$

To apply the *maximum principle*, we need to check the condition on $\mathbb{R}^n \backslash D$. Obviously,

$$t\psi(x) - v_k(x) > 0, \quad x \in \mathbb{R}^n \backslash \Omega_k.$$

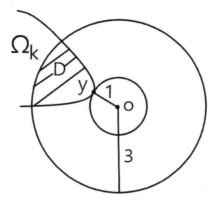

Fig. 9.1

Meanwhile, since $\Omega_k \to \mathbb{R}^n_+$ as $k \to \infty$, $t\psi(x)$ may not be able to control $v_k(x)$ in Ω_k when $|x|$ large. To overcome this difficulty, we may add a positive constant, for example 1, to $t\psi(x)$. It's easy to see that $t\psi(x)+1-v_k(x)$ satisfies both conditions required in the maximum principle, and consequently

$$t\psi(x) + 1 \geq v_k(x), \quad x \in D. \tag{9.16}$$

However, $t\psi(x)+1$ does not vanish at $y \in \partial\Omega_k$. Only when $v_k(x)$ is controlled by a Hölder continuous function that *vanishes* at the point y, will we be able to derive Hölder continuity for $v_k(x)$ itself at y. To this end, we modify the auxiliary function by adding a cutoff function to $t\psi(x)$. Let $\xi(x)$ be a smooth cutoff function such that $\xi(x) \in [0,1]$ in $\mathbb{R}^n$ and $\xi(x) = 0$ in $\overline{B_1(0)}$ and $\xi(x) = 1$ in $B_3^c(0)$. Then it is easy to see that $t\psi(y) + \xi(y) = 0$. We also have

$$(-\triangle)^{\alpha/2}\xi(x) = \frac{C_{n,\alpha}}{2} \int_{\mathbb{R}^n} \frac{2\xi(x) - \xi(x+y) - \xi(x-y)}{|y|^{n+\alpha}} dy$$
$$\geq -C_1.$$

Let

$$\varphi_1(x) = t\psi(x) + \xi(x),$$

then for t sufficiently large and for all $x \in D$,

$$(-\triangle)^{\alpha/2}\varphi_1(x) = t(-\triangle)^{\alpha/2}\psi(x) + (-\triangle)^{\alpha/2}\xi(x)$$
$$\geq \frac{tC}{|x|^{n+\alpha}} - C_1$$
$$\geq 1.$$

Thus

$$\begin{cases} (-\triangle)^{\alpha/2}(\varphi_1 - v_k) \geq 0, \ x \in D, \\ \varphi_1(x) - v_k(x) > 0, \quad x \in D^c. \end{cases} \tag{9.17}$$

Applying the *maximum principle* to (9.17), it gives

$$\varphi_1(x) \geq v_k(x), \quad x \in D.$$

Hence for $x \in D$,

$$0 \leq v_k(x) - v_k(y) = v_k(x) \leq \varphi_1(x) = \varphi_1(x) - \varphi_1(y) \leq C|x-y|^{\alpha/2}.$$

This shows that v_k is Hölder continuous near p_k indeed, and hence completes the proof of *Case iii.*

9.1.2 A Priori Estimates for More General Problems

We consider Dirichlet problems involving more general nonlinearities and more general uniformly elliptic nonlocal operators:

$$\begin{cases} (-\triangle)_a^{\alpha/2} u(x) = f(x, u), \ x \in \Omega, \\ u(x) \equiv 0, \qquad\qquad\quad x \in \mathbb{R}^n \backslash \Omega, \end{cases} \tag{9.18}$$

where

$$(-\triangle)_a^{\alpha/2} u(x) = \frac{C_{n,\alpha}}{2} \int_{\mathbb{R}^n} \frac{2u(x) - u(x+y) - u(x-y)}{|y|^{n+\alpha}} a(y) dy, \tag{9.19}$$

and

$$0 < c_o \leq a(x) \leq C_o, \ x \in \Omega.$$

Under mainly a subcritical growth condition on f, we obtain an a priori estimate on the positive solutions.

Theorem 9.1.2 *Consider Dirichlet problem (9.18) with*

$$\lim_{|x| \to 0} a(x) = 1. \tag{9.20}$$

Assume that

1.

$$\lim_{s \to \infty} \frac{f(x, s)}{s^p} = g(x) \ uniformly \ for \ all \ x \ in \ \Omega, \tag{9.21}$$

for some $1 < p < \frac{n+\alpha}{n-\alpha}$ and for some positive function $g(x)$ which is bounded away from zero in Ω;
there are some $\tau_1, \tau_2 \in (0, 1)$, such that

2. $|f(x, s) - f(y, s)| \leq C s^p |x - y|^{\tau_1}, \ \forall s \in (0, \infty)$; and

3. $|f(x, s) - f(x, t)| \leq C |s - t|^{\tau_2}, \ \forall x \in \Omega.$

Let $u \in L_\alpha \cap C_{loc}^{1,1}(\Omega)$ be a positive solution of (9.18) that is upper semi-continuous on $\bar{\Omega}$. Then

$$\|u\|_{L^\infty(\Omega)} \leq C, \tag{9.22}$$

for some positive constant C independent of u.

Proof. Suppose (9.22) does not hold, then there exists a sequence of solutions $\{u_k\}$ to (9.18) and a sequence of points $\{x^k\} \subset \Omega$ such that

$$u_k(x^k) = \max_\Omega u_k := m_k \to \infty. \tag{9.23}$$

Let

$$\lambda_k = m_k^{\frac{1-p}{\alpha}} \text{ and } v_k(x) = \frac{1}{m_k} u_k(\lambda_k x + x^k). \tag{9.24}$$

Through an elementary calculation, we have

$$(-\triangle)_{a_k}^{\alpha/2} v_k(x) = F_k(x), \ x \in \Omega_k := \{x \in \mathbb{R}^n \mid x = \frac{y - x^k}{\lambda_k}, y \in \Omega\}, \tag{9.25}$$

with

$$(-\triangle)_{a_k}^{\alpha/2} v_k(x) = \frac{C_{n,\alpha}}{2} \int_{\mathbb{R}^n} \frac{2v_k(x) - v_k(x+y) - v_k(x-y)}{|y|^{n+\alpha}} a(\lambda_k y) dy,$$

and

$$F_k(x) = m_k^{-p} f(\lambda_k x + x^k, m_k v_k(x)).$$

Let $d_k = dist(x^k, \partial\Omega)$.

Similar to the previous section, we derive contradictions in the following three possible cases.

Case i. $\lim\limits_{k \to \infty} \frac{d_k}{\lambda_k} = \infty.$

Case ii. $\lim\limits_{k \to \infty} \frac{d_k}{\lambda_k} = C > 0.$

Case iii. $\lim\limits_{k \to \infty} \frac{d_k}{\lambda_k} = 0.$

As can be seen previously, the key ingredient in the proofs for the first two cases is to obtain a uniform regularity estimate for v_k.

Given any point $x^o \in \mathbb{R}^n$, for k sufficiently large, we have $B_3(x^o) \subset \Omega_k$. Condition (9.21) guarantees that $F_k(x)$ is uniformly bounded in Ω_k, and it follows from Theorem 11.1 in [CS1] that, for some $0 < \sigma < 1$,

$$\|v_k\|_{C^\sigma(B_2(x^o))} \le C(\sup_{\mathbb{R}^n} |v_k| + \sup_{B_3(x^o)} |F_k(x)|) \le C_1, \text{independent of } k \text{ and } x^o.$$

Then from conditions 2 and 3 in Theorem 9.1.2, we derive that for some $\sigma \in (0, 1)$,

$$\|F_k\|_{C^\sigma(B_2(x^o))} \le C, \text{ with } C \text{ independent of } k \text{ and } x^o.$$

Indeed, for any $x, y \in B_2(x^o)$,

$$|F_k(x) - F_k(y)|$$
$$= m_k^{-p}|f(\lambda_k x + x^k, m_k v_k(x)) - f(\lambda_k y + x^k, m_k v_k(y))|$$
$$\leq m_k^{-p}|f(\lambda_k x + x^k, m_k v_k(x)) - f(\lambda_k x + x^k, m_k v_k(y))|$$
$$\quad + m_k^{-p}|f(\lambda_k x + x^k, m_k v_k(y)) - f(\lambda_k y + x^k, m_k v_k(y))|$$
$$\leq m_k^{-p} C m_k^{\tau_2}|v_k(x) - v_k(y)|^{\tau_2}$$
$$\quad + m_k^{-p} C[m_k v_k(x)]^p |\lambda_k x + x^k - (\lambda_k y + x^k)|^{\tau_1}$$
$$\leq C|x - y|^{s\tau_2} + C|x - y|^{\tau_1}$$
$$\leq C|x - y|^\sigma,$$

with $\sigma \in (0, \min\{\tau_1, s\tau_2\})$.

Applying the Schauder estimate to the solutions of (9.25) (see Theorem 1.2 in [DK]), we arrive at

$$\|(-\triangle)^{\alpha/2} v_k\|_{C^\sigma(B_1(x^o))} \leq C(\sup_{\mathbb{R}^n} |v_k| + \|F_k\|_{C^\sigma(B_2(x^o))}) \leq C_1.$$

Here C_1 is independent of k and x^o. Then similarly, we arrive at uniform estimates (9.10).

Now through a similar argument as in the previous section, we can rule out *Case i* and *Case ii*.

To derive a contradiction in *Case iii*, we need the following maximum principle for a general uniformly elliptic operator $(-\triangle)_a^{\alpha/2}$.

Lemma 9.1.1 *Assume Ω is bounded in $\mathbb{R}^n$, $a(x)$ is nonnegative and bounded, $u \in L_\alpha \cap C_{loc}^{1,1}$ and is lower semi-continuous on $\bar{\Omega}$. If*

$$\begin{cases} (-\triangle)_a^{\alpha/2} u(x) \geq 0, & x \in \Omega, \\ u(x) \geq 0, & x \in \mathbb{R}^n \backslash \Omega, \end{cases} \qquad (9.26)$$

then

$$u(x) \geq 0, \quad x \in \Omega. \qquad (9.27)$$

Proof. Suppose that (9.27) does not hold, then there exists some $x^o \in \Omega$ such that

$$u(x^o) = \min_\Omega u(x) < 0.$$

Therefore,

$$(-\triangle)_a^{\alpha/2} u(x^o)$$
$$= C_{n,\alpha} \, PV \int_{\mathbb{R}^n} \frac{u(x^o) - u(z)}{|x^o - z|^{n+\alpha}} a(x^o - z)\, dz$$
$$= C_{n,\alpha} \, PV \int_\Omega \frac{u(x^o) - u(z)}{|x^o - z|^{n+\alpha}} a(x^o - z)\, dz$$
$$\quad + C_{n,\alpha} \int_{\mathbb{R}^n \backslash \Omega} \frac{u(x^o) - u(z)}{|x^o - z|^{n+\alpha}} a(x^o - z)\, dz$$
$$< 0.$$

This is a contradiction with (9.26) and thus proves (9.27).

To prove that *Case iii* will not happen, similar to the simple case, we will derive a contradiction between the uniformly Hölder regularity of v_k and its jump discontinuity near $\partial\Omega_k$, which are (9.14) and (9.13). Eq. (9.14) is obviously true so we only need to prove (9.13). Due to the singularity caused by the fractional Laplacian near the boundary, this time we need to construct a super solution $\varphi(x)$ for (9.25), which is different from the one in the previous section. The following lemma plays an important role in the construction, and it may be applied widely to study other problems involving nonlocal operators.

Lemma 9.1.2 *For $t \in R$, let*

$$t_+ = \begin{cases} t, & \text{if } t \geq 0, \\ 0, & \text{if } t < 0. \end{cases}$$

Then for $0 < \gamma < \alpha$,

$$(-\triangle)^{\alpha/2}(t_+^\gamma) = C_\gamma t_+^{\gamma-\alpha}, \quad t > 0, \tag{9.28}$$

where

$$C_\gamma = C_{1,\alpha} \int_0^1 \frac{(\tau^{\alpha/2} - \tau^\gamma)(1 - \tau^{\alpha/2-\gamma-1})}{(1-\tau)^{1+\alpha}} d\tau. \tag{9.29}$$

Remark 9.1.1 *i. One can easily see that*

$$C_\gamma \text{ is } \begin{cases} > 0, & \text{if } 0 < \gamma < \alpha/2, \\ = 0, & \text{if } \gamma = \alpha/2, \\ < 0, & \text{if } \alpha > \gamma > \alpha/2. \end{cases} \tag{9.30}$$

ii. This fact can be used to construct super- or sub-solutions (or barrier functions) near the boundary as we will see in Lemma 9.1.3.

Proof. Since for $t > 0$, $t_+ = t > 0$,

$$t_+^{\alpha-\gamma}(-\triangle)^{\alpha/2}(t_+^\gamma)$$
$$= t^{\alpha-\gamma}CPV \int_{-\infty}^\infty \frac{t^\gamma - s_+^\gamma}{|t-s|^{1+\alpha}} ds$$
$$= CPV \int_{-\infty}^\infty \frac{t^\alpha - t^{\alpha-\gamma}s_+^\gamma}{|t-s|^{1+\alpha}} ds$$
$$= CPV \int_{-\infty}^\infty \frac{t^{\alpha/2}(t^{\alpha/2} - s_+^{\alpha/2})}{|t-s|^{1+\alpha}} ds + CPV \int_{-\infty}^\infty \frac{t^{\alpha/2}s_+^{\alpha/2} - t^{\alpha-\gamma}s_+^\gamma}{|t-s|^{1+\alpha}} ds$$
$$= t^{\alpha/2}(-\triangle)^{\alpha/2}(t_+^{\alpha/2}) + CPV \int_{-\infty}^\infty \frac{t^{\alpha/2}s_+^{\alpha/2} - t^{\alpha-\gamma}s_+^\gamma}{|t-s|^{1+\alpha}} ds$$
$$= I_1 + I_2.$$

It's well known that

$$(-\triangle)^{\alpha/2}(t_+^{\alpha/2}) = 0, \quad t > 0.$$

Therefore we only need to calculate I_2.

$$CPV \int_0^\infty \frac{t^{\alpha/2}s^{\alpha/2} - t^{\alpha-\gamma}s^\gamma}{|t-s|^{1+\alpha}} ds$$

$$= CPV \int_0^\infty \frac{\tau^{\alpha/2} - \tau^\gamma}{|1-\tau|^{1+\alpha}} d\tau, \text{ where } s = \tau t$$

$$= C \int_0^1 \frac{\tau^{\alpha/2} - \tau^\gamma}{|1-\tau|^{1+\alpha}} d\tau + C \int_1^\infty \frac{\tau^{\alpha/2} - \tau^\gamma}{|1-\tau|^{1+\alpha}} d\tau$$

$$= C \int_0^1 \frac{\tau^{\alpha/2} - \tau^\gamma}{|1-\tau|^{1+\alpha}} d\tau + C \int_0^1 \frac{\eta^{\alpha/2-1} - \eta^{\alpha-\gamma-1}}{|1-\eta|^{1+\alpha}} d\eta, \text{ where } \tau = \frac{1}{\eta}$$

$$= C \int_0^1 \frac{(\tau^{\alpha/2} - \tau^\gamma)(1 - \tau^{\alpha/2-\gamma-1})}{|1-\tau|^{1+\alpha}} d\tau$$

$$= C_\gamma. \quad \square$$

One can generalize Lemma 9.1.2 to n-dimensions.

Corollary 9.1.1 *Let $x = (x_1, x_2, \cdots, x_n)$. Then for $0 < \gamma < \alpha$,*

$$(-\triangle)^{\alpha/2}(x_n)_+^\gamma = C_{\gamma,\alpha,n}(x_n)_+^{\gamma-\alpha}, \quad x_n > 0.$$

Proof. Let $x = (x', x_n) \in \mathbb{R}^n$, then

$$(-\triangle)^{\alpha/2}(x_n)_+^\gamma$$

$$= C_{n,\alpha}PV \int_{\mathbb{R}^n} \frac{(x_n)_+^\gamma - (y_n)_+^\gamma}{|x-y|^{n+\alpha}} dy$$

$$= C_{n,\alpha}PV \int_{-\infty}^\infty \int_{\mathbb{R}^{n-1}} \frac{(x_n)_+^\gamma - (y_n)_+^\gamma}{|(x_n - y_n)^2 + (x' - y')^2|^{\frac{n+\alpha}{2}}} dy' \, dy_n.$$

Let $s = |x_n - y_n|$, $r = |x' - y'|$, by Lemma 9.1.2, we have

$$C_{n,\alpha}PV \int_{-\infty}^\infty ((x_n)_+^\gamma - (y_n)_+^\gamma) \int_0^\infty \frac{\omega_{n-2}r^{n-2}}{(r^2 + s^2)^{\frac{n+\alpha}{2}}} dr \, dy_n$$

$$= C_{n,\alpha}PV \int_{-\infty}^\infty \frac{(x_n)_+^\gamma - (y_n)_+^\gamma}{s^{1+\alpha}} \int_0^\infty \frac{\omega_{n-2}t^{n-2}}{(t^2 + 1)^{\frac{n+\alpha}{2}}} dt \, dy_n, \text{ where } r = st$$

$$= C_{n,\alpha}PV \int_{-\infty}^\infty \frac{(x_n)_+^\gamma - (y_n)_+^\gamma}{|x_n - y_n|^{1+\alpha}} dy_n$$

$$= C_{n,\alpha}C_\gamma(x_n)_+^{\gamma-\alpha}$$

$$= C_{\gamma,n,\alpha}(x_n)_+^{\gamma-\alpha}. \quad \square$$

Now we can construct the barrier function.

Lemma 9.1.3 *Let $\psi(x) = (|x|^2 - 1)_+^\gamma$ with $0 < \gamma < \alpha/2$. Then there exists some $\delta > 0$ small and $C_0 > 0$ such that*

$$(-\triangle)_a^{\alpha/2}\psi(x) \geq C_0(|x| - 1)^{\gamma-\alpha}, \quad \forall 1 < |x| < 1 + \delta.$$

Proof. Suppose in contrary, there exists a sequence $\{x^k\}$ and $|x^k| \to 1$ such that

$$(|x^k| - 1)^{\alpha - \gamma}(-\triangle)_a^{\alpha/2}\psi(x^k) \to 0. \tag{9.31}$$

Without loss of generality, we may assume that $x^k = (0', 1 + d_k)$. Then

$$d_k = |x^k| - 1 \to 0.$$

$$(|x^k| - 1)^{\alpha - \gamma}(-\triangle)_a^{\alpha/2}\psi(x^k)$$

$$= \frac{C_{n,\alpha}}{2} \int_{\mathbb{R}^n} \frac{d_k^{\alpha-\gamma}\left[2(|x^k|^2 - 1)_+^\gamma - (|x^k + y|^2 - 1)_+^\gamma - (|x^k - y|^2 - 1)_+^\gamma\right]}{|y|^{n+\alpha}}$$

$$\cdot a(y)\, dy$$

$$= \frac{C_{n,\alpha}}{2} \int_{\mathbb{R}^n} \frac{a(y)}{|y|^{n+\alpha}} d_k^{\alpha-\gamma}\left[2(2d_k + d_k^2)_+^\gamma - (2d_k + d_k^2 + 2(1 + d_k)y_n + |y|^2)_+^\gamma\right.$$

$$\left. - (2d_k + d_k^2 - 2(1 + d_k)y_n + |y|^2)_+^\gamma\right] dy$$

$$= \frac{C_{n,\alpha}}{2} \int_{\mathbb{R}^n} \frac{a(d_k z)}{|z|^{n+\alpha}}\left[2(2 + d_k)^\gamma - (2 + d_k + 2(1 + d_k)z_n + d_k|z|^2)_+^\gamma\right.$$

$$\left. - (2 + d_k - 2(1 + d_k)z_n + d_k|z|^2)_+^\gamma\right] dz \quad \text{let } z = \frac{y}{d_k}$$

$$= I.$$

As $k \to \infty$, $d_k \to 0$ and $a(d_k z) \to 1$. By the *Lebesgue's dominated convergence theorem*, we have

$$I \to 2^\gamma \frac{C_{n,\alpha}}{2} \int_{\mathbb{R}^n} \frac{2 - (1 + z_n)_+^\gamma - (1 - z_n)_+^\gamma}{|z|^{n+\alpha}} dz$$

$$= C_{\gamma,n,\alpha} > 0.$$

The last equality can be obtained by choosing $x_n = 1$ in Corollary 9.1.1. This contradicts (9.31) and proves the lemma.

Corollary 9.1.2 *For any* $0 < \gamma < \alpha/2$, *let* $\psi(x) = (|x|^2 - 1)_+^\gamma$ *and* $a_k(y) = a(\lambda_k y)$ *for* $k \in \mathbb{N}$. *Then there exists some* $\delta > 0$ *small and* $C_0 > 0$ *(independent of* k*) such that*

$$(-\triangle)_{a_k}^{\alpha/2}\psi(x) \geq C_0(|x| - 1)^{\gamma-\alpha}, \quad \forall 1 < |x| < 1 + \delta.$$

With the condition that

$$\lim_{s \to \infty} \frac{f(x,s)}{s^p} = g(x), \quad \text{uniformly for all } x \in \Omega,$$

where $g(x)$ is bounded away from zero in Ω, for k large we have

$$|(-\triangle)_{a_k}^{\alpha/2} v_k(x)| = |m_k^{-p} f(\lambda_k x + x^k, m_k v_k(x))|$$
$$\leq C m_k^{-p} (m_k v_k(x))^p |g(x)|$$
$$\leq C_o. \qquad (9.32)$$

Let $\xi(x)$ be a smooth cutoff function such that $0 \leq \xi(x) \leq 1$ in $\mathbb{R}^n$ and $\xi(x) = 0$ in $\overline{B_1(0)}$ and $\xi(x) = 1$ in $B_{1+\delta}^c(0)$. Then for all k,

$$(-\triangle)_{a_k}^{\alpha/2} \xi(x) = \frac{C_{n,\alpha}}{2} \int_{\mathbb{R}^n} \frac{2\xi(x) - \xi(x+y) - \xi(x-y)}{|y|^{n+\alpha}} a(\lambda_k y) dy$$
$$\geq -C_\delta.$$

Let

$$\varphi(x) = t\psi(x) + \xi(x), \ t \text{ is a positive constant,}$$

and $D = (B_{1+\delta}(0) \backslash B_1(0)) \cap \Omega_k$. For t sufficiently large and $x \in D$,

$$(-\triangle)_{a_k}^{\alpha/2} \varphi(x) = t(-\triangle)_{a_k}^{\alpha/2} \psi(x) + (-\triangle)_{a_k}^{\alpha/2} \xi(x)$$
$$\geq t C_0 (|x| - 1)^{\gamma - \alpha} - C_\delta$$
$$\geq t C_0 \delta^{\gamma - \alpha} - C_\delta$$
$$> C_o. \qquad (9.33)$$

Combining (9.32) and (9.33), it yields that

$$\begin{cases} (-\triangle)_{a_k}^{\alpha/2}(\varphi - v_k) > 0, \ x \in D, \\ \varphi(x) - v_k(x) > 0, \qquad x \in \mathbb{R}^n \backslash D. \end{cases} \qquad (9.34)$$

Applying Lemma 9.1.1 to (9.34), we have

$$\varphi(x) - v_k(x) > 0, \quad x \in D.$$

Let $\lim_{k \to \infty} x^k = \tilde{x}^o \in \partial \Omega_k$. Then

$$0 \leq v_k(x) - v_k(\tilde{x}^o) = v_k(x) \leq \varphi(x) = \varphi(x) - \varphi(\tilde{x}^o) \leq C|x - \tilde{x}^o|^\gamma.$$

This proves (9.22) and completes the proof in *Case iii*. $\square$

9.2 The Brouwer Degree

In this section, we introduce the Brouwer degree on finite dimensional spaces, then in the next section, we generalize it to the Leray-Schauder degree on infinite dimensional Banach spaces, which is used to establish existence of solutions based on the a priori estimates. More details concerning the degree theory can be found, for example, in K. C. Chang's book [C].

We denote $\mathbb{R}^n$ and $\mathbb{R}^m$ by X and Y respectively. Assume that Ω is an open set in X, $f \in C^p(\Omega, Y)$ for $p \geq 1$. For each $x_o \in \Omega$, define linear operator $f'(x_o) : X \to Y$ as

$$f'(x_o) = (\frac{\partial f_j}{\partial x_i})_{\substack{0 \leq i \leq n \\ 0 \leq j \leq m}} (x_o),$$

where $f(x) = (f_1(x), f_2(x), \cdots, f_m(x))$ and $x = (x_1, x_2, \cdots, x_n)$.

Definition 9.2.1 *We say $x_o \in \Omega$ is a regular point of f if $f'(x_o)$ is surjective. And $y = f(x_o)$ is a regular value of f. Otherwise, x_o is a critical point of f and $y = f(x_o)$ is a critical value of f.*

Example 9.2.1 *1. When $n = m = 1$, f' is the derivative of f. If $f'(x_o) \neq 0$, then x_o is a regular point of f, and $f(x_o)$ is a regular value. If $f'(x_o) = 0$, then x_o is a critical point of f, and $f(x_o)$ is a critical value.*

2. When $n = m \geq 2$, $f'(x_o) = J_f(x_o)$, the Jacobian matrix of f evaluated at x_o. If the Jacobian determinant $|J_f(x_o)| \neq 0$, then x_o is a regular point and $f(x_o)$ is a regular value. If $|J_f(x_o)| = 0$, then x_o is a critical point and $f(x_o)$ is a critical value.

Roughly speaking, taking $n = m = 1$ as an example, the Brouwer degree $deg(f, \Omega, p)$ (we will define it precisely later) indicates algebraically how many times f maps points in Ω into p by summing up the signs of $f'(x_i)$ on all the pre-images of p, i.e. on the points $x_i \in \Omega$, such that $f(x_i) = p$. It is meaningful to first define it on a regular value p of f where $f' \neq 0$, and then by approximation, we can define such a degree on a critical value. To make this kind of approximation possible, we need to show that the set of regular values is dense in the image of f. A more precise statement is given by the Sard Theorem in the next subsection. Then we define the Brouwer degree for differentiable mappings and derive its properties, among which an important one is its invariance under a small perturbation of a function f, which enables us to extend the degree to continuous mapping by the density of the differentiable mappings.

9.2.1 The Sard Theorem

Theorem 9.2.1 (Sard Theorem) *Let $\Omega \subset \mathbb{R}^n$ be an open set. If $f \in C^p(\Omega, \mathbb{R}^m)$ for $p > \max\{n - m, 0\}$ and E is the set of critical points for f, then $f(E)$ is Lebesgue measure zero, i.e., $m(f(E)) = 0$.*

To better illustrate the idea, we prove the Sard Theorem only for $n = m = 1$. For more general cases, please see [ZF] and [C] for the detailed proofs.

Theorem 9.2.2 *Let $\Omega \subset R$ be an open set, $f \in C^1(\Omega, R)$, and $E = \{x \in \Omega \mid f'(x)\} = 0$. Then $m(f(E)) = 0$.*

Proof. For any $\Omega \subset R$, there exist countable closed intervals C_i such that $\Omega \subset \bigcup_{i=1}^{\infty} C_i$. Thus

$$f(E) = f(E \cap (\bigcup_{i=1}^{\infty} C_i)) = \bigcup_{i=1}^{\infty} f(E \cap C_i).$$

In order to prove $m(f(E)) = 0$, it suffices to show for each i, $m(f(E \cap C_i)) = 0$.

Without loss of generality, we take the length of C_i to be 1 and divide C_i into n subintervals K_j of equal length $\frac{1}{n}$ with $j = 1, 2, \cdots, n$.

Suppose that for some j_0, $x_0 \in K_{j_0}$ is a critical point, i.e., $f'(x_0) = 0$. It follows from the continuity of f' that for any $\varepsilon > 0$ and $x \in K_{j_0}$, there exists an N such that for $n > N$,

$$|f'(x)| = |f'(x) - f'(x_0)| < \varepsilon.$$

Therefore,

$$|f(x) - f(x_0)| = |f'(\xi)(x - x_0)| < \frac{\varepsilon}{n}.$$

This implies that $m(f(K_{j_0})) < \frac{2\varepsilon}{n}$.

If C_i has M subintervals containing critical points, without loss of generality, we may assume that they are K_j with $j = 1, \cdots, M$, then

$$m(f(E \cap C_i)) \le m\left(\bigcup_{j=1}^{M} f(K_j)\right) \le \sum_{j=1}^{M} m(f(K_j)) \le M\frac{2\varepsilon}{n} < 2\varepsilon.$$

Due to the arbitrariness of ε, we have

$$m(f(E \cap C_i)) = 0.$$

This proves the theorem.

We present more lemmas on the Brouwer degree in $\mathbb{R}^n$ for the rest of Section 9.2. Here we only give the proofs in R^1 so that the readers can quick grasp the essence in a straight forward way. Readers who are interested in these proofs in $\mathbb{R}^n$ may refer to Chapter 3 in [ZF].

9.2.2 The Brouwer Degree for Differentiable Mappings

Let $\Omega = (a, b)$ and $f \in C(\bar{\Omega}, R)$. Consider

$$f(x) = p, \quad x \in \Omega. \tag{9.35}$$

Assume that $f \in C^1(\Omega)$ and $p \notin f(\partial\Omega)$. We want to define an integer-valued function $deg(f, \Omega, p)$, read as degree, to roughly estimate the number of solutions to (9.35) in (a, b). We expect that degree does not change when a small perturbation of f occurs.

Ideally, we hope to define $deg(f, \Omega, p)$ as the number of solutions to (9.35). Unfortunately, in such case, a small change in f may lead to a big change in the degree. For example, let $f(x) = \sin x - 1 - \varepsilon$ and $p = 0$ in (9.35). From Figs. 9.2(a)-(c), it's easy to see that:

1. when ε is small positive, (9.35) has no solution in $(0, \pi)$,
2. when ε is small negative, (9.35) has two solutions in $(0, \pi)$,
3. when $\varepsilon = 0$, (9.35) has one solution in $(0, \pi)$.

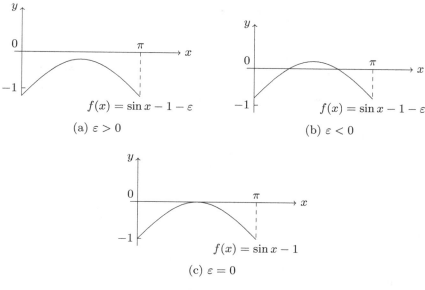

Fig. 9.2

The figures show how a small change of ϵ results in abrupt change in the number of solutions. However, one may notice that if we consider the sum of the signs of f' in the pre-image of 0, i.e., $\sum\limits_{x \in f^{-1}(p)} sgn f'(x)$, then in the second case, it is $1 + (-1) = 0$ and so is it in the third case. It is invariant under small perturbation.

Since Ω is bounded and p is a regular value of f, it follows from the cluster point argument that there are at most finitely many solutions to (9.35). Hence we define

$$deg(f, (a, b), p) = \sum_{x \in f^{-1}(p)} sgn\, f'(x). \tag{9.36}$$

Notice that the degree defined above is always an integer, but may be positive or negative depending on the orientations.

By the Sard Theorem, we know that the regular values are dense in $f(\Omega)$. Therefore, if p is a critical value of f, for any $\delta > 0$, there exist regular values in $(p - \delta, p + \delta)$. Later, we will show that for any $p_1, p_2 \in (p - \delta, p + \delta)$, it holds that

$$deg(f, (a, b), p_1) = deg(f, (a, b), p_2). \tag{9.37}$$

Thus for any $p \in R$, we can define

$$deg(f, (a, b), p) = deg(f, (a, b), p_1). \tag{9.38}$$

Let's go back to the example:

$$f(x) = \sin x - 1 - \varepsilon = 0, \quad x \in \Omega = (0, \pi),$$

to check if the degree defined in (9.36) is invariant or not under small perturbations.

1. For $\varepsilon > 0$ small, since $0 \notin f(\Omega)$, we have

$$deg(f, \Omega, 0) = 0.$$

2. For $\varepsilon < 0$ small, 0 is a regular value. From the case above we know that

$$deg(f, \Omega, 0) = \underset{x \in f^{-1}(0)}{sgn} \; f'(x) = 0.$$

3. For $\varepsilon = 0$, the equation becomes $f(x) = \sin x - 1 = 0$ and 0 is a critical value. Suppose that $p_1 < 0 < p_2$ are any two regular values of f close to 0. Then there exist $x_1 < x_2$ in $(0, \pi)$ such that $f(x_1) = f(x_2) = p_1$, and

$$deg(f, \Omega, p_1) = sgn \, f'(x_1) + sgn \, f'(x_2) = 1 + (-1) = 0.$$

On the other hand, since $p_2 \notin f(\Omega)$,

$$deg(f, \Omega, p_2) = 0.$$

Hence

$$deg(f, \Omega, 0) = deg(f, \Omega, p_i) = 0, i = 1, 2.$$

To summarize, $deg(f, \Omega, 0) = 0$ for any small $\varepsilon \in R$. The degree stays the same under small perturbations of the function f indeed, we prove this in the next subsection.

As a generalization of (9.36) to mappings from $\mathbb{R}^n$ to $\mathbb{R}^n$, we define the degree as below:

- If p is a regular value of f, let

$$deg(f, \Omega, p) = \sum_{x \in f^{-1}(p)} sgn \, |J_f(x)|. \tag{9.39}$$

- If p is a critical value of f, let

$$deg(f, \Omega, p) = deg(f, \Omega, p_1), \tag{9.40}$$

where p_1 is any regular value in $B_\delta(p)$ for δ small.

Now we prove (9.37). First, we need to rewrite (9.36) as an integral through mollifiers. Assume that $\varphi \in C^\infty(R, R)$ is a nonnegative function satisfying

$$\varphi(x) = 0, \quad |x| \geq 1,$$

and

$$\int_R \varphi(x) dx = 1,$$

then φ is a mollifier. For example, let

$$\varphi(x) = \begin{cases} C \, exp(\frac{1}{|x|^2 - 1}), & |x| < 1, \\ 0, & |x| \geq 1. \end{cases} \tag{9.41}$$

For $\varepsilon > 0$, let

$$\varphi_\varepsilon(x) = \frac{1}{\varepsilon}\varphi(\frac{x}{\varepsilon}).$$

Then

$$\varphi_\varepsilon(x) = 0, \quad |x| \geq \varepsilon,$$

and

$$\int_R \varphi_\varepsilon(x)dx = 1.$$

Lemma 9.2.1 *Assume that $\Omega \subset \mathbb{R}^n$ is open and bounded, $f \in C^1(\bar{\Omega}, \mathbb{R}^n)$ and $p \notin f(\partial\Omega)$ is a regular value of f. Then there exists some $\varepsilon_0 = \varepsilon_0(p, f)$ such that $\forall \varepsilon \in (0, \varepsilon_0)$,*

$$deg(f, \Omega, p) = \int_\Omega \varphi_\varepsilon(f(x) - p)|J_f(x)|dx. \tag{9.42}$$

Proof. Here we only show the proof in R^1, i.e.,

$$deg(f, \Omega, p) = \int_\Omega \varphi_\varepsilon(f(x) - p)f'(x)dx.$$

- If $f^{-1}(p) = \emptyset$, take $\varepsilon_0 = dist(p, f(\bar{\Omega}))$. Then for $0 < \varepsilon < \varepsilon_0$, $\varphi_\varepsilon = 0$.

- If $f^{-1}(p) \neq \emptyset$, because p is a regular value of f, $f^{-1}(p)$ consists of at most finitely many points, denoted as $f^{-1}(p) = \{x_1, x_2, \cdots, x_k\}$. It follows from the *inverse function theorem* that there exists some ε_0 such that $f^{-1}(p - \varepsilon_0, p + \varepsilon_0) = \bigcup_{i=1}^{k} U_i$, where $U_i \in \Omega$ is a neighborhood of x_i, and $U_j \cap U_i = \emptyset$ for $i \neq j$ and f' does not change sign in each U_i. For $x \in \Omega \backslash \bigcup_{i=1}^{k} U_i$, $|f(x) - p| \geq \varepsilon_0$. Hence for $0 < \varepsilon < \varepsilon_0$, we have

$$\int_\Omega \varphi_\varepsilon(f(x) - p)f'(x)dx$$
$$= \sum_{i=1}^{k} sgn f'(x) \int_{U_i} \varphi_\varepsilon(f(x) - p)|f'(x)|dx. \tag{9.43}$$

Let $y = f(x)$, then $dy = |f'(x)|dx$ and

$$\int_{U_i} \varphi_\varepsilon(f(x) - p)|f'(x)|dx$$
$$= \int_{B_{\varepsilon_0}(p)} \varphi_\varepsilon(y - p)dy$$
$$= \int_{B_{\varepsilon_0}(0)} \varphi_\varepsilon(y)dy = 1.$$

Together with (9.43) we obtain

$$deg(f, \Omega, p) = \int_\Omega \varphi_\varepsilon(f(x) - p)f'(x)dx.$$

Next we prove (9.37).

Proof. Assume $f \in C^1(\bar{\Omega}, R)$, $p \notin f(\partial\Omega)$ and $\rho = dist(p, f(\partial\Omega))$. Let p_1 and p_2 be any two regular values of f satisfying $|p_i - p| < \rho$, $i = 1, 2$. It follows from Lemma 9.2.1 that there exists some $\varepsilon_0 = \varepsilon_0(p_1, p_2, f)$ such that for any $0 < \varepsilon < \varepsilon_0$, we have

$$deg(f, \Omega, p_1) = \int_\Omega \varphi_\varepsilon(f(x) - p_1)f'(x)dx,$$

$$deg(f, \Omega, p_2) = \int_\Omega \varphi_\varepsilon(f(x) - p_2)f'(x)dx.$$

In order to prove (9.37), we only need to show that

$$\int_\Omega [\varphi_\varepsilon(f(x) - p_1) - \varphi_\varepsilon(f(x) - p_2)]f'(x)dx = 0. \tag{9.44}$$

Let $\varepsilon \in (0, \min\{\varepsilon_0, \rho - |p - p_i|\})$ for $i = 1, 2$. Then

$$\varphi_\varepsilon(f(x) - p_i) = 0, \quad x \in \partial\Omega, \; i = 1, 2. \tag{9.45}$$

Notice that

$$\varphi_\varepsilon(f(x) - p_1) - \varphi_\varepsilon(f(x) - p_2) = \int_0^1 \varphi'_\varepsilon(f(x) - p_1 - t(p_2 - p_1))(p_2 - p_1)dt,$$

thus

$$[\varphi_\varepsilon(f(x) - p_1) - \varphi_\varepsilon(f(x) - p_2)]f'(x)$$

$$= \int_0^1 \varphi'_\varepsilon(f(x) - p_1 - t(p_2 - p_1))f'(x)(p_2 - p_1)dt$$

$$= \frac{d}{dx}\left(\int_0^1 \varphi_\varepsilon(f(x) - p_1 - t(p_2 - p_1))(p_2 - p_1)dt\right).$$

Integrating the equation above in Ω, it follows from the *divergence theorem* and (9.45) that

$$\int_\Omega [\varphi_\varepsilon(f(x) - p_1) - \varphi_\varepsilon(f(x) - p_2)]f'(x)dx$$

$$= \int_\Omega \left[\frac{d}{dx}\left(\int_0^1 \varphi_\varepsilon(f(x) - p_1 - t(p_2 - p_1))(p_2 - p_1)dt\right)\right]dx$$

$$= \int_0^1 \left\{\int_{\partial\Omega} (\varphi_\varepsilon(f(x) - p_1 - t(p_2 - p_1))(p_2 - p_1))\, d\sigma_x\right\}dt$$

$$= 0.$$

This proves (9.37).

9.2.3 The Brouwer Degree for Continuous Mappings

In the previous section, we defined the Brouwer degree for differentiable mappings. Now we extend such a definition to C^0 mappings. A natural idea is to approximate a continuous function by differentiable functions. Therefore we need to show that the degree is invariant under a small perturbation of the function.

Lemma 9.2.2 *Assume that $\Omega \subset \mathbb{R}^n$ is open and bounded, $f \in C^2(\bar{\Omega}, \mathbb{R}^n)$, $p \in \mathbb{R}^n$ and $p \notin f(\partial\Omega)$. Then for any $g \in C^2(\bar{\Omega}, \mathbb{R}^n)$, there exists a $\delta = \delta(f, p, g) > 0$ such that for $|t| < \delta$, we have $p \notin (f + tg)(\partial\Omega)$ and*

$$deg(f + tg, \Omega, p) = deg(f, \Omega, p).$$

Proof. Just to illustrate the idea, we prove the lemma in R^1. Let $M = \max_{x \in \bar{\Omega}} |g(x)|$, we carry out the proof in three cases.

Case i. $f^{-1}(p) = \emptyset$.
Let $\delta = dist(p, f(\bar{\Omega}))M^{-1}$. For any $|t| < \delta$, $x \in \bar{\Omega}$,

$$
\begin{aligned}
|f(x) + tg(x) - p| &\geq |f(x) - p| - |t||g(x)| \\
&\geq dist(p, f(\bar{\Omega})) - |t|M \\
&> dist(p, f(\bar{\Omega})) - \delta M \\
&= 0.
\end{aligned}
$$

Hence $(f + tg)^{-1}(p) = \emptyset$.

Case ii. $f^{-1}(p) \neq \emptyset$ and p is a regular value of f.
Let $f^{-1}(p) := \{x_1, x_2, \cdots, x_k\}$, $\rho = dist(p, f(\partial\Omega))$ and

$$F(x, t) = f(x) + tg(x).$$

Take $\delta_1 = \rho M^{-1}$, then for $|t| < \delta_1$, $p \notin F(\partial\Omega)$.

Since $\frac{\partial F}{\partial x}(x_i, 0) = f'(x_i) \neq 0$ and $F(x_i, 0) = p$, it follows from the *implicit function theorem* that there exists a δ_2 and disjoint intervals $U_i(x_i)$, such that for $t \in (-\delta_2, \delta_2)$, the equation

$$F(x, t) = p \tag{9.46}$$

has a uniquely continuous solution $x_i(t)$ in $U_i(x_i)$ and $x_i(0) = x_i$. Due to the continuity of $\frac{\partial F}{\partial x}$ about both x and t, it holds that for $|t| < \delta_2$,

$$sgn \frac{\partial F}{\partial x}(x_i(t), t) = sgn \frac{\partial F}{\partial x}(x_i(0), 0) = sgn \frac{\partial f}{\partial x}(x_i). \tag{9.47}$$

This shows that p is also a regular value of $F(x, t)$ for all small t.

Let $V = \bigcup_{i=1}^{k} U_i(x_i)$. Since $f(x) = p$ has no solution in $\overline{\Omega \backslash V}$, there exists a $\beta > 0$ such that

$$dist(f(\overline{\Omega\backslash V}), p) = \beta.$$

Let $\delta_3 = \frac{\beta}{M}$, then (9.46) has no solution in $\overline{\Omega\backslash V}$ for $|t| < \delta_3$. Therefore, when $|t| < \delta = \min\{\delta_1, \delta_2, \delta_3\}$, the only solutions for (9.46) are $x_1(t)$, $x_2(t), \cdots, x_k(t)$. Hence

$$deg(F, \Omega, p) = \sum_{i=1}^{k} sgn \frac{\partial F}{\partial x}(x_i(t), t)$$

$$= \sum_{i=1}^{k} sgn f'(x_i)$$

$$= deg(f, \Omega, p).$$

Case iii. $f^{-1}(p) \neq \emptyset$ and p is a critical value of f.

The Sard Theorem implies that f has a regular value p_1 satisfying

$$|p - p_1| < \rho := dist(p, f(\partial\Omega)).$$

From *Case ii* we know that there exists a $\delta > 0$ such that for $|t| < \delta$, it holds that $dist(p, F(\partial\Omega)) > 0$,

$$deg(F, \Omega, p_1) = deg(f, \Omega, p_1).$$

and p_1 is a regular value of F. Thus

$$deg(F, \Omega, p) = deg(F, \Omega, p_1)$$

$$= deg(f, \Omega, p_1)$$

$$= deg(f, \Omega, p).$$

This proves that the Brouwer degree is homotopy invariant in a "small" scale, or, with t small. Next we show that the homotopy invariance is also true in a "larger" scale with $t \in [0, 1]$.

Lemma 9.2.3 *Suppose $\Omega \subset \mathbb{R}^n$ is open and bounded, $f \in C(\bar{\Omega}, \mathbb{R}^n)$, $p \in R\backslash f(\partial\Omega)$, and $f_1, f_2 \in C^2(\bar{\Omega}, \mathbb{R}^n)$. Let $\rho = dist(p, f(\partial\Omega))$. If for some $\varepsilon_0 > 0$ small,*

$$\sup_{x\in\bar{\Omega}} |f_i(x) - f(x)| \leq \rho - \varepsilon_0, \quad i = 1, 2.$$

Then

$$deg(f_1, \Omega, p) = deg(f_2, \Omega, p).$$

Proof. Let $h(x, t) = t f_1(x) + (1 - t) f_2(x)$. For any fixed $t \in [0, 1]$ and $x \in \bar{\Omega}$, we have

$$|h(x, t) - f(x)| \leq t|f_1(x) - f(x)| + (1 - t)|f_2(x) - f(x)|$$

$$\leq \rho - \varepsilon_0,$$

which implies that $p \notin h(\partial\Omega, t)$. Let $\varphi(t) = deg(h(\cdot, t), \Omega, p)$, from Lemma 9.2.2 we know that for any $t_0 \in [0, 1]$, there exists a $\delta = \frac{\varepsilon_0}{M}$, such that for $|t - t_0| < \delta$, $\varphi(t) = \varphi(t_0)$. Because t_0 is arbitrary, $\varphi(t)$ is constant on $[0, 1]$. Moreover $\varphi(0) = \varphi(1)$, i.e.,

$$deg(f_1, \Omega, p) = deg(f_2, \Omega, p).$$

Remark 9.2.1 *With Lemma 9.2.3, we can define the degree for a continuous mapping f in Ω at a given point p by the degree of a C^2 function g close to f, on condition that $p \notin g(\partial\Omega)$. This is reasonable because Lemma 9.2.3 shows that the degree is invariant for any two C^2 functions f_i that are close to f, as long as $p \notin f_i(\partial\Omega)$.*

Definition 9.2.2 *Suppose $\Omega \subset \mathbb{R}^n$ is open and bounded, $f \in C(\bar{\Omega}, \mathbb{R}^n)$, $p \in \mathbb{R}^n \backslash f(\partial\Omega)$ and $f_1 \in C^2(\bar{\Omega}, \mathbb{R}^n)$. Let $\rho = dist(p, f(\partial\Omega))$. If for some $\varepsilon_0 > 0$,*

$$\sup_{x \in \bar{\Omega}} |f_1(x) - f(x)| \le \rho - \varepsilon_0,$$

then we define

$$deg(f, \Omega, p) = deg(f_1, \Omega, p).$$

From Lemma 9.2.3, we know that $deg(f, \Omega, p)$ is independent of f_1. For a continuous function, we can prove its degree is homotopy invariant with a similar argument to that for the C^2 function.

Theorem 9.2.3 *Assume that $\Omega \in \mathbb{R}^n$ is open and bounded, $f \in C(\bar{\Omega}, \mathbb{R}^n)$, $p \in \mathbb{R}^n$ and $p \notin f(\partial\Omega)$. Then for any given $g \in C(\bar{\Omega}, \mathbb{R}^n)$, if*

$$p \notin (f + tg)(\partial\Omega), \quad \forall t \in [0, 1],$$

then

$$deg(f + tg, \Omega, p) = deg(f, \Omega, p), \quad \forall t \in [0, 1].$$

From the proof of Lemma 9.2.3, one can see that in order to guarantee the homotopy invariance of the degree, we require that

$$p \notin h(\partial\Omega, t), \quad \forall t \in [0, 1]. \tag{9.48}$$

Figure 9.3(a)-(c) shed some light on why this condition is needed. Here $\Omega = (-1, 1)$, $\partial\Omega = \{-1, 1\}$ and $p = 0$. In figure (a) and (b), condition (9.48) is satisfied, hence

$$deg(f_1, \Omega, 0) = deg(f_2, \Omega, 0) = \begin{cases} 1 & \text{in case (a)} \\ 0 & \text{in case (b)}. \end{cases}$$

However in case (c), at $x = 1$, $tf_1(1) + (1 - t)f_2(1) = 0$ for some t, and $deg(f_1, \Omega, 0) = 1$, $deg(f_2, \Omega, 0) = 0$.

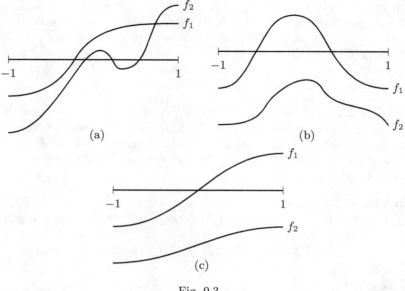

Fig. 9.3

9.3 The Leray-Schauder Degree

In this section, we extend the definition of the Brouwer degree from finite dimensional spaces to infinite dimensional Banach spaces. However, the generalization may not be suitable to arbitrary continuous mappings.

To give an example, let B be the unit open ball at the origin in an infinite dimensional real Banach spaces. Let $\phi \in C(\bar{B}, \bar{B})$, and I be the identity operator. Assume that for any $t \in [0, 1]$, $0 \notin (I - t\phi)(\partial B)$. If the degree were well-defined for the C function in X, then it would still possess fundamental properties such as the homotopy invariance. From this invariance, we would have

$$deg(I - t\phi, B, 0) = deg(I, B, 0) = 1,$$

and hence there exists an $x \in B$, such that $(I - \phi)(x) = 0$, that is, ϕ has a fixed point:

$$\phi(x) = x.$$

We show that this may not be valid for all continuous mappings. Let

$$X = l^2 = \{x = (x_1, x_2, \cdots, x_n, \cdots) \mid \|x\|^2 = \sum_{n=1}^{\infty} x_n^2 < \infty\},$$

and

$$\phi : x \mapsto (\sqrt{1 - \|x\|^2}, x_1, x_2, \cdots).$$

Then $\phi : \bar{B} \to \bar{B}$ is continuous. But ϕ has no fixed point on $\bar{B}$. Indeed, if for some $x^o \in \bar{B}$, $\phi(x^o) = x^o$, then

$$\sqrt{1 - \|x^o\|^2} = x_1^o,$$
$$x_1^o = x_2^o,$$
$$\cdots,$$
$$x_n^o = x_{n+1}^o, \qquad (9.49)$$
$$\cdots,$$

and

$$\|x^o\| = \|\phi(x^o)\| = 1. \qquad (9.50)$$

On the other hand, since $x^o \in X$, it must be true that $\lim_{n\to\infty} x_n^o = 0$. Together with (9.49), it gives that $x_i^o = 0$, $i = 1, 2, \cdots, n, \cdots$. This is a contradiction to (9.50).

Hence we have to "narrow down" to a smaller and better group from the big family of the continuous functions, and as we will see that, this is a group of mappings of the form $I - K$, where I is the identity map and K is any compact map.

In the following we always assume that Ω is a bounded open set in a real Banach space X.

Theorem 9.3.1 *Let $K : \bar{\Omega} \to X$ be a compact map. Then for any $\varepsilon > 0$, there is a continuous operator K_ε taking values in a finite dimensional linear subspace E_{m_ε} such that*

$$|K(x) - K_\varepsilon(x)| \leq \varepsilon, \quad \forall x \in \bar{\Omega}. \qquad (9.51)$$

Proof. Let $B_\varepsilon(y^i)$, $i = 1, \cdots, m_\varepsilon$, be finitely many ε-balls covering $K(\bar{\Omega})$. Set

$$\psi_i(x) = (\varepsilon - |x - y^i|)_+$$

where

$$\beta_+ = \begin{cases} \beta, & \beta \geq 0, \\ 0, & \beta < 0. \end{cases}$$

Let

$$\varphi_i(x) = \frac{\psi_i(x)}{\sum_{j=1}^{m_\varepsilon} \psi_j(x)}, i = 1, 2, \cdots, m_\varepsilon,$$

and

$$K_\varepsilon(x) = \sum_{i=1}^{m_\varepsilon} \varphi_i(K(x))y^i,$$

then $K_\varepsilon(x) \in span\{y^1, y^2, \cdots, y^{m_\varepsilon}\}$ and

$$|K(x) - K_\varepsilon(x)| \leq \sum_{i=1}^{m_\varepsilon} \varphi_i(K(x))|K(x) - y^i|$$

$$\leq \varepsilon \sum_{i=1}^{m_\varepsilon} \varphi_i(K(x)) = \varepsilon.$$

The above theorem enables us to approximate any operator of the form $f := I - K$ by a family of $f_\varepsilon := I - K_\varepsilon$, and hence provides a possibility to define the degree of f via the Brouwer degree of f_ε, which is well-defined, since it is an operator on a finite dimensional space. However before doing so, we must make sure of two things

(a) if $y_0 \in X$ and $y_0 \notin f(\partial\Omega)$, then for $\varepsilon > 0$ small, $y_0 \notin f_\varepsilon(\partial\Omega)$,

(b) for any two f_{ε_1} and f_{ε_2} satisfying (9.51), their Brouwer degrees are equal.

To prove part (a), we need the following lemma.

Lemma 9.3.1 *If $K : \bar\Omega \to X$ is compact, $y_o \notin (I - K)(\partial\Omega)$, then*

$$dist(y_o, (I - K)(\partial\Omega)) > 0.$$

Proof. Assume in the contrary, there exists $\{y_i\} \subset (I - K)(\partial\Omega)$, such that $y_i \to y_o$. Then there exists $\{x_i\} \subset \partial\Omega$ satisfying $(I - K)(x_i) = y_i$. Therefore,

$$x_i - K(x_i) \to y_o. \tag{9.52}$$

Since K is compact and $\{x_i\}$ is bounded, there exists a subsequence of $\{x_i\}$, still denoted by $\{x_i\}$, and $z_o \in X$ such that

$$K(x_i) \to z_o. \tag{9.53}$$

Combining (9.52) and (9.53), it yields that

$$x_i \to y_o + z_o. \tag{9.54}$$

Since $\partial\Omega$ is closed, we must have $y_o + z_o \in \partial\Omega$. Taking the limit of (9.52) yields that

$$y_o + z_o - K(y_o + z_o) = y_o. \tag{9.55}$$

This means that $y_o \in (I - K)(\partial\Omega)$, which contradicts the assumption. Hence we complete the proof.

Lemma 9.3.1 enables us to choose a proper $0 < \varepsilon < dist(y_o, (I - K)(\partial\Omega))$ such that

$$dist(y_o, f_\varepsilon(\partial\Omega)) > 0.$$

Now, we consider part (b).

Let

$$f_\varepsilon : \bar\Omega \cap \mathbb{R}^{m_\varepsilon} \to \mathbb{R}^{m_\varepsilon}$$

where $m_\varepsilon = dim\{K_\varepsilon(\Omega)\}$. The Brouwer degree $deg(f_\varepsilon, \Omega \cap \mathbb{R}^{m_\varepsilon}, y_0)$ is well defined. Since m_ε depends on ε, we should compare all the degrees of f_ε.

Lemma 9.3.2 *Let $\Omega \subset \mathbb{R}^n$ be a bounded open set. For $m < n$, let $i : \mathbb{R}^m \to \mathbb{R}^n$ be the canonical immersion:*

$$x = (x_1, x_2, \cdots, x_m) \to \hat{x} = (x_1, x_2, \cdots, x_m, 0, \cdots, 0).$$

Suppose that $K : \bar\Omega \to \mathbb{R}^m$ is continuous, and $p \in \mathbb{R}^m$ satisfying $\hat{p} \notin f(\partial\Omega)$ with $f = I - K$. Then

$$deg(f, \Omega, \hat{p}) = deg(f|_{\bar\Omega \cap \mathbb{R}^m}, \mathbb{R}^m \cap \Omega, p).$$

Proof. We prove the lemma when $m = 1$ and $n = 2$. For more general proofs, please refer to [C].

Let Ω be a bounded open set in R^2, $F \in C(\bar{\Omega}, R)$ and $\varepsilon = dist(\hat{p}, f(\partial\Omega)) > 0$. There exists an $\hat{F} \in C^1(\bar{\Omega}, R)$ satisfying

$$|F - \hat{F}|_{C(\bar{\Omega})} < \frac{\varepsilon}{2}.$$

Let $g = I - \hat{F}$. Then

$$|J_g(y)| = \begin{vmatrix} 1 - \frac{\partial \hat{F}}{\partial x_1} & -\frac{\partial \hat{F}}{\partial x_2} \\ 0 & 1 \end{vmatrix} = |1 - \frac{\partial \hat{F}}{\partial x_1}| = |J_{g|_{R \cap \bar{\Omega}}}(y)|. \tag{9.56}$$

Therefore, if q is a regular value of g, it must be true that

$$deg(g, \Omega, \hat{q}) = \sum_{y \in g^{-1}(\hat{q})} sgn|J_g(y)|$$

$$= \sum_{y \in g^{-1}|_{R \cap \bar{\Omega}}(q)} sgn|J_g(y)|$$

$$= deg(g|_{R \cap \bar{\Omega}}, R \cap \bar{\Omega}, q). \tag{9.57}$$

By the Sard theorem, given $p \in g(\Omega)$, for any $\varepsilon > 0$, there exists a regular value q of g such that $|p - q| < \frac{\varepsilon}{2}$ and $\hat{q} \notin g(\partial\Omega)$. Thus

$$deg(f, \Omega, \hat{p}) = deg(g, \Omega, \hat{p})$$
$$= deg(g, \Omega, \hat{q}) \quad (homotopy\ invariance)$$
$$= deg(g|_{R \cap \bar{\Omega}}, R \cap \Omega, q)$$
$$= deg(g|_{R \cap \bar{\Omega}}, R \cap \Omega, p)$$
$$= deg(f|_{R \cap \bar{\Omega}}, R \cap \Omega, p).$$

Now we are ready to define the topological degree for $f = I - K$ when K is compact.

Definition 9.3.1 (Leray-Schauder Degree) *Let X be a real Banach space and $\Omega \subset X$ be open and bounded. Assume that $K : \bar{\Omega} \to X$ is compact, $p \in X \backslash f(\partial\Omega)$ with $f = I - K$. Define*

$$deg(f, \Omega, p) = deg(f_\varepsilon, \Omega \cap E_\varepsilon, p),$$

where

$$0 < \varepsilon < dist(p, f(\partial\Omega)),\ f_\varepsilon = I - K_\varepsilon,$$

K_ε is a continuous operator defined in a finite dimensional space E_ε with $p \in E_\varepsilon$ and satisfies

$$|K(x) - K_\varepsilon(x)| \le \varepsilon, \forall x \in \bar{\Omega}.$$

The definition makes sense for the following reasons.

1. Lemma 9.3.1 guarantees that $dist(p, f(\partial\Omega)) > 0$, and Theorem 9.3.1 ensures the existence of such K_ε and E_ε.
2. We verify that $deg(f_\varepsilon, \Omega \cap E_\varepsilon, p)$ is well-defined for all such f_ε and E_ε and is independent of ε.

It's easy to see that

$$dist(p, f_\varepsilon(\partial\Omega \cap E_\varepsilon)) \geq dist(p, f(\partial\Omega)) - \varepsilon > 0.$$

Let $(f_{\varepsilon_0}, E_{\varepsilon_0})$ and $(f_{\varepsilon_1}, E_{\varepsilon_1})$ be two arbitrary pairs of mappings and finite dimensional linear subspaces satisfying the hypotheses of the definition. Let

$$\hat{E} = span\{E_{\varepsilon_0}, E_{\varepsilon_1}\}, \quad \hat{f}_{\varepsilon_i} = I - \hat{K}_{\varepsilon_i}, i = 0, 1,$$

where $\hat{K}_{\varepsilon_i}(x) = (K_{\varepsilon_i}(x), 0)$ and 0 is the zero element in $\hat{E}\backslash E_{\varepsilon_i}$. Let us define $\phi : (\Omega \cap \hat{E}) \times [0, 1] \to \hat{E}$ by

$$\phi(x, t) = t\hat{f}_{\varepsilon_0}(x) + (1 - t)\hat{f}_{\varepsilon_1}(x), \quad \forall t \in [0, 1].$$

It follows from the homotopy invariance of the Brouwer degree that

$$deg(\hat{f}_{\varepsilon_0}, \hat{E} \cap \Omega, \hat{p}) = deg(\hat{f}_{\varepsilon_1}, \hat{E} \cap \Omega, \hat{p}).$$

By Lemma 9.3.2, we have

$$deg(f_{\varepsilon_i}, E_{\varepsilon_i} \cap \Omega, p) = deg(\hat{f}_{\varepsilon_i}, \hat{E} \cap \Omega, \hat{p}), \quad i = 0, 1.$$

This proves

$$deg(f_{\varepsilon_0}, E_{\varepsilon_0} \cap \Omega, p) = deg(f_{\varepsilon_1}, E_{\varepsilon_1} \cap \Omega, p).$$

9.4 Applications

One of major applications of the Leray-Schauder degree is to prove the existence of solutions for nonlinear partial differential equations. In this process, one needs to make sure that the solution so obtained is non-trivial, and in many cases, one wish to find a positive solution. Notice that, the set of all positive functions in a certain Banach space, say in the space of continuous functions, is a convex set. Hence, we extend the Leray-Schauder degree theory to closed convex sets, then apply it to show the existence of a positive solution for a Dirichlet problem involving the fractional Laplacian.

9.4.1 The Degree Theory on Closed Convex Sets

From the previous sections, one can see that when applying the Leray-Schauder degree theory to the operator $I - K$, we need to pay special attention to the boundary conditions, to make sure that there is no fixed point of the compact operator K on the boundary. However, as one will see from the following discussions, for mappings with images contained in a closed convex set, we can reduce the notion of the boundary.

Assume that C is a closed convex subset of a real Banach space X and that $U \subset C$ is a bounded relatively open set. Assume that $K : U \to C$ is compact and that $0 \notin (I - K)(\partial U)$, where ∂U is the relative boundary of U to C. Let r be a retraction of C, that is, r is a continuous map from X to C with

$$r(x) = x \quad \forall x \in C.$$

Chose $R > 0$ sufficiently large such that $U \subset B_R(0)$. Since $B_R(0) \cap r^{-1}(U)$ is an open set in X, on which the degree of the map $I - K \circ r$ is well-defined. Now we can define the degree of the map $I - K$ on U as

$$\deg_C(I - K, U, 0) = \deg(I - K \circ r, B_R \cap r^{-1}(U), 0). \tag{9.58}$$

We first show that the definition makes sense, because

$$0 \notin (I - K \circ r)(\partial(B_R(0) \cap r^{-1}(U))).$$

Indeed, if otherwise, there exists a $p \in \partial(B_R(0) \cap r^{-1}(U))$ such that

$$p = K \circ r(p) \in C. \tag{9.59}$$

Since r is a retraction of C, it implies that $r(p) = p$. Hence $p = K(p)$, which implies $p \in C \cap \partial(B_R(0) \cap r^{-1}(U))$, i.e. $p \in \partial U$. This contradicts the assumption that

$$0 \notin (I - K)(\partial U).$$

Next, we show that the degree defined in (9.58) is independent of the radius R and the retraction r.

Indeed, let $R_1 < R_2$ such that $U \subset int(B_{R_1}(0))$, then all fixed points of K in U must be in $B_{R_1}(0)$, and hence

$$deg(I - K \circ r, r^{-1}(U) \cap B_{R_2}(0), 0)$$
$$= deg(I - K \circ r, r^{-1}(U) \cap B_{R_1}(0), 0).$$

Let r_1, r_2 any two retractions from X to C, then

$$deg(I - K \circ r_i, r_i^{-1}(U) \cap B_R(0), 0)$$
$$= deg(I - K \circ r_i, r_1^{-1}(U) \cap r_2^{-1}(U) \cap B_R(0), 0), \quad i = 1, 2.$$

Let

$$F(x, t) := I - (tK \circ r_1 + (1 - t)K \circ r_2), \quad \forall t \in [0, 1].$$

The homotopy invariance implies that the value of $deg_C(I - K, U, 0)$ does not rely on the choice of r_i, $i = 1, 2$.

In particular, if X is a real ordered Banach space (OBS), i.e. X is a real Banach space with a closed positive cone P, which is a closed convex subset, then $deg_P(I - K, U, 0)$ is well defined.

Now we give a few examples for degree computations for special boundary conditions in an ordered Banach space.

Lemma 9.4.1 *Suppose that (X, P) is an OBS and that $U \subset P$ is bounded and open. Assume that $K : \bar{U} \mapsto P$ is compact, satisfying*

$$\exists y \in P \backslash \{0\} \text{ such that } x - Kx \neq ty, \ \forall t \geq 0 \text{ and } \forall x \in \partial U,$$

then

$$deg_P(I - K, U, 0) = 0.$$

Proof. From the homotopy invariance, we know that

$$deg(I - K - ty, U, 0) = constant, \quad \forall t \geq 0.$$

Since K is compact, there exists $C > 0$ such that

$$\|x - Kx\| \leq C, \quad \forall x \in U.$$

Let $t > \frac{C}{\|y\|}$, then for any $x \in U$, $x - Kx \neq ty$. Therefore $deg_P(I - K, U, 0) = 0$.

Theorem 9.4.1 *Suppose that (X, P) is an OBS, $U \subset P$ is bounded open and contains 0. Assume that there exists $\rho > 0$ such that $B_\rho(0) \cap P \subset U$ and that $K : \bar{U} \mapsto P$ is compact and satisfies:*

1. *for any $x \in P$ with $|x| = \rho$, and $\lambda \in [0, 1)$, $x \neq \lambda Kx$,*
2. *there exists some $y \in P \backslash \{0\}$, such that $x - Kx \neq ty$ for any $t \geq 0$ and $x \in \partial U$.*

Then K possesses a fixed point on $\bar{U}_\rho$, where $U_\rho = U \backslash B_\rho(0)$.

Proof. If K possesses a fixed point on $\partial B_\rho(0)$, then we are done. Next, we suppose that K has no fixed point on $\partial B_\rho(0)$, then

$$deg(I - K, U, 0) = deg(I - K, U_\rho, 0) + deg(I - K, B_\rho(0) \cap P, 0).$$

By Lemma 9.4.1 and assumption 2, the left-hand side equals 0. From assumption 1 and the homotopy invariance, we know that the second term on the right-hand side is 1. Hence $deg(I - K, U_\rho, 0) = -1$. Therefore, there must be an $x \in U_\rho$, such that

$$(I - K)(x) = 0.$$

To illustrate the idea, we consider the following simple example.

Example 9.4.1 *Let $X = (-\infty, \infty)$, $P = [0, \infty)$, $U = [0, 1)$ and $r : X \mapsto P$ such that*

$$r(x) = \begin{cases} x, & x \in P \\ -x, & x \notin P. \end{cases}$$

Assume that $x - Kx \neq 0$, for any $x \in \partial U$, i.e., $1 - K1 \neq 0$.
  By definition,

$$deg(I - K, [0, 1), 0) = deg(I - K \circ r, (-1, 1), 0).$$

In order to meet the first condition in Theorem 9.4.1, we should have

$$(I - K)(\rho) > 0,$$

since for $\lambda = 0$, we have $I(\rho) > 0$.

In order to satisfy the second condition, that is, for any $t \geq 0$, $x - Kx - tx_o \neq 0$ for some $x_o > 0$, we must have

$$x - Kx < 0, x \in \partial U, \text{ i.e., } (I - K)(1) < 0.$$

Then by the intermediate value theorem, there exists an $x \in [\rho, 1)$, such that

$$(I - K)(x) = 0.$$

This coincides with the result obtained via the degree theory.

A rough sketch of the graph of $(I - K)(x)$ can be seen in Fig. 9.4.

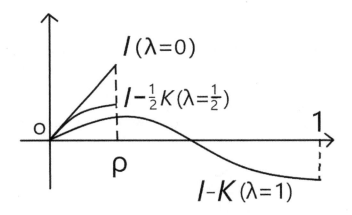

Fig. 9.4

The figure shows a typical example of such maps on xy plane. For $\lambda = 0$, the graph of $y = (I - \lambda K)(x)$ is the line $y = x$, which is above x-axis. Condition 1 in Theorem 9.4.1 guarantees that for all values of λ, the points $(\rho, (I - \lambda K)(\rho))$ are above the x-axis, in particular, this is the case for the point $(\rho, (I - K)(\rho))$. While condition 2 in the theorem implies the point $(1, (I - K)(1))$ must lay below the x-axis. Hence the graph of $y = (I - K)(x)$ must intersect with the x-axis, and

$$deg(I - K, (\rho, 1), 0) = -1.$$

9.4.2 Existence of Positive Solutions

One significant application of Theorem 9.4.1 is the study of the existence of positive solutions for nonlinear equations. Here we consider Dirichlet problems involving nonlocal elliptic operators.

For the Fractional Laplacian

We first consider a semi-linear equation involving the fractional Laplacian.

Let $\Omega \subset \mathbb{R}^n$ be a bounded domain with smooth boundary. We study the following equation:

$$\begin{cases} (-\triangle)^{\alpha/2} u = u^p, \ x \in \Omega, \\ u = 0, \qquad\qquad x \notin \Omega. \end{cases} \tag{9.60}$$

In the previous chapter, we establish a priori estimates for positive solutions to a family of such equations, and as a consequence, we have

Proposition 9.4.1 *Let* $1 < p < \frac{n+\alpha}{n-\alpha}$ *and* A *be any constant. Then all positive solutions of*

$$\begin{cases} (-\triangle)^{\alpha/2} u \leq u^p + A, \ x \in \Omega, \\ u = 0, \qquad\qquad x \notin \Omega \end{cases} \tag{9.61}$$

are uniformly bounded, i.e. there exists some constant M, *independent of* u, *such that*

$$\|u\|_{C_0^0(\Omega)} \leq M. \tag{9.62}$$

The proof is similar to that of Theorem 9.1.1. Hence we omit it. Combining this a priori estimate and the degree theory, we are able to prove the existence of positive solutions.

Let

$$T(u) = \int_\Omega G(x,y) u^p(y) dy,$$

where $G(x,y)$ is the Green's function for $(-\triangle)^{\alpha/2}$ on Ω.

Lemma 9.4.2 *The operator* T *is compact.*

Proof. We first show that T is continuous. Indeed,

$$|T(u_n) - T(u)| = |\int_\Omega G(x,y)(u_n^p(y) - u^p(y)) dy|$$

$$\leq \int_\Omega G(x,y)|u_n^p(y) - u^p(y)| dy$$

$$\leq \int_\Omega G(x,y) dy \|u_n^p(y) - u^p(y)\|_{C^0(\Omega)}$$

$$\leq C \|u_n^p(y) - u^p(y)\|_{C^0(\Omega)}.$$

Next we show T is compact. Let $\{u_n\}$ be bounded in $C_0^0(\Omega)$, then $\{T(u_n)\}$ is bounded in $C_0^0(\Omega)$. Let $v_n = T(u_n)$, then $(-\triangle)^{\alpha/2} v_n = u_n^p$. Hence, $\{v_n\}$ is bounded in $W_0^{\alpha,q}(\Omega)$. Because $W^{\alpha,q}$ is compactly embedded into C^0, we derive that $\{v_n\}$ has a convergent subsequence in C^0. This completes the proof.

Theorem 9.4.2 *Assume* $1 < p < \frac{n+\alpha}{n-\alpha}$, *then problem (9.60) possesses a positive solution.*

Proof. Let $X = C_0^0(\Omega) = \{u \in C^0(\Omega) \mid u \equiv 0 \text{ in } \Omega^c\}$ and $P = \{u \in X \mid u \geq 0\}$ in Theorem 9.4.1.

Applying $(-\triangle)^{-\alpha/2}$ to both sides of the first equation in (9.60), we have

$$u = (-\triangle)^{-\alpha/2}(u^p) = T(u).$$

(1) First we show that there exists a $\rho > 0$ small such that for any $u \in \partial B_\rho(0)$ and $0 \leq \lambda < 1$,

$$u - \lambda T(u) \neq 0.$$

Since $\|u\|_{C_0^0(\Omega)} \leq C$, it holds that

$$|Tu(x)| = |\int_\Omega G(x,y)u^p(y)dy|$$
$$\leq \int_\Omega G(x,y)|u|^p dy$$
$$\leq \|u\|_{C_0^0(\Omega)}^p \int_\Omega G(x,y)dy$$
$$\leq C\|u\|_{C_0^0(\Omega)}^p.$$

Hence

$$\|T(u)\|_{C_0^0(\Omega)} \leq C\|u\|_{C_0^0(\Omega)}^p. \tag{9.63}$$

On the other hand, since $p > 1$, for $\rho = \|u\|_{C_0^0(\Omega)}$ sufficiently small, it holds that

$$C\|u\|_{C_0^0(\Omega)}^p < \|u\|_{C_0^0(\Omega)}. \tag{9.64}$$

Combining (9.63) and (9.64), it shows that there exists $\rho > 0$ small such that

$$u - \lambda T(u) \neq 0, \ u \in \partial B_\rho(0) \cap P, \ 0 \leq \lambda < 1.$$

(2) We claim that there exists a $\psi \in P$ and $\psi \neq 0$ such that

$$u - T(u) \neq t\psi, \ t \geq 0, \ u \in \partial U,$$

where U will be chosen as $B_M(0) \cap P$ with sufficiently large M.

Let λ_1 be the first eigenvalue of the corresponding Dirichlet problem, i.e.

$$\lambda_1 = \sup\{\lambda \in R \mid (-\triangle)^{\alpha/2}u \geq \lambda u, u > 0 \text{ in } \Omega, u = 0 \text{ in } \Omega^c\}.$$

Let ψ be the unique positive solution of

$$\begin{cases} (-\triangle)^{\alpha/2}\psi = 1, \ x \in \Omega, \\ \psi = 0, \qquad\quad x \notin \Omega. \end{cases}$$

Then if $u - T(u) = t\psi$, we have

$$\begin{cases} (-\triangle)^{\alpha/2}u = u^p + t, & x \in \Omega, \\ u = 0, & x \notin \Omega. \end{cases} \tag{9.65}$$

Choose a constant $C_1 > \lambda_1$. Since $u > 0$ and $p > 1$, there exists another constant $C_2 > 0$, such that

$$u^p(x) \geq C_1 u(x) - C_2.$$

Indeed, if $u^{p-1}(x) \geq C_1$, then

$$u^p(x) \geq C_1 u(x) \geq C_1 u(x) - C_2;$$

and if $u^{p-1}(x) \leq C_1$, by choosing $C_2 \geq C_1^{\frac{p}{p-1}}$, we have

$$C_1 u(x) - C_2 \leq 0,$$

and hence

$$u^p(x) \geq C_1 u(x) - C_2.$$

Therefore, for such a choice of C_1 and C_2, we have

$$(-\triangle)^{\alpha/2}u = u^p + t \geq C_1 u - C_2 + t.$$

If $t - C_2 \geq 0$, we have $(-\triangle)^{\alpha/2}u \geq C_1 u$. This contradicts the definition of λ_1. Hence we must have $t - C_2 < 0$, then

$$(-\triangle)^{\alpha/2}u \leq u^p + C_2.$$

It follows from the Proposition 9.4.1 that the positive solutions of (9.65) are a priori bounded, i.e., there exists $M > 0$ such that $\|u\|_{C^0(\Omega)} < M$. Let

$$U = B_M(0) \cap P,$$

then it is obvious that $u - Tu = t\psi$ has no solution on ∂U for all $t > 0$. This verifies the claim.

Combining (1), (2), and Theorem 9.4.1, we prove the theorem.   □

For General Uniformly Elliptic Nonlocal Operators

Let Ω be a bounded domain in $\mathbb{R}^n$ with C^2 boundary. We study the following equation:

$$\begin{cases} (-\triangle)_a^{\alpha/2}u = u^p, & x \in \Omega, \\ u = 0, & x \notin \Omega. \end{cases} \tag{9.66}$$

In the previous chapter, we obtain a priori estimates for positive solutions to a family of such equations, and as a consequence, we have

Proposition 9.4.2 *Let* $1 < p < \frac{n+\alpha}{n-\alpha}$ *and* A *be any constant. Then all positive solutions of*

$$
\begin{cases}
(-\triangle)_a^{\alpha/2} u \leq u^p + A, & x \in \Omega, \\
u = 0, & x \notin \Omega,
\end{cases}
\tag{9.67}
$$

are uniformly bounded, i.e., there exists some constant M, *independent of* u, *such that*

$$
\|u\|_{C_0^0(\Omega)} \leq M.
\tag{9.68}
$$

Unlike $(-\triangle)^{\alpha/2}$, whose corresponding inverse operator can sometimes be explicitly expressed as an integral via the Green's functions, little is known about such expression for the more general operator $(-\triangle)_a^{\alpha/2}$. Fortunately, we can validate its existence with the following lemma.

Lemma 9.4.3 *For a bounded domain* Ω, *assume that* $f \in C(\Omega)$ *and* u *is a solution of*

$$
\begin{cases}
(-\triangle)_a^{\alpha/2} u(x) = f(x), & x \in \Omega, \\
u \equiv 0, & x \in \mathbb{R}^n \backslash \Omega.
\end{cases}
\tag{9.69}
$$

Then

$$
\|u\|_{C_0^0(\Omega)} \leq C \|f\|_{C^0(\Omega)}.
$$

Proof. Let ψ be the unique positive solution of

$$
\begin{cases}
(-\triangle)_a^{\alpha/2} \psi = 1, & x \in \Omega, \\
\psi = 0, & x \notin \Omega.
\end{cases}
$$

Let

$$
v = \|f\|_{C^0(\Omega)} \psi.
$$

Then

$$
\begin{cases}
(-\triangle)_a^{\alpha/2} v = \|f\|_{C^0(\Omega)}, & x \in \Omega, \\
v = 0, & x \notin \Omega,
\end{cases}
$$

and

$$
\begin{cases}
(-\triangle)_a^{\alpha/2}(v - u) = \|f\|_{C^0(\Omega)} - f \geq 0, & x \in \Omega, \\
v - u = 0, & x \notin \Omega,
\end{cases}
$$

By the maximum principle, we have

$$
v \geq u, \quad x \in \Omega.
\tag{9.70}
$$

On the other hand,

$$
\begin{cases}
(-\triangle)_a^{\alpha/2}(v + u) = \|f\|_{C^0(\Omega)} + f \geq 0, & x \in \Omega, \\
v + u = 0, & x \notin \Omega.
\end{cases}
$$

It follows from the maximum principle that

$$u \geq -v, \quad x \in \Omega. \tag{9.71}$$

Hence

$$|u| \leq v \leq C\|f\|_{C^0(\Omega)}, \quad x \in \Omega,$$

or

$$\|u\|_{C^0(\Omega)} \leq C\|f\|_{C^0(\Omega)}. \qquad \square$$

Let $X = C_0^0(\Omega)$ and $P = \{u \in X \mid u \geq 0\}$ in Theorem 9.4.1. This lemma guarantees that for $f = v^p \in P$, we can define an operator $T : P \to P$ by means of $u = T(v)$. Thus the nonnegative solutions of (9.66) are the fixed points for T.

Lemma 9.4.4 *The operator $T : P \to P$ is compact.*

Proof. Let $\{u_n\} \subset P$ be the solutions for (9.66). We start with the continuity. Suppose $u_n \to u$ in X. Then

$$(-\triangle)_a^{\alpha/2} u_n = u_n^p, \text{ and } u_n = T(u_n),$$

$$(-\triangle)_a^{\alpha/2} u = u^p, \text{ and } u = T(u).$$

Due to the linearity of $(-\triangle)_a^{\alpha/2}$, it holds that

$$(-\triangle)_a^{\alpha/2}(u_n - u) = u_n^p - u^p. \tag{9.72}$$

Applying Lemma 9.4.3 to (9.72), we have

$$\|u_n - u\|_{C^0(\Omega)} \leq C\|u_n^p - u^p\|_{C^0(\Omega)}.$$

Recall that

$$u_n - u = T(u_n) - T(u),$$

thus

$$\|T(u_n) - T(u)\|_{C^0(\Omega)} \leq C\|u_n^p - u^p\|_{C^0(\Omega)} \to 0, \quad n \to \infty.$$

This proves the continuity. The proof of the compactness of T is similar to that in Lemma 9.4.2, we omit the details. $\qquad \square$

Now we are ready to claim that the existence of solutions.

Theorem 9.4.3 *Assume $1 < p < \frac{n+\alpha}{n-\alpha}$, then problem (9.66) possesses a positive solution.*

Proof. Obviously, the nonnegative solutions of (9.66) in X are the fixed points of T.

In order to obtain the desired existence through Theorem 9.4.1, we only need to check the conditions.

(a) Suppose we have $u = \lambda T(u)$ for some $\lambda \in [0, 1)$ and $u \in P$. This is equivalent to

$$\begin{cases} (-\triangle)_a^{\alpha/2}u = \lambda u^p, \ x \in \Omega, \\ u = 0, \qquad\qquad x \notin \Omega. \end{cases}$$

By Lemma 9.4.3, we have

$$\|u\|_{C^0(\Omega)} \le C\|\lambda u^p\|_{C^0(\Omega)} \le C\|\lambda u\|_{C^0(\Omega)}^p.$$

Since $p > 1$, there must exist some constant C such that

$$\|u\|_{C^0(\Omega)} \ge C > 0.$$

This implies that $u = \lambda T(u)$ has no positive solution on $\|u\|_{C^0(\Omega)} = \rho$ for ρ sufficiently small.

(b) We claim that there exists a $\psi \in P$ and $\psi \neq 0$ such that

$$u - T(u) \neq t\psi, \ \forall t \ge 0, \ \forall u \in \partial U,$$

where U will be chosen as $B_M(0) \cap P$ with sufficiently large M.

Let λ_1 be the first eigenvalue of the corresponding Dirichlet problem, i.e.

$$\lambda_1 = \sup\{\lambda \in R \mid (-\triangle)_a^{\alpha/2}u \ge \lambda u, u > 0 \text{ in } \Omega, u = 0 \text{ in } \Omega^c\}.$$

Let ψ be the unique positive solution of

$$\begin{cases} (-\triangle)_a^{\alpha/2}\psi = 1, \ x \in \Omega, \\ \psi = 0, \qquad\qquad x \notin \Omega. \end{cases}$$

Then $u - T(u) = t\psi$ is equivalent to

$$\begin{cases} (-\triangle)_a^{\alpha/2}u = u^p + t, \ x \in \Omega, \\ u = 0, \qquad\qquad\quad x \notin \Omega. \end{cases} \tag{9.73}$$

We claim that for $t \ge C_1^{\frac{p}{p-1}}$, Eq. (9.73) does not have any positive solution. Choose a constant $C_1 > \lambda_1$. Indeed, if $u^{p-1}(x) \ge C_1$, then

$$u^p(x) \ge C_1 u(x). \tag{9.74}$$

And if $u^{p-1}(x) \le C_1$, then for $t \ge C_1^{\frac{p}{p-1}}$, we have

$$u^p + t > t \ge C_1 u. \tag{9.75}$$

In any case, we have $u^p + t \ge C_1 u$ for $t > C_1^{\frac{p}{p-1}}$. This contradicts the definition of λ_1. Therefore, a necessary condition for (9.73) to have a positive solution is that

$$t \le C_1^{\frac{p}{p-1}}.$$

In this case, $u^p + t$ is bounded. By Proposition 9.4.2, we deduce that positive solutions of (9.73) must be bounded. Let

$$U = B_M(0) \cap P,$$

then it is obvious that $u - Tu = t\psi$ has no solution on ∂U for all $t > 0$ when M is sufficiently large.

Combining (a), (b), and Theorem 9.4.1, it completes the proof. $\quad\square$

10

Variational Methods and Pohozaev Identities

In this chapter, we continue to consider the Dirichlet problem for fractional order semilinear equations on bounded domains in $\mathbb{R}^n$:

$$\begin{cases} (-\triangle)^s u = f(x,u), \ x \in \Omega, \\ u(x) = 0, \qquad\quad x \in \mathbb{R}^n \backslash \Omega. \end{cases} \tag{10.1}$$

We employ a variational method to show the existence of a weak solution in the subcritical case and use a Pohozaev identity to derive non-existence of solutions in the critical and supercritical cases when Ω is a star-shaped domain.

Let $H_0^s(\mathbb{R}^n)$ be the usual fractional Sobolev space endowed with the norm

$$\|u\|_{H_0^s(\mathbb{R}^n)} = \left(\int_{\mathbb{R}^n} |\xi|^{2s} |\hat{u}(\xi)|^2 d\xi \right)^{\frac{1}{2}},$$

where $\hat{u}$ is the Fourier transform of u. Denote by $H_0^s(\Omega)$ the completion of $C_0^\infty(\Omega)$ under this norm.

Definition 10.0.1 *We say that $u \in H_0^s(\Omega)$ is a weak solution of (10.1) if*

$$\int_{\mathbb{R}^n} |\xi|^{2s} \hat{u}(\xi) \overline{\hat{v}(\xi)} d\xi = \int_\Omega f(x, u(x)) v(x) dx, \ \forall v \in H_0^s(\Omega).$$

10.1 The Variational Method

10.1.1 The Mountain Pass Lemma

To seek a weak solution of (10.1), we consider the functional

$$I(u) = \frac{1}{2} \|u\|_{H_0^s}^2 - \int_\Omega F(x, u) dx$$

in the Hilbert space $H_0^s(\Omega)$. It is easy to show that a critical point of $I(u)$ in this space is a weak solution of (10.1). Since the functional is not bounded from below, we look for saddle points by using the well-known *mountain pass lemma*. To this end, we require that the functional satisfies some compactness property as described in the following

Definition 10.1.1 (Palais-Smale Condition) *A functional I defined on a Banach space X satisfies the Palais-Smale condition (abbreviated as (PS)), if every sequence $\{u_k\}_{k=1}^\infty \subset X$ possesses a convergent subsequence on condition that*

1. $\{I(u_k)\}_{k=1}^\infty$ is bounded, and
2. $I'(u_k) \to 0$ as $k \to \infty$.

Theorem 10.1.1 (Mountain Pass Theorem, see [R]) *Let X be a real Banach space and $I \in C^1(X, \mathbb{R})$. Assume that I satisfies (PS), $I(0) = 0$, and*

i. there exist positive constants r and C such that $I(u) \geq C$ if $\|u\|_X = r$,
ii. there exists an $e \in X \setminus \overline{B_r(0)}$ such that $I(e) \leq 0$.

Then I has a critical value c characterized by

$$c = \inf_{g \in \Gamma} \max_{0 \leq t \leq 1} I(g(t)),$$

where

$$\Gamma = \{g \in C([0,1], X) \mid g(0) = 0, g(1) = e\}.$$

The proof of *mountain pass theorem* is based on the following

Theorem 10.1.2 (Deformation Theorem, see [R]) *Let X be a real Banach space. Assume that $I \in C^1(X, \mathbb{R})$ and satisfies the Palais-Smale condition. Suppose that c is not a critical value of I. Then for each sufficiently small $\varepsilon > 0$, there exists a constant $0 < \delta < \varepsilon$ and a continuous mapping $\eta(t, u)$ from $[0, 1] \times X$ to X, such that*

i. $\eta(0, u) = u$, $\forall u \in X$,
ii. $\eta(1, u) = u$, $\forall u \notin I^{-1}[c - \varepsilon, c + \varepsilon]$,
iii. $I(\eta(t, u)) \leq I(u)$, $\forall u \in X$, $t \in [0, 1]$,
iv. $\eta(1, I_{c+\delta}) \subset I_{c-\delta}$.

Here

$$I_c = \{u \in X \mid I(u) \leq c\}.$$

10.1.2 Existence of Weak Solutions

Now we apply the mountain pass theorem to seek weak solutions of the following problem

$$\begin{cases} (-\triangle)^s u = f(x,u),\ x \in \Omega, \\ u(x) = 0, \qquad\qquad x \notin \Omega, \end{cases} \tag{10.2}$$

where $\Omega \subset \mathbb{R}^n$ is a bounded domain, $0 < s < 1$ and $n \geq 2$.
    Assume that $f(x,t)$ satisfies

(f_1) $f \in C(\bar{\Omega} \times R, R)$, $f(x,t) \geq 0$ if $t > 0$ and $f(x,t) = 0$ if $t \leq 0$.

(f_2) $\lim\limits_{t \to 0} \frac{f(x,t)}{t} = 0$ uniformly in $x \in \bar{\Omega}$.

(f_3) There exists some $1 < p < \frac{n+2s}{n-2s}$, so that

$$f(x,t) \leq C(1 + |t|)^p \text{ for all } t \in \mathbb{R} \text{ and } x \in \bar{\Omega}.$$

(f_4) There exists a $\theta > 2$ such that for all $t > 0$ and a.e. $x \in \bar{\Omega}$,

$$0 < \theta F(x,t) \leq t f(x,t), \ \ F(x,t) = \int_0^t f(x,\tau)d\tau.$$

Lemma 10.1.1 *Let* $2 \leq q \leq 2^*_s := \frac{2n}{n-2s}$, *then we have*

$$\|u\|_{L^q(\mathbb{R}^n)} \leq C\|u\|_{H^s(\mathbb{R}^n)}, \ \ \forall u \in H^s(\mathbb{R}^n).$$

If $2 \leq q < 2^*_s$, $\Omega \subset \mathbb{R}^n$ *is a bounded domain, then every bounded sequence* $\{u_k\} \subset H^s_0(\Omega)$ *has a converging subsequence in* $L^q(\Omega)$.

Definition 10.1.2 *We say that* $u \in H^s_0(\Omega)$ *is a weak solution of (10.2) if*

$$\int_{\mathbb{R}^n} |\xi|^{2s}\hat{u}\hat{v}d\xi = \int_\Omega fvdx, \ \forall v \in H^s_0(\Omega).$$

Using (f_1) and (f_2) and Lemma 10.1.1, it is easy to see $I(u) \in C^1(H^s_0(\Omega), R)$ and

$$\langle I'(u), v \rangle = (u,v)_{H^s_0(\Omega)} - \int_\Omega f(x,u)vdx \ \ \forall v \in H^s_0(\Omega).$$

Moreover, the critical points of I are weak solutions to problem (10.2).

Lemma 10.1.2 *Assume* $f : \bar{\Omega} \times \mathbb{R} \to \mathbb{R}$ *is a function satisfying conditions* (f_2)-(f_4). *Then for any* $\varepsilon > 0$ *there exists a* $\delta = \delta(\varepsilon)$ *such that for any* $x \in \Omega$ *and* $t \in \mathbb{R}$

$$|f(x,t)| \leq 2\varepsilon|t| + (p+1)\delta(\varepsilon)|t|^p. \tag{10.3}$$

As a result, we have

$$|F(x,t)| \leq \varepsilon|t|^2 + \delta(\varepsilon)|t|^{p+1}, \tag{10.4}$$

and for some constants C_1,

$$F(x,t) \geq C_1|t|^\theta, \tag{10.5}$$

where F *is defined as in* (f_3).

Proof. From (f_2), we know for any $\varepsilon > 0$ there exists a $C_\varepsilon > 0$ such that when $|t| < C_\varepsilon$,

$$|f(x,t)| \leq 2\varepsilon|t|. \tag{10.6}$$

From (f_3) we have when $|t| \geq C_\varepsilon$, then $1 < \frac{|t|}{C_\varepsilon}$,

$$\begin{aligned}|f(x,t)| &\leq C(1+|t|)^p \\ &\leq C(\frac{|t|}{C_\varepsilon} + |t|)^p \\ &= C(1+\frac{1}{C_\varepsilon})^p|t|^p \\ &:= (p+1)\delta(\varepsilon)|t|^p.\end{aligned} \tag{10.7}$$

Then (10.3) follows from (10.6) and (10.7), and (10.4) is a direct consequence of (10.3). From (f_4), we have

$$\frac{f(x,t)}{F(x,t)} \geq \frac{\theta}{t}.$$

Integrating both sides, it yields

$$\ln F(x,t) \geq \theta \ln t + C.$$

Thus it's trivial that

$$F(x,t) \geq C_1|t|^\theta.$$

Theorem 10.1.3 *If f satisfies (f_1)-(f_4), then problem (10.2) has a weak solution.*

Proof. By (10.4),

$$\begin{aligned}I(u) &= \frac{1}{2}\|u\|_{H_0^s}^2 - \int_\Omega F(x,u)dx \\ &\geq \frac{1}{2}\|u\|_{H_0^s}^2 - \int_\Omega (\varepsilon|u|^2 dx + \delta(\varepsilon)|u|^{p+1})dx \\ &= \frac{1}{2}\|u\|_{H_0^s}^2 - \varepsilon\|u\|_{L^2}^2 - \delta(\varepsilon)\|u\|_{L^{p+1}}^{p+1} \\ &\geq \frac{1}{2}\|u\|_{H_0^s}^2 - C_1\varepsilon\|u\|_{H_0^s}^2 - C_2\delta(\varepsilon)\|u\|_{H_0^s}^{p+1}.\end{aligned}$$

Let $\|u\|_{H_0^s(\Omega)} = \rho$ be sufficiently small, we have $I|_{\partial B_\rho(0)}(u) \geq C > 0$.

For any $t > 0$, from (10.5),

$$\begin{aligned}I(tu) &= \frac{1}{2}t^2\|u\|_{H_0^s}^2 - \int_\Omega F(x,tu)dx \\ &\leq \frac{1}{2}t^2\|u\|_{H_0^s}^2 - C_1\|tu\|_{L^\theta}^\theta \\ &\leq \frac{1}{2}t^2\|u\|_{H_0^s}^2 - C_3 t^\theta\|u\|_{L^\theta}^\theta \\ &\to -\infty \quad t \to \infty.\end{aligned}$$

Notice that $\theta > 2$, there exists t_0 large enough, such that $I(t_0 u) < 0$. Let $e = t_0 u$, we have $I(e) < 0$.

Next we verify I satisfies the (PS) condition. Let $\{u_k\}$ be a (PS) sequence, such that

$$I(u_k) \text{ is bounded, } I'(u_k) \to 0, \quad \text{as } k \to \infty. \qquad (10.8)$$

First we prove the boundedness of the (PS) sequence. We have

$$M + \|u_k\|_{H_0^s} \geq I(u_k) - \frac{1}{\theta}\langle I'(u_k), u_k \rangle$$

$$= (\frac{1}{2} - \frac{1}{\theta})\|u_k\|_{H_0^s}^2 + \int_\Omega [\theta^{-1} f(x, u_k) u_k - F(x, u_k)] dx$$

$$\geq (\frac{1}{2} - \frac{1}{\theta})\|u_k\|_{H_0^s}^2.$$

Notice that $\theta > 2$, hence $\frac{1}{2} - \frac{1}{\theta} > 0$, and therefore $\{u_k\}$ is bounded. Consequently, it follows from Lemma 10.1.1,

$$u_k \rightharpoonup u, \text{ in } H_0^s(\Omega), \qquad (10.9)$$

$$u_k \to u, \text{ in } L^q(\Omega), \quad 2 \leq q < \frac{2n}{n - 2s}, \qquad (10.10)$$

$$u_k \to u, \text{ a.e. in } \mathbb{R}^n. \qquad (10.11)$$

Then, by (10.8) and (10.9)

$$|\langle I'(u_k) - I'(u), u_k - u \rangle|$$

$$\leq |\langle I'(u_k), u_k - u \rangle| + |\langle I'(u), u_k - u \rangle|$$

$$\leq \|I'(u_k)\|_{H_0^s}\|u_k - u\|_{H_0^s} + |\langle I'(u), u_k - u \rangle|$$

$$\to 0, \ k \to \infty. \qquad (10.12)$$

We know

$$o(1) = \langle I'(u_k) - I'(u), u_k - u \rangle$$

$$= \|u_k - u\|_{H_0^s}^2 - \int_\Omega (u_k - u)(f(x, u_k) - f(x, u)) dx.$$

Hence

$$\|u_k - u\|_{H_0^s}^2 = \langle I'(u_k) - I'(u), u_k - u \rangle + \int_\Omega (u_k - u)(f(x, u_k) - f(x, u)) dx$$

$$\leq |\langle I'(u_k) - I'(u), u_k - u \rangle| + \int_\Omega |u_k - u||f(x, u_k) - f(x, u)| dx. \qquad (10.13)$$

Using the Hölder inequality and (f_3), we arrive at

$$\int_\Omega |u_k - u||f(x, u_k) - f(x, u)| dx$$

$$\leq \|u_k - u\|_{L^\nu}\|f(x, u_k) - f(x, u)\|_{L^{\frac{\nu}{\nu-1}}}$$

$$\leq \|u_k - u\|_{L^\nu}(\|f(x, u_k)\|_{L^{\frac{\nu}{\nu-1}}} + \|f(x, u)\|_{L^{\frac{\nu}{\nu-1}}})$$

$$\leq \|u_k - u\|_{L^\nu}(\|C(1 + |u_k|)^p\|_{L^{\frac{\nu}{\nu-1}}} + \|C(1 + |u|)^p\|_{L^{\frac{\nu}{\nu-1}}})$$

$$\leq \|u_k - u\|_{L^\nu}(C + \|u_k\|_{L^{\frac{p\nu}{\nu-1}}}^p + \|u\|_{L^{\frac{p\nu}{\nu-1}}}^p). \qquad (10.14)$$

For $\varepsilon > 0$ sufficiently small, let $p = \frac{n+2s-\varepsilon}{n-2s}$. We want $\frac{p\nu}{\nu-1} \le \frac{2n}{n-2s}$ to satisfy Lemma 10.1.1, then $\nu \ge \frac{2n}{n-2s+\varepsilon}$.

Let $p = \frac{n+2s-\varepsilon}{n-2s}$ and $\frac{2n}{n-2s+\varepsilon} \le \nu < \frac{2n}{n-2s}$. Then by (10.10) and Lemma 10.1.1, it holds

$$\|u_k - u\|_{L^\nu} \to 0 \text{ and } \|u_k\|_{L^{\frac{p\nu}{\nu-1}}}^p \le C\|u_k\|_{H_0^s}.$$

Combining this with (10.12), (10.13), and (10.14), we derive that

$$\|u_k - u\|_{H_0^s}^2 \to 0.$$

This proves

$$u_k \to u \text{ in } H_0^s.$$

We verify I satisfies (PS) condition. Therefore by Mountain Pass Theorem, I has a critical point, that is problem (10.2) has a weak solution.

10.2 The Pohozaev Identity

The well-known Pohozaev identity is a powerful tool in proving the non-existence of solutions for semi-linear second order equations. For equations involving the fractional Laplacian, one can derive similar identities and use them to prove non-existence of solutions in star-shaped domains.

10.2.1 The Pohozaev Identity for Single Equations

Let $s \in (0,1)$. Consider the fractional elliptic problem

$$\begin{cases} (-\triangle)^s u(x) = f(u), & x \in \Omega, \\ u \ge 0, & x \in \mathbb{R}^n \backslash \Omega \end{cases} \tag{10.15}$$

in a bounded domain $\Omega \subset \mathbb{R}^n$.

Let

$$H^s(\mathbb{R}^n) =: \{u : \mathbb{R}^n \to \mathbb{R} \mid \int_{\mathbb{R}^n} |\xi|^{2s} |\hat{u}(\xi)|^2 d\xi < +\infty\},$$

and

$$H_0^s(\Omega) =: \{u \in H^s(\mathbb{R}^n) \mid u(x) = 0, x \in \mathbb{R}^n \backslash \Omega\}.$$

We say that $u \in H_0^s(\Omega)$ is a weak solution of

$$\begin{cases} (-\triangle)^s u = f(x,u), & x \in \Omega, \\ u = 0, & x \in \mathbb{R}^n \backslash \Omega, \end{cases}$$

if for all $\varphi \in C_0^\infty(\Omega)$,

$$\int_{\mathbb{R}^n} (-\triangle)^{\frac{s}{2}} u \, (-\triangle)^{\frac{s}{2}} \varphi \, dx = \int_\Omega f(x,u) \, \varphi \, dx,$$

where the left hand side is defined via Fourier transform

$$\int_{\mathbb{R}^n} (-\triangle)^{\frac{s}{2}} u \, (-\triangle)^{\frac{s}{2}} \varphi \, dx = \int_{\mathbb{R}^n} |\xi|^{2s} \hat{u}(\xi) \hat{\varphi}(\xi) d\xi.$$

In [RS], the authors established the following Pohozaev identity.

Theorem 10.2.1 *Let Ω be a bounded and $C^{1,1}$ domain, f be a locally Lipschitz function, u be a bounded solution of (10.15), and*

$$\delta(x) = dist(x, \partial\Omega).$$

Then,

$$u/\delta^s|_\Omega \in C^\alpha(\bar{\Omega}) \text{ for some } \alpha \in (0,1),$$

meaning that $u/\delta^s|_\Omega$ has a continuous extension to $\bar{\Omega}$ which is $C^\alpha(\bar{\Omega})$, and the following identity holds

$$(2s-n) \int_\Omega uf(u)dx + 2n \int_\Omega F(u)dx = \Gamma^2(1+s) \int_{\partial\Omega} \left(\frac{u}{\delta^s}\right)^2 (x \cdot \nu)d\sigma, \quad (10.16)$$

where $F(t) = \int_0^t f(s)ds$, ν is the unit outward normal vector to $\partial\Omega$ at x, and Γ is the Gamma function.

From the above identity, they obtained non-existence of solutions if Ω is a star-shaped domain.

Corollary 10.2.1 *Let Ω be a bounded, $C^{1,1}$, and star-shaped domain, and f be a locally Lipschitz function. If*

$$\frac{n-2s}{2n} uf(u) \geq \int_0^u f(t)dt \text{ for all } u \in R, \quad (10.17)$$

then problem (10.15) admits no positive bounded solution. Moreover, if the inequality in (10.17) is strict, then problem (10.15) admits no nontrivial bounded solution.

In particular, for the pure power nonlinearity, the non-existence result reads as follows.

Corollary 10.2.2 *Let Ω be a bounded, $C^{1,1}$, and star-shaped domain. If $p \geq \frac{n+2s}{n-2s}$, then problem*

$$\begin{cases} (-\triangle)^s u(x) = |u|^{p-1}u, & x \in \Omega, \\ u \geq 0, & x \in \mathbb{R}^n \backslash \Omega, \end{cases} \quad (10.18)$$

admits no positive bounded solution. Moreover, if $p > \frac{n+2s}{n-2s}$, then (10.18) admits no nontrivial bounded solution.

Identity (10.16) was derived from the following more general proposition in which one does not assume that u is a solution of the problem.

Proposition 10.2.1 *([RS]) Let Ω be a bounded and $C^{1,1}$ domain. Assume $u \in H_0^s(\Omega)$ and satisfies*

(a) $u \in C^s(\mathbb{R}^n)$ and, for every $\beta \in [s, 1+2s)$, u is of class $C^\beta(\Omega)$ and

$$[u]_{C^\beta(\{x \in \Omega : \delta(x) \geq \rho\})} \leq C\rho^{s-\beta} \ \text{for all } \rho \in (0,1),$$

(b) the function $u/\delta^s|_\Omega$ can be continuously extended to $\bar{\Omega}$. Moreover, there exists $\alpha \in (0,1)$ such that $u/\delta^s \in C^\alpha(\bar{\Omega})$. In addition, for all $\beta \in [\alpha, s+\alpha]$, it holds the estimate

$$[u/\delta^s]_{C^\beta(\{x \in \Omega : \delta(x) \geq \rho\})} \leq C\rho^{s-\beta} \ \text{for all } \rho \in (0,1),$$

(c) $(-\triangle)^s u$ is pointwise bounded in Ω.

Then the following identity holds

$$\int_\Omega (x \cdot \nabla u)(-\triangle)^s u \, dx = \frac{2s-n}{2} \int_\Omega u(-\triangle)^s u \, dx - \frac{\Gamma^2(1+s)}{2} \int_{\partial\Omega} \left(\frac{u}{\delta^s}\right)^2 (x \cdot \nu) d\sigma,$$

where ν is the unit outward normal vector to $\partial\Omega$ at x, and Γ is the Gamma function, and $[u]_{C^\beta}$ is the semi-norm of u in C^β.

When u is a solution of the problem, they showed that the above conditions (a), (b), and (c) are satisfied.

Theorem 10.2.2 *Let Ω be a bounded $C^{1,1}$ domain, $f \in C_{loc}^{0,1}(\bar{\Omega} \times R)$, u be a bounded solution of*

$$\begin{cases} (-\triangle)^s u(x) = f(x,u), \ x \in \Omega, \\ u \geq 0, \qquad\qquad\quad x \in \mathbb{R}^n \backslash \Omega \end{cases} \tag{10.19}$$

and $\delta(x) = dist(x, \partial\Omega)$. Then

(a) $u \in C^s(\mathbb{R}^n)$ and, for every $\beta \in [s, 1+2s)$, u is of class $C^\beta(\Omega)$ and

$$[u]_{C^\beta(\{x \in \Omega : \delta(x) \geq \rho\})} \leq C\rho^{s-\beta} \ \text{for all } \rho \in (0,1),$$

(b) the function $u/\delta^s|_\Omega$ can be continuously extended to $\bar{\Omega}$. Moreover, there exists an $\alpha \in (0,1)$ such that $u/\delta^s \in C^\alpha(\bar{\Omega})$. In addition, for all $\beta \in [\alpha, s+\alpha]$, it holds the estimate

$$[u/\delta^s]_{C^\beta(\{x \in \Omega : \delta(x) \geq \rho\})} \leq C\rho^{s-\beta} \ \text{for all } \rho \in (0,1).$$

The constants α and C depend only on Ω, s, f, $\|u\|_{L^\infty(\mathbb{R}^n)}$, and β.

Now Pohozaev identity (10.16) is a direct consequence of Proposition 10.2.1 and Theorem 10.2.2. We prove Proposition 10.2.1, and to this end, we need the following estimates, which reveal the nonlocal nature that distinguishes the fractional Laplacian from the regular Laplacian.

Proposition 10.2.2 *Let Ω be a bounded $C^{1,1}$ domain, and u be a function such that $u \equiv 0$ in $\mathbb{R}^n \backslash \Omega$ and that u satisfies (b) in Proposition 10.2.1. Then, there exists a $C^\alpha(\mathbb{R}^n)$ extension v of $u/\delta^s|_\Omega$ such that*

$$(-\triangle)^{s/2}u(x) = c_1\{\log^- \delta(x) + c_2\chi_\Omega(x)\}v(x) + h(x), \ x \in \mathbb{R}^n, \qquad (10.20)$$

where $h \in C^\alpha(\mathbb{R}^n)$ and $\log^- t = \min\{\log t, 0\}$,

$$c_1 = \frac{\Gamma(1+s)\sin(\frac{\pi s}{2})}{\pi} \ and \ c_2 = \frac{\pi}{\tan(\frac{\pi s}{2})}. \qquad (10.21)$$

Moreover, if u also satisfies (a) in Proposition 10.2.1, then for all $\beta \in (0, 1+s)$

$$[(-\triangle)^{s/2}u]_{C^\beta(\{x \in \Omega : \delta(x) \geq \rho\})} \leq C\rho^{-\beta} \ for \ all \ \rho \in (0,1) \qquad (10.22)$$

for some constant C which does not depend on ρ.

Proposition 10.2.3 *Let A and B be real numbers, and*

$$\varphi(t) = A\log^-|t-1| + B\chi_{[0,1]}(t) + h(t),$$

where $\log^- t = \min\{\log t, 0\}$ and h is a function satisfying, for some constants α and γ in $(0,1)$, and $C_0 > 0$, the following conditions:

(i) $\|h\|_{C^\alpha([0,\infty])} \leq C_0$.

(ii) For all $\beta \in [\gamma, 1+\gamma]$

$$\|h\|_{C^\beta((0,1-\rho)\cup(1+\rho,2))} \leq C_0\rho^{-\beta}, \ for \ all \ \rho \in (0,1).$$

(iii) $|h'(t)| \leq C_0 t^{-2-\gamma}$ and $|h''(t)| \leq C_0 t^{-3-\gamma}$ for all $t > 2$.

Then,

$$-\frac{d}{d\lambda}\bigg|_{\lambda=1+}\int_0^\infty \varphi(\lambda t)\varphi(\frac{t}{\lambda})dt = A^2\pi^2 + B^2.$$

Proof of Proposition 10.2.1 for Star-Shaped Domains

Without loss of generality, we assume that Ω is strictly star-shaped with respect to the origin.

We first show that

$$\int_\Omega (x \cdot \nabla u)(-\triangle)^s u dx = \frac{d}{d\lambda}\bigg|_{\lambda=1+}\int_\Omega u_\lambda(-\triangle)^s u dx, \qquad (10.23)$$

where $u_\lambda = u(\lambda x)$ and $\frac{d}{d\lambda}|_{\lambda=1+}$ is the right hand side derivative at $\lambda = 1$. Indeed, let $g = (-\triangle)^s u$. By assumption (a), g is defined pointwise in Ω, and by assumption (c), $g \in L^\infty(\Omega)$. Then, making the change of variables $y = \lambda x$ and using that supp $u_\lambda = \frac{1}{\lambda}\Omega \subset \Omega$ for $\lambda > 1$, we obtain

$$\frac{d}{d\lambda}\Big|_{\lambda=1+} \int_{\Omega} u_{\lambda}(x)\, g(x)dx$$

$$= \lim_{\lambda\downarrow 1} \int_{\Omega} \frac{u(\lambda x) - u(x)}{\lambda - 1} g(x)dx$$

$$= \lim_{\lambda\downarrow 1} \lambda^{-n} \int_{\lambda\Omega} \frac{u(y) - u(y/\lambda)}{\lambda - 1} g(y/\lambda)dy$$

$$= \lim_{\lambda\downarrow 1} \int_{\Omega} \frac{u(y) - u(y/\lambda)}{\lambda - 1} g(y/\lambda)dy + \lim_{(\lambda\downarrow 1)} \int_{(\lambda\Omega)\backslash\Omega} \frac{-u(y/\lambda)}{\lambda - 1} g(y/\lambda)dy.$$

By the dominated convergence theorem, we can change the order of integration and take limit to obtain

$$\lim_{\lambda\downarrow 1} \int_{\Omega} \frac{u(y) - u(y/\lambda)}{\lambda - 1} g(y/\lambda)dy = \int_{\Omega} (y \cdot \nabla u)g(y)dy.$$

Since $g \in L^{\infty}(\Omega)$, we have

$$|\nabla u(\xi)| \le C\delta(\xi)^{s-1} \le C\lambda^{1-s}\delta(y)^{s-1}$$

for all ξ in the segment joining y and y/λ. Because $s > 1$, δ^{s-1} is integrable. The gradient bound $|\nabla u(\xi)| \le C\delta(\xi)^{s-1}$ follows from assumption (a) with $\beta = 1$. Hence, to prove (10.23) it remains only to show that

$$\lim_{(\lambda\downarrow 1)} \int_{(\lambda\Omega)\backslash\Omega} \frac{-u(y/\lambda)}{\lambda - 1} g(y/\lambda)dy = 0.$$

Indeed, $|(\lambda\Omega)\backslash\Omega| \le C(\lambda - 1)$, and by (a), $u \in C^s(\mathbb{R}^n)$ and $u \equiv 0$ outside Ω. Hence, $\|u\|_{L^{\infty}((\lambda\Omega)\backslash\Omega)} \to 0$ as $\lambda \downarrow 1$ and (10.23) holds.

Recall that the integration by parts formula states: If u and v are $H^s(\mathbb{R}^n)$ functions and $u \equiv v \equiv 0$ in $\mathbb{R}^n\backslash\Omega$, then

$$\int_{\Omega} v(-\triangle)^s u dx = \int_{\mathbb{R}^n} (-\triangle)^{s/2}v(-\triangle)^{s/2}u dx. \qquad (10.24)$$

Let $v = u_{\lambda}$ in (10.24), then

$$\int_{\Omega} u_{\lambda}(-\triangle)^s u dx = \int_{\mathbb{R}^n} u_{\lambda}(-\triangle)^s u dx$$

$$= \int_{\mathbb{R}^n} (-\triangle)^{s/2}u_{\lambda}(-\triangle)^{s/2}u dx$$

$$= \lambda^s \int_{\mathbb{R}^n} [(-\triangle)^{s/2}u](\lambda x)(-\triangle)^{s/2}u(x)dx$$

$$= \lambda^s \int_{\mathbb{R}^n} w_{\lambda}w dx,$$

where

$$w(x) = (-\triangle)^{s/2}u(x) \ and \ w_{\lambda}(x) = w(\lambda x).$$

By the change of variables $y = \sqrt{\lambda}\, x$ this integral becomes

$$\lambda^s \int_{\mathbb{R}^n} w_\lambda w dx = \lambda^{\frac{2s-n}{2}} \int_{\mathbb{R}^n} w_{\sqrt{\lambda}} w_{1/\sqrt{\lambda}} dy,$$

and thus

$$\int_\Omega u_\lambda (-\triangle)^s u dx = \frac{2s-n}{2} \int_{\mathbb{R}^n} w_{\sqrt{\lambda}} w_{1/\sqrt{\lambda}} dy.$$

Furthermore, this leads to

$$\int_\Omega (\nabla u \cdot x)(-\triangle)^s u dx$$

$$= \frac{d}{d\lambda}\Big|_{\lambda=1^+} \left(\lambda^{\frac{2s-n}{2}} \int_{\mathbb{R}^n} w_{\sqrt{\lambda}} w_{1/\sqrt{\lambda}} dy \right)$$

$$= \frac{2s-n}{2} \int_{\mathbb{R}^n} |(-\triangle)^{s/2} u|^2 dx + \frac{d}{d\lambda}\Big|_{\lambda=1^+} \int_{\mathbb{R}^n} w_{\sqrt{\lambda}} w_{1/\sqrt{\lambda}} dy$$

$$= \frac{2s-n}{2} \int_{\mathbb{R}^n} u(-\triangle)^s u dx + \frac{1}{2}\frac{d}{d\lambda}\Big|_{\lambda=1^+} \int_{\mathbb{R}^n} w_\lambda w_{1/\lambda} dy. \qquad (10.25)$$

Hence, what remains is to prove that

$$-\frac{d}{d\lambda}\Big|_{\lambda=1^+} I_\lambda = \Gamma^2(1+s) \int_{\partial\Omega} (\frac{u}{\delta^s})^2 (x \cdot \nu) d\sigma, \qquad (10.26)$$

where

$$I_\lambda = \int_{\mathbb{R}^n} w_\lambda w_{1/\lambda} dy. \qquad (10.27)$$

Now for each $\theta \in S^{n-1}$ there exists a unique $r_\theta > 0$ such that $r_\theta \theta \in \partial\Omega$. Writing the integral (10.27) in spherical coordinates and using the change of variables $t = r/r_\theta$, we have

$$\frac{d}{d\lambda}\Big|_{\lambda\to 1^+} I_\lambda$$

$$= \frac{d}{d\lambda}\Big|_{\lambda\to 1^+} \int_{S^{n-1}} d\theta \int_0^\infty r^{n-1} w(\lambda r\theta) w(\frac{r}{\lambda}\theta) dr$$

$$= \frac{d}{d\lambda}\Big|_{\lambda\to 1^+} \int_{S^{n-1}} r_\theta d\theta \int_0^\infty (r_\theta t)^{n-1} w(\lambda r_\theta t\theta) w(\frac{r_\theta t}{\lambda}\theta) dt$$

$$= \frac{d}{d\lambda}\Big|_{\lambda\to 1^+} \int_{\partial\Omega} (x \cdot \nu) d\sigma \int_0^\infty t^{n-1} w(\lambda tx) w(\frac{tx}{\lambda}) dt,$$

where

$$r_\theta^{n-1} d\theta = (\frac{x}{|x|} \cdot \nu) d\sigma = \frac{1}{r_\theta}(x \cdot \nu) d\sigma.$$

Note that in the change of variables, every point in S^{n-1} is mapped to its radial projection on $\partial\Omega$, and it is unique because of the strict star-shapedness of Ω.

Fix $x_0 \in \partial\Omega$ and define

$$\varphi(t) = t^{\frac{n-1}{2}}[(-\triangle)^{s/2} u](tx_0).$$

By Proposition 10.2.2,

$$\varphi(t) = c_1\{\log^- \delta(tx_0) + c_2\chi_{[0,1]}\}v(tx_0) + h_0(t), \quad t \in [0,\infty),$$

where v is a $C^\alpha(\mathbb{R}^n)$ extension of $u/\delta^s|_\Omega$, and h_0 is a $C^\alpha([0,\infty))$ function. Next we modify this expression in order to apply Proposition 10.2.2.

Since Ω is $C^{1,1}$ and strictly star-shaped, it is not difficult to see that $\frac{|r-r_\theta|}{\delta(r\theta)}$ is a Lipschitz function of r in $[0,\infty)$ and is bounded below by a positive constant (independent of x_0). Similarly, $\frac{|t-1|}{\delta(tx_0)}$ and $\frac{\min\{|t-1|,1\}}{\min\{\delta(tx_0),1\}}$ are positive and Lipschitz functions of t in $[0,\infty)$. Therefore,

$$\log^- |t-1| - \log^- \delta(tx_0)$$

is Lipschitz in $[0,\infty)$ as a function of t.

Hence, for $t \in [0,\infty)$,

$$\varphi(t) = c_1\{\log^- |t-1| + c_2\chi_{[0,1]}(t)\}v(tx_0) + h_1(t),$$

where h_1 is a C^α function on the same interval.

Moreover, note that the difference

$$v(tx_0) - v(x_0)$$

is C^α and vanishes at $t = 1$. Thus, for $t \in [0,\infty)$

$$\varphi(t) = c_1\{\log^- |t-1| + c_2\chi_{[0,1]}(t)\}v(x_0) + h(t),$$

and

$$h(t) = t^{\frac{n-1}{2}}(-\Delta)^{s/2}u(tx_0) - c_1\{\log^- |t-1| + c_2\chi_{[0,1]}(t)\}v(x_0),$$

where h is C^α in $[0,\infty)$. Therefore,

$$\frac{d}{d\lambda}\Big|_{\lambda\to 1^+} I_\lambda = \frac{d}{d\lambda}\Big|_{\lambda\to 1^+} \int_{\partial\Omega} (x\cdot\nu)d\sigma \int_0^\infty \varphi(\lambda t)\varphi(\frac{t}{\lambda})dt,$$

and from Proposition 10.2.3, we know

$$\frac{d}{d\lambda}\Big|_{\lambda\to 1^+} \int_0^\infty \varphi(\lambda t)\varphi(\frac{t}{\lambda})dt = -c_1^2(\pi^2 + c_2^2)v^2(x_0)$$

and

$$c_1 = \frac{\Gamma(1+s)\sin(\frac{\pi s}{2})}{\pi} \quad \text{and} \quad c_2 = \frac{\pi}{\tan(\frac{\pi s}{2})}.$$

Therefore

$$c_1^2(\pi^2 + c_2^2) = \frac{\Gamma^2(1+s)\sin^2(\frac{\pi s}{2})}{\pi^2}\left(\pi^2 + \frac{\pi^2}{\tan^2(\frac{\pi s}{2})}\right)$$

$$= \Gamma^2(1+s).$$

Now we can express $\frac{d}{d\lambda}\big|_{\lambda\to1+}I_\lambda$ in terms of u, that is,

$$\frac{d}{d\lambda}\Big|_{\lambda\to1+}I_\lambda = \frac{d}{d\lambda}\Big|_{\lambda\to1+}\int_{\partial\Omega}(x\cdot\nu)d\sigma\int_0^\infty\varphi(\lambda t)\psi(\frac{t}{\lambda})dt$$

$$= -\Gamma^2(1+s)\int_{\partial\Omega}(\frac{u}{\delta^s})^2(x\cdot\nu)d\sigma.$$

We complete the proof. $\square$

10.2.2 The Pohozaev Identity for Systems

In this section we present a Pohozaev identity associated with a fractional system proved in [MLL]. Consider

$$\begin{cases}(-\triangle)^su(x)=|x|^av^p, & x\in\Omega,\\ (-\triangle)^sv(x)=|x|^bu^q, & x\in\Omega,\\ u=0,v=0, & x\in\mathbb{R}^n\backslash\Omega,\end{cases} \tag{10.28}$$

in a bounded star-sharped domain $\Omega\subset\mathbb{R}^n$ with $C^{1,1}$ boundary. Assume that $s\in(0,1)$, $pq>1$, $p,q,a,b\geq0$, $n\geq1$. We say that the system is in subcritical case if $\frac{n+a}{p+1}+\frac{n+b}{q+1}>n-2s$; in critical case if $\frac{n+a}{p+1}+\frac{n+b}{q+1}=n-2s$; and in supercritical case if $\frac{n+a}{p+1}+\frac{n+b}{q+1}<n-2s$.

Theorem 10.2.3 *Let Ω be a bounded $C^{1,1}$ domain, $\delta(x)=dist(x,\partial\Omega)$. Assume that $(u,v)\in H^s(\mathbb{R}^n)\times H^s(\mathbb{R}^n)$ is a pair of solutions of (10.28), then the following identity holds*

$$\int_\Omega(x\cdot\nabla v)(-\triangle)^su\,dx$$

$$=\frac{2s-n}{2}\int_\Omega v\,(-\triangle)^su\,dx-\frac{\Gamma^2(1+s)}{2}\int_{\partial\Omega}\frac{u}{\delta^s}\frac{v}{\delta^s}(x\cdot\nu)d\sigma$$

$$-\frac{1}{2}\int_{\Omega\cup(\mathbb{R}^n\backslash\bar\Omega)}\Big(x\nabla U(x)V(x)-xU(x)\nabla V(x)\Big)dx,$$

$$\tag{10.29}$$

where $U(x)=(-\triangle)^{s/2}u(x)$ and $V(x)=(-\triangle)^{s/2}v(x)$.

Remark 10.2.1 (i). Identity (10.29) is far less handy than its analogue for single equations due to the uncertainty of the signs for the last integral.
(ii). Notice that in (10.29),

$$\int_{\Omega\cup(\mathbb{R}^n\backslash\bar\Omega)}\Big(x\nabla U(x)V(x)-xU(x)\nabla V(x)\Big)dx$$

$$\neq\int_{\mathbb{R}^n}\Big(x\nabla U(x)V(x)-xU(x)\nabla V(x)\Big)dx,$$

because the integrand is highly singular on $\partial\Omega$, in a sense similar to the Delta measure. To illustrate this, we consider a simple example,

$$f(t) = \begin{cases} t+1, & t \geq 0, \\ t, & t < 0. \end{cases}$$

Obviously $f'(t) = 1 + \delta(t)$, where δ is the Delta function. Since $\int_{-1}^{0} f'(t)dt = \lim_{\varepsilon\to 0}\int_{-1}^{-\varepsilon} f'(t)dt = 1$ and $\int_{0}^{1} f'(t)dt = \lim_{\varepsilon\to 0}\int_{\varepsilon}^{1} f'(t)dt = 1$, we have

$$\int_{-1}^{0} f'(t)dt + \int_{0}^{1} f'(t)dt = 2 \neq \int_{-1}^{1} f'(t)dt = 3.$$

Let $\Omega_\varepsilon = \{x \in \mathbb{R}^n \mid \text{dist}(x, \Omega) < \varepsilon\}$. We point out that

$$\int_{\mathbb{R}^n\setminus\Omega} \left(x\nabla U(x)V(x) - xU(x)\nabla V(x) \right) dx$$

$$=: \lim_{\varepsilon\to 0} \int_{\mathbb{R}^n\setminus\Omega_\varepsilon} \left(x\nabla U(x)V(x) - xU(x)\nabla V(x) \right) dx.$$

The same definition applies to

$$\int_{\Omega} (x\nabla U(x)V(x) - xU(x)\nabla V(x))dx.$$

Remark 10.2.2 In the fractional Pohozaev identity, the functions $u/\delta^s|_{\partial\Omega}$ and $v/\delta^s|_{\partial\Omega}$ play the roles of $\partial u/\partial\nu$ and $\partial v/\partial\nu$ in the classical Pohozaev identity. Surprisingly, from a nonlocal problem we obtain an identity with a boundary term (an integral over $\partial\Omega$) which is completely local.

From this Pohozaev identity, we derive the non-existence of solutions.

Theorem 10.2.4 *Let Ω be a bounded star-shaped domain with $C^{1,1}$ boundary. If $\frac{n+a}{p+1} + \frac{n+b}{q+1} \leq n - 2s$, then problem (10.28) admits no positive bounded solution.*

Proof. From Theorem 10.2.3, we know

$$\int_{\Omega}(x\cdot\nabla v)(-\triangle)^s u = \frac{2s-n}{2}\int_{\Omega} v(-\triangle)^s u dx - \frac{\Gamma^2(1+s)}{2}\int_{\partial\Omega} \frac{u}{\delta^s}\frac{v}{\delta^s}(x\cdot\nu)d\sigma$$

$$-\frac{1}{2}\int_{\Omega\cup(\mathbb{R}^n\setminus\bar\Omega)} \left(x\nabla U(x)V(x) - xU(x)\nabla V(x) \right) dx.$$

Using integration by parts, we obtain

$$\int_{\Omega} |x|^a v^p(x\cdot\nabla v)dx = -\frac{n+a}{p+1}\int_{\Omega} |x|^a v^{p+1}dx.$$

Therefore,

$$(s - \frac{n}{2}) \int_\Omega v(-\triangle)^s u dx - \frac{\Gamma^2(1+s)}{2} \int_{\partial\Omega} \frac{u}{\delta^s} \frac{v}{\delta^s} (x \cdot \nu) d\sigma$$

$$- \frac{1}{2} \int_{\Omega \cup (\mathbb{R}^n \setminus \bar{\Omega})} \Big(x \nabla U(x) V(x) - x U(x) \nabla V(x) \Big) dx$$

$$= -\frac{n+a}{p+1} \int_\Omega |x|^a v^{p+1} dx. \tag{10.30}$$

Likewise, for the second equation in problem (10.28), we derive

$$(s - \frac{n}{2}) \int_\Omega u (-\triangle)^s v \, dx - \frac{\Gamma^2(1+s)}{2} \int_{\partial\Omega} \frac{v}{\delta^s} \frac{u}{\delta^s} (x \cdot \nu) d\sigma$$

$$- \frac{1}{2} \int_{\Omega \cup (\mathbb{R}^n \setminus \bar{\Omega})} \Big(x \nabla V(x) U(x) - x V(x) \nabla U(x) \Big) dx$$

$$= -\frac{n+b}{q+1} \int_\Omega |x|^b u^{q+1} dx. \tag{10.31}$$

Combining (10.30) and (10.31), we arrive at the Pohozaev identity associated with (10.28),

$$(s - \frac{n}{2}) \int_\Omega u (-\triangle)^s v \, dx + (s - \frac{n}{2}) \int_\Omega v (-\triangle)^s u \, dx$$

$$- \Gamma^2(1+s) \int_{\partial\Omega} \frac{v}{\delta^s} \frac{u}{\delta^s} (x \cdot \nu) d\sigma$$

$$= -\frac{n+b}{q+1} \int_\Omega |x|^b u^{q+1} dx - \frac{n+a}{p+1} \int_\Omega |x|^a v^{p+1} dx. \tag{10.32}$$

Since Ω is a star-shaped domain, we must have $x \cdot \nu > 0$. For $u, v > 0$, it holds that

$$-\Gamma^2(1+s) \int_{\partial\Omega} \frac{v}{\delta^s} \frac{u}{\delta^s} (x \cdot \nu) d\sigma < 0.$$

Hence from (10.28) and (10.32), it yields

$$\Gamma^2(1+s) \int_{\partial\Omega} \frac{v}{\delta^s} \frac{u}{\delta^s} (x \cdot \nu) d\sigma$$

$$= (s - \frac{n}{2}) \int_\Omega u(-\triangle)^s v dx + (s - \frac{n}{2}) \int_\Omega v(-\triangle)^s u dx$$

$$= (s - \frac{n}{2}) \int_\Omega |x|^b u^{q+1} dx + (s - \frac{n}{2}) \int_\Omega |x|^a v^{p+1} dx$$

$$+ \frac{n+b}{q+1} \int_\Omega |x|^b u^{q+1} dx + \frac{n+a}{p+1} \int_\Omega |x|^a v^{p+1} dx$$

$$> 0. \tag{10.33}$$

From $u, v \in H_0^s(\Omega)$ it follows that

$$\int_\Omega |x|^b u^{q+1} dx = \int_{\mathbb{R}^n} |x|^b u^{q+1} dx = \int_{\mathbb{R}^n} u(-\triangle)^s v dx = \int_{\mathbb{R}^n} (-\triangle)^{\frac{s}{2}} u(-\triangle)^{\frac{s}{2}} v dx,$$

and

$$\int_\Omega |x|^a v^{p+1} dx = \int_{\mathbb{R}^n} |x|^a v^{p+1} dx = \int_{\mathbb{R}^n} v(-\triangle)^s u\, dx = \int_{\mathbb{R}^n} (-\triangle)^{\frac{s}{2}} u(-\triangle)^{\frac{s}{2}} v\, dx,$$

thus

$$\int_\Omega |x|^a v^{p+1} dx = \int_\Omega |x|^b u^{q+1} dx. \qquad (10.34)$$

If $\frac{n+a}{p+1} + \frac{n+b}{q+1} \leq n - 2s$, then

$$\Gamma^2(1+s) \int_{\partial\Omega} \frac{v}{\delta^s} \frac{u}{\delta^s} (x \cdot \nu) d\sigma$$

$$= (s - \frac{n}{2} + \frac{n+b}{q+1}) \int_\Omega |x|^b u^{q+1} dx + (s - \frac{n}{2} + \frac{n+a}{p+1}) \int_\Omega |x|^a v^{p+1} dx$$

$$= (2s - n + \frac{n+b}{q+1} + \frac{n+a}{p+1}) \int_\Omega |x|^b u^{q+1} dx$$

$$\leq 0.$$

This is a contradiction with (10.33).
    We complete the proof. $\square$

11

Higher Order Fractional Laplacians

Let m be an even integer greater than or equal to 4. In this chapter, we discuss the higher-order fractional Laplacian $(-\triangle)^{\alpha/2}$ for $m - 2 < \alpha < m$.

We first introduce two kinds of definitions, one via an integral in the Cauchy principal value, and the other via an integral of difference quotients. We also show that, in the class of smooth functions, it can be defined equivalently via the Fourier transform.

Then we consider semi-linear equations involving higher order fractional Laplacians. To prove symmetry or non-existence of solutions, we introduce two approaches. One is to write it as a system of lower order equations, and the other is to consider the equivalent higher order integral equation.

11.1 Definition

Recall that, for $0 < \alpha < 2$, the fractional Laplacian is defined as

$$(-\triangle)^{\alpha/2}u(x) = C_{n,\alpha} \lim_{\varepsilon \to 0} \int_{\mathbb{R}^n \setminus B_\varepsilon(x)} \frac{u(x) - u(y)}{|x - y|^{n+\alpha}} dy.$$

The condition $u \in L_\alpha$ guarantees the convergence of the integral near infinity, and $u \in C_{loc}^{1,1}$ ensures the existence of the limit as $\epsilon \to 0$. This integral is in the Cauchy principal value sense, or equivalently, we can evaluate it as the following

$$\int_0^\infty \int_{S_r(x)} \frac{u(x) - u(y)}{|x - y|^{n+\alpha}} d\sigma_y dr,$$

where $S_r(x)$ is the sphere of radius r centered at x.

For r sufficiently small and for $y \in S_r(x)$, by Taylor expansion, we have

$$u(x) - u(y) = -\nabla u(x) \cdot (y - x) + O(|y - x|^2).$$

Due to symmetry,

$$\int_{S_r(x)} \frac{-\nabla u(x) \cdot (y - x)}{|y - x|^{n+\alpha}} d\sigma_y = 0.$$

Then for the remaining term, it's integral on $B_1(x)$ is

$$\int_0^1 \int_{S_r(x)} \frac{O(|y - x|^2)}{|x - y|^{n+\alpha}} d\sigma_y dr = c \int_0^1 \frac{r^2 r^{n-1}}{r^{n+\alpha}} dr = c \int_0^1 r^{1-\alpha} dr. \qquad (11.1)$$

Obviously, this integral converges for $\alpha < 2$. However, it diverges when $\alpha > 2$.

From the above analysis, one can see that in the case $2 < \alpha < 4$, we need to have $O(|y - x|^4)$ in the numerator of the integrand in order for the integral to converge. For this reason, we add a term in the definition:

$$(-\triangle)^{\alpha/2} u(x) = C_{n,\alpha} \lim_{\varepsilon \to 0} \int_{\mathbb{R}^n \backslash B_\varepsilon(x)} \frac{u(x) - u(y) + \frac{1}{2n} \triangle u(x)|x - y|^2}{|x - y|^{n+\alpha}} dy. \qquad (11.2)$$

Then due to symmetry, for r small, one has

$$\int_{S_r(x)} \frac{u(x) - u(y) + \frac{1}{2n} \triangle u(x)|x - y|^2}{|x - y|^{n+\alpha}} d\sigma_y = \int_{S_r(x)} \frac{O(|x - y|^4)}{|x - y|^{n+\alpha}} d\sigma_y,$$

which converges for $\alpha < 4$ when is integrated with respect to r on a finite interval containing 0.

More generally, given any even integer m, for $m - 2 < \alpha < m$, the higher-order fractional Laplacian $(-\triangle)^{\alpha/2}$ can be defined in the sense of P.V., the Cauchy principal value, as follows

$$(-\triangle)^{\alpha/2} u(x) = C_{n,\alpha} P.V. \int_{\mathbb{R}^n} \frac{\sum\limits_{k=0}^{\frac{m-2}{2}} H_k \triangle^k u(x)|y - x|^{2k} - u(y)}{|x - y|^{n+\alpha}} dy$$

$$= C_{n,\alpha} \lim_{\varepsilon \to 0} \int_{\mathbb{R}^n \backslash B_\varepsilon(x)} \frac{\sum\limits_{k=0}^{\frac{m-2}{2}} H_k \triangle^k u(x)|y - x|^{2k} - u(y)}{|x - y|^{n+\alpha}} dy,$$

where

$$H_k = \frac{1}{2^k k! n(n + 2) \cdots (n + 2k - 2)}, \quad k \geq 1 \text{ and } H_0 = 1.$$

An equivalent approach to define it without using P.V. is via the difference quotient:

$$(-\triangle)^{\alpha/2} u(x) = C_{n,\alpha} \int_{\mathbb{R}^n} \frac{\sum\limits_{k=0}^m (-1)^{k - \frac{m}{2}} C_m^k u(x + (k - \frac{m}{2})y)}{|y|^{n+\alpha}} dy, \qquad (11.3)$$

with

$$C_m^k = \frac{m!}{k!(m - k)!}.$$

For example, when $m = 4$ and $2 < \alpha < 4$,

$$(-\triangle)^{\alpha/2}u(x)$$
$$= C_{n,\alpha} \int_{\mathbb{R}^n} \frac{u(x+2y) - 4u(x+y) + 6u(x) - 4u(x-y) + u(x-2y)}{|y|^{n+\alpha}} dy.$$

$$(11.4)$$

For $m = 6$ and $4 < \alpha < 6$,

$$(-\triangle)^{\alpha/2}u(x) = C_{n,\alpha} P.V. \int_{\mathbb{R}^n} \frac{u(x) - u(y) + \frac{1}{2n}\triangle u(x)|x-y|^2 + \frac{1}{8n(n+2)}\triangle^2 u(x)|x-y|^4}{|x-y|^{n+\alpha}} dy.$$

Or

$$(-\triangle)^{\alpha/2}u(x) = C_{n,\alpha} \int_{\mathbb{R}^n} \frac{-u(x+3y) + 6u(x+2y) - 15u(x+y) + 20u(x)}{|y|^{n+\alpha}}$$
$$\frac{-15u(x-y) + 6u(x-2y) - u(x-3y)}{} dy.$$

When the function u is smooth enough, similar to the lower order fractional Laplacian, we can also define the higher one via the Fourier transform in $\mathbb{R}^n$:

Lemma 11.1.1

$$\mathcal{F}((-\triangle)^{\alpha/2}u(x)) = |\xi|^\alpha \hat{u}(\xi). \tag{11.5}$$

Proof. We show (11.5) for $2 < \alpha < 4$ through (11.2) and (11.4) respectively.

Via (11.2).

Let $|x - y| = z$ in (11.2), then

$$(-\triangle)^{\alpha/2}u(x) = C_{n,\alpha} \lim_{\varepsilon \to 0} \int_{\mathbb{R}^n \backslash B_\varepsilon(0)} \frac{u(x) - u(x-z) + \triangle u(x) \cdot \frac{|z|^2}{2n}}{|z|^{n+\alpha}} dz.$$

Let

$$I(x) = \int_{\mathbb{R}^n \backslash B_\varepsilon(0)} \frac{u(x) - u(x-z) + \triangle u(x) \cdot \frac{|z|^2}{2n}}{|z|^{n+\alpha}} dz.$$

$$\hat{I}(\xi) = \int_{\mathbb{R}^n} e^{-2\pi i x \cdot \xi} I(x) dx$$

$$= \int_{\mathbb{R}^n \backslash B_\varepsilon(0)} \int_{\mathbb{R}^n} e^{-2\pi i x \cdot \xi} \left(u(x) - u(x-z) + \triangle u(x) \cdot \frac{|z|^2}{2n} \right) dx \frac{dz}{|z|^{n+\alpha}}$$

$$= \int_{\mathbb{R}^n \backslash B_\varepsilon(0)} \hat{u}(\xi) \left(1 - e^{-2\pi i z \cdot \xi} - \frac{2\pi^2 |\xi|^2 |z|^2}{n} \right) \frac{dz}{|z|^{n+\alpha}}$$

$$= |\xi|^\alpha \hat{u}(\xi) \int_{\mathbb{R}^n \backslash B_{\varepsilon|\xi|}(0)} \left(1 - e^{-2\pi i y \cdot \frac{\xi}{|\xi|}} - \frac{2\pi^2 |y|^2}{n} \right) \frac{dy}{|y|^{n+\alpha}}, \qquad z = \frac{y}{|\xi|}.$$

To prove the convergence of the integral above, we use the symmetry of the integrating domain. Without loss of generality, let

$$\frac{\xi}{|\xi|} = \vec{e} = (1, 0, \cdots, 0).$$

Then

$$\int_{\mathbb{R}^n \backslash B_{\epsilon|\xi|}(0)} e^{-2\pi i y \cdot \frac{\xi}{|\xi|}} \frac{dy}{|y|^{n+\alpha}} = \int_{\mathbb{R}^n \backslash B_{\epsilon|\xi|}(0)} e^{-2\pi i y_1} \frac{dy}{|y|^{n+\alpha}}$$

$$= \int_{\mathbb{R}^n \backslash B_{\epsilon|\xi|}(0)} (\cos 2\pi y_1 - i \sin 2\pi y_1) \frac{dy}{|y|^{n+\alpha}}$$

$$= \int_{\mathbb{R}^n \backslash B_{\epsilon|\xi|}(0)} \cos 2\pi y_1 \frac{dy}{|y|^{n+\alpha}}.$$

The last equality is a consequence of the symmetry of $\mathbb{R}^n \backslash B_{\epsilon|\xi|}(0)$ and the oddness of the sine function.

It's trivial that for R large,

$$\int_{\mathbb{R}^n \backslash B_R(0)} \left(1 - \cos 2\pi y_1 - \frac{2\pi^2 |y|^2}{n}\right) \frac{dy}{|y|^{n+\alpha}} < \infty. \tag{11.6}$$

What remains is to show that the integral converges in $B_1(0) \backslash B_{\epsilon|\xi|}(0)$. Indeed, by the Taylor expansion,

$$\int_{B_1(0) \backslash B_{\epsilon|\xi|}(0)} \left(1 - \cos 2\pi y_1 - \frac{2\pi^2 |y|^2}{n}\right) \frac{dy}{|y|^{n+\alpha}}$$

$$= \int_{B_1(0) \backslash B_{\epsilon|\xi|}(0)} \left(2\pi^2 y_1^2 + O(y_1^4) - \frac{2\pi^2 |y|^2}{n}\right) \frac{dy}{|y|^{n+\alpha}}$$

$$= \int_{B_1(0) \backslash B_{\epsilon|\xi|}(0)} \frac{O(y_1^4)}{|y|^{n+\alpha}} dy < \infty.$$

Together with (11.6), this proves the convergence.

Via (11.4).

By (11.4),

$$\mathcal{F}((-\triangle)^{\frac{\alpha}{2}} u(\xi))$$

$$= \int_{\mathbb{R}^n} e^{-2\pi i x \cdot \xi} \int_{\mathbb{R}^n} \frac{u(x+2y) - 4u(x+y) + 6u(x) - 4u(x-y)}{|y|^{n+\alpha}}$$

$$\frac{+u(x-2y)}{} dy dx$$

$$= \int_{\mathbb{R}^n} \frac{\mathcal{F}(u(x+2y)) - 4\mathcal{F}(u(x+y)) + 6\mathcal{F}(u(x)) - 4\mathcal{F}(u(x-y))}{|y|^{n+\alpha}}$$

$$\frac{+\mathcal{F}(u(x-2y))}{} dy.$$

Similar to

$$\mathcal{F}(u(x+2y)) = \int_{\mathbb{R}^n} e^{-2\pi i x \cdot \xi} u(x+2y) dx$$

$$= \int_{\mathbb{R}^n} e^{-2\pi i (z-2y)\cdot\xi} u(z) dz, \quad z = x+2y$$

$$= e^{4\pi i y \cdot \xi} \int_{\mathbb{R}^n} e^{-2\pi i z \cdot \xi} u(z) dz$$

$$= e^{4\pi i y \cdot \xi} \hat{u}(\xi),$$

we derive

$$\mathcal{F}(u(x-2y)) = e^{-4\pi i y \cdot \xi} \hat{u}(\xi),$$
$$\mathcal{F}(u(x+y)) = e^{2\pi i y \cdot \xi} \hat{u}(\xi),$$
$$\mathcal{F}(u(x-y)) = e^{-2\pi i y \cdot \xi} \hat{u}(\xi).$$

Therefore,

$$\mathcal{F}((-\triangle)^{\frac{\alpha}{2}} u(\xi))$$
$$= \int_{\mathbb{R}^n} \frac{e^{4\pi i y \cdot \xi} \hat{u}(\xi) - 4e^{2\pi i y \cdot \xi}\hat{u}(\xi) + 6\hat{u}(\xi) - 4e^{-2\pi i y \cdot \xi}\hat{u}(\xi) + e^{-4\pi i y \cdot \xi}\hat{u}(\xi)}{|y|^{n+\alpha}} dy$$
$$= \hat{u}(\xi) \int_{\mathbb{R}^n} \frac{e^{4\pi i y \cdot \xi} - 4e^{2\pi i y \cdot \xi} + 6 - 4e^{-2\pi i y \cdot \xi} + e^{-4\pi i y \cdot \xi}}{|y|^{n+\alpha}} dy$$
$$= \hat{u}(\xi) \int_{\mathbb{R}^n} \frac{2\cos(4\pi y \cdot \xi) + 6 - 8\cos(2\pi y \cdot \xi)}{|y|^{n+\alpha}} dy$$
$$= \hat{u}(\xi) \int_{\mathbb{R}^n} \frac{2(2\cos^2(2\pi y \cdot \xi) - 1) + 6 - 8\cos(2\pi y \cdot \xi)}{|y|^{n+\alpha}} dy$$
$$= 4\hat{u}(\xi) \int_{\mathbb{R}^n} \frac{(\cos(2\pi y \cdot \xi) - 1)^2}{|y|^{n+\alpha}} dy$$
$$= 4(2\pi)^{\alpha}|\xi|^{\alpha}\hat{u}(\xi) \int_{\mathbb{R}^n} \frac{(\cos(z \cdot \frac{\xi}{|\xi|}) - 1)^2}{|z|^{n+\alpha}} dz, \quad z = 2\pi|\xi|y.$$

It remains to show the convergence of the integral above. Let $t = z \cdot \frac{\xi}{|\xi|}$. For z close to the origin, we have $t \sim 0$. It follows from the Taylor expansion that

$$1 - \cos t = 1 - (1 - \frac{t^2}{2} + O(t^4))$$
$$= \frac{t^2}{2} + O(t^4)$$
$$\sim (z \cdot \frac{\xi}{|\xi|})^2$$
$$= |z|^2.$$

Thus

$$\int_{B_1(0)} \frac{(\cos(z \cdot \frac{\xi}{|\xi|}) - 1)^2}{|z|^{n+\alpha}} dz \le C \int_{B_1(0)} \frac{O(|z|^4)}{|z|^{n+\alpha}} dz < \infty. \tag{11.7}$$

On the other hand, for $|z|$ large, it's easy to see that

$$\int_{\mathbb{R}^n \setminus B_1(0)} \frac{(\cos(z \cdot \frac{\xi}{|\xi|}) - 1)^2}{|z|^{n+\alpha}} dz < \int_{\mathbb{R}^n \setminus B_1(0)} \frac{4}{|z|^{n+\alpha}} dz < \infty.$$

Together with (11.7), it yields the expected convergence. $\square$

11.2 The System Method

To prove symmetry of solutions, a common practice is to apply the method of moving planes. So far, there has not been success in deploying the method directly on higher-order equations. However, we can split them into systems of lower order equations, or we can show that they are equivalent to integral equations, then we will be able to carry out the method of moving planes on the equivalents.

We first introduce the system method.

Lemma 11.2.1 (Narrow Region Principle) *Assume that Ω is a bounded narrow region in $\Sigma_\lambda = \{x \in \mathbb{R}^n \mid x_1 < \lambda\}$. Without loss of generality, we may assume that Ω is contained in the slab $\{x \in \mathbb{R}^n \mid \lambda - l < x_1 < \lambda\}$ with $l > 0$ small.*

Consider

$$\begin{cases} -\triangle U(x) + c_1(x)V(x) \geq 0, & x \in \Omega, \\ (-\triangle)^{\alpha/2}V(x) + c_2(x)U(x) \geq 0, & x \in \Omega, \\ U(x^\lambda) = -U(x) \\ V(x^\lambda) = -V(x), & x \in \Sigma_\lambda, \\ U(x) \geq 0, & x \in \partial\Omega, \\ V(x) \geq 0, & x \in \Sigma_\lambda \setminus \Omega, \end{cases} \tag{11.8}$$

where $c_i(x) \leq 0$, $i = 1, 2$ and are bounded in Ω. Assume that $U \in C^2(\Omega)$, $V \in C^{1,1}_{loc}(\Omega) \cap L_\alpha$ and are lower semi-continuous on $\bar{\Omega}$.

Then for l sufficiently small, we have

$$U(x), V(x) \geq 0, \; x \in \Omega. \tag{11.9}$$

For an unbounded domain Ω, (11.9) still holds on condition that

$$U(x), V(x) \to 0, \quad |x| \to \infty.$$

Further, under the assumption that $c_1(x) \neq 0$ in Ω, if either $U(x)$ or $V(x)$ equals 0 at some point in Ω, then

$$U(x) \equiv 0, x \in \Omega \; and \; V(x) \equiv 0, \; x \in \mathbb{R}^n. \tag{11.10}$$

Proof. Suppose otherwise, then at least one of $U(x)$ and $V(x)$ is negative somewhere in Ω. Without loss of generality, we may assume that $V(x)$ is negative somewhere. Then there must exist some $x^0 \in \Omega$ such that

$$V(x^0) = \min_{\Sigma_\lambda} V(x) < 0.$$

Then for $0 < l \ll 1$, we have

$$(-\triangle)^{\alpha/2}V(x^0) = C_{n,\alpha}PV \int_{\mathbb{R}^n} \frac{V(x^0) - V(y)}{|x^0 - y|^{n+\alpha}} dy$$

$$= C_{n,\alpha}PV \left\{ \int_{\Sigma_\lambda} \frac{V(x^0) - V(y)}{|x^0 - y|^{n+\alpha}} dy + \int_{\mathbb{R}^n \setminus \Sigma_\lambda} \frac{V(x^0) - V(y)}{|x^0 - y|^{n+\alpha}} dy \right\}$$

$$= C_{n,\alpha}PV \left\{ \int_{\Sigma_\lambda} \frac{V(x^0) - V(y)}{|x^0 - y|^{n+\alpha}} dy + \int_{\Sigma_\lambda} \frac{V(x^0) + V(y)}{|x^0 - y^\lambda|^{n+\alpha}} dy \right\}$$

$$\leq C_{n,\alpha} \int_{\Sigma_\lambda} \frac{2V(x^0)}{|x^0 - y^\lambda|^{n+\alpha}} dy \tag{11.11}$$

$$\leq CV(x^0) \int_{B_1(x^o) \setminus B_l(x^o)} \frac{1}{|x^o - y|^{n+\alpha}} dy$$

$$\leq \frac{CV(x^0)}{l^\alpha} < 0. \tag{11.12}$$

Together with

$$(-\triangle)^{\alpha/2}V(x) + c_2(x)U(x) \geq 0, \quad x \in \Omega, \tag{11.13}$$

it's easy to see that $c_2(x^0) \neq 0$. Further, with $c_2(x) \leq 0$ in Ω, from (11.13) we deduce that

$$U(x^0) < 0.$$

It thus implies that there exists some $\bar{x} \in \Sigma_\lambda \cap B_1(0)$ such that

$$U(\bar{x}) = \min_{\Sigma_\lambda} U(x) < 0. \tag{11.14}$$

For $\delta > 0$ small, let

$$\phi(x) = \sin \left(\frac{x_1 - \lambda + l}{l} + \delta \right).$$

Then $\phi(x)$ is positively bounded away from 0 on Ω and satisfies

$$\triangle\phi(x) = -\frac{\phi(x)}{l^2}.$$

Let $w(x) = \frac{U(x)}{\phi(x)}$. It follows from (11.14) that there exists some ξ such that

$$w(\xi) = \min_{\Sigma_\lambda} w < 0.$$

At the negative minimum point of w, we have

$$\triangle U(\xi) = \triangle w(\xi)\phi(\xi) + 2\nabla w(\xi) \cdot \nabla\phi(\xi) + w(\xi)\triangle\phi(\xi)$$

$$= \triangle w(\xi)\phi(\xi) - w(\xi)\frac{\phi(\xi)}{l^2}$$

$$\geq -w(\xi)\frac{\phi(\xi)}{l^2} \geq -w(\bar{x})\frac{\phi(\xi)}{l^2} \geq -U(\bar{x})\frac{\phi(\xi)}{\phi(\bar{x})l^2}. \tag{11.15}$$

On the other hand, by (11.12) and

$$-\triangle U(x) + c_1(x)V(x) \geq 0, \tag{11.16}$$

we know

$$\triangle U(\xi) \leq c_1(\xi)\, V(\xi) \tag{11.17}$$
$$\leq c_1(\xi)\, V(x^0)$$
$$\leq -c_1(\xi)\, c_2(x^0)\, U(x^0)\, l^\alpha. \tag{11.18}$$

Combining (11.15) and (11.17), it implies that $c_1(\xi) \neq 0$. And by $c_1(x) \leq 0$ in Ω and (11.18) we arrive at

$$l^{2+\alpha}\, \frac{\phi(\bar{x})\, c_1(\xi)\, c_2(x^0)}{\phi(\xi)} \geq 1.$$

The inequality above does not hold for l sufficiently small. Hence we obtain that $V(x)$ must be non-negative. Then from (11.16) it gives

$$\begin{cases} -\triangle U(x) \geq 0, \ x \in \Omega, \\ U(x) \geq 0, \qquad x \in \partial\Omega. \end{cases}$$

Applying the *maximum principle*, it yields

$$U(x) \geq 0, \quad x \in \Omega.$$

What remains to be shown is (11.10), also known as the strong maximum principle.

Suppose there exists some $\eta \in \Omega$ such that $U(\eta) = 0$. Then η is the minimum point of U. Together with (11.16), we have

$$0 \leq \triangle U(\eta) \leq c_1(\eta)V(\eta) \leq 0.$$

Since $c_1(x) < 0$ in Ω, it's easy to see that

$$V(\eta) = 0 = \min_{\Sigma_\lambda} V.$$

Therefore,

$$(-\triangle)^{\alpha/2} V(\eta)$$
$$= C_{n,\alpha} PV \int_{\mathbb{R}^n} \frac{-V(y)}{|\eta - y|^{n+\alpha}} dy$$
$$= C_{n,\alpha} PV \int_{\Sigma_\lambda} \frac{-V(y)}{|\eta - y|^{n+\alpha}} dy + \int_{\Sigma_\lambda} \frac{-V(y^\lambda)}{|\eta - y^\lambda|^{n+\alpha}} dy$$
$$= C_{n,\alpha} PV \int_{\Sigma_\lambda} \left(\frac{1}{|\eta - y^\lambda|^{n+\alpha}} - \frac{1}{|\eta - y|^{n+\alpha}} \right) V(y)\, dy. \tag{11.19}$$

If

$$V(x) \not\equiv 0, \quad x \in \Sigma_\lambda,$$

then (11.19) implies that

$$(-\triangle)^{\alpha/2}V(\eta) < 0.$$

Together with (11.13), it shows that

$$U(\eta) < 0.$$

However, previously we already proved that $U(x) \geq 0$ in Ω. This is a contradiction.

It thus shows that $V(x)$ must be identically 0 in Σ_λ. By

$$V(x^\lambda) = -V(x), \quad x \in \Sigma_\lambda,$$

we deduce that

$$V(x) \equiv 0, \quad x \in \mathbb{R}^n.$$

Together with (11.13), we obtain $U(x) \leq 0, x \in \Omega$. Formerly we already prove that $U(x) \geq 0, x \in \Omega$. Therefore, $U(x) = 0, x \in \Omega$. Thus we prove

$$U(x) \equiv 0, \quad x \in \Omega, \text{ and } V(x) \equiv 0, \quad x \in \mathbb{R}^n. \tag{11.20}$$

Suppose there exists a $\xi \in \Omega$ such that $V(\xi) = 0$. If

$$V(x) \equiv 0, x \in \Sigma_\lambda,$$

then repeating the argument right above we can obtain (11.20). If

$$V(x) \not\equiv 0, x \in \Sigma_\lambda,$$

then from (11.13) and (11.19) it follows

$$-c_2(\xi)U(\xi) \leq (-\triangle)^{\alpha/2}V(\xi) < 0.$$

Thus $U(\xi) = 0$. The rest is the same as the previous argument. And we complete the proof of the lemma. $\square$

Theorem 11.2.1 *Let $s = 1 + \frac{\alpha}{2}$ for $0 < \alpha < 2$. Assume that $u \in C^{3,1}_{loc}(B_1(0))$ is lower semi-continuous on $\overline{B_1(0)}$. Let u be a positive solution of*

$$\begin{cases} (-\triangle)^s u = f(u), x \in B_1, \\ u = \triangle u = 0, \quad x \in \mathbb{R}^n \backslash B_1, \end{cases} \tag{11.21}$$

where $f(t)$ is Lipschitz continuous and increasing in t. If $f(t) \geq 0$ for $t > 0$, then u must be symmetric about the origin.

Remark 11.2.1 *Here and in the sequel, we define the higher order operator as the composition of two lower order ones as the following:*

$$(-\triangle)^s u = (-\triangle)^{\alpha/2} \circ (-\triangle)u.$$

For domains with boundaries, the order of $(-\triangle)^{\alpha/2}$ and $(-\triangle)$ may not be exchanged in general.

Proof. Let $-\triangle u = v$. Then (11.21) can be split into two equations:

$$\begin{cases} (-\triangle)^{\alpha/2} v = f(u), \ x \in B_1, \\ v = 0, \qquad\qquad x \in \mathbb{R}^n \backslash B_1, \end{cases} \tag{11.22}$$

and

$$\begin{cases} -\triangle u = v, \ x \in B_1, \\ u = 0, \qquad x \in \mathbb{R}^n \backslash B_1. \end{cases} \tag{11.23}$$

From (11.22) and (11.11), it's easy to see that

$$v(x) \geq 0, \quad x \in B_1. \tag{11.24}$$

We carry out the proof via the *method of moving planes*.
To start with, we choose an arbitrary direction to be the x_1-direction. Let

$$T_\lambda = \{x \in \mathbb{R}^n \mid x_1 = \lambda\}, \ \ \Sigma_\lambda = \{x \in \mathbb{R}^n \mid x_1 < \lambda\},$$

and the reflection of x about the plane T_λ be

$$x^\lambda = \{(2\lambda - x_1, x') \mid x = (x_1, x') \in \mathbb{R}^n\}.$$

Let

$$u_\lambda(x) = u(x^\lambda), \quad v_\lambda(x) = v(x^\lambda),$$
$$U_\lambda(x) = u_\lambda(x) - u(x), \qquad V_\lambda(x) = v_\lambda(x) - v(x).$$

Obviously,

$$U_\lambda(x) \geq 0, \quad x \in \partial(\Sigma_\lambda \cap B_1(0)).$$

From (11.24) we can also see that

$$V_\lambda(x) \geq 0, \quad x \in \Sigma_\lambda \backslash (\Sigma_\lambda \cap B_1(0)).$$

Thus we have

$$\begin{cases} -\triangle U_\lambda(x) = V_\lambda(x), & x \in \Sigma_\lambda \cap B_1(0), \\ (-\triangle)^{\alpha/2} V_\lambda(x) = f(u_\lambda) - f(u), & x \in \Sigma_\lambda \cap B_1(0), \\ U_\lambda(x) \geq 0, & x \in \partial(\Sigma_\lambda \cap B_1(0)), \\ V_\lambda(x) \geq 0, & x \in \Sigma_\lambda \backslash (\Sigma_\lambda \cap B_1(0)). \end{cases} \tag{11.25}$$

Step 1. Moving the plane T_λ from -1 to the right along the x_1-axis.
For λ near -1, we show that

$$U_\lambda(x), V_\lambda(x) \geq 0, \quad x \in \Sigma_\lambda. \tag{11.26}$$

Notice that

$$f(u_\lambda) - f(u) = \frac{f(u_\lambda) - f(u)}{u_\lambda(x) - u(x)} U_\lambda(x).$$

This allows us to rewrite (11.25) into the same form as in (11.8) with

$$c_1(x) = -1, \quad c_2(x) = -\frac{f(u_\lambda) - f(u)}{u_\lambda(x) - u(x)}.$$

When λ is close to -1, $\Sigma_\lambda \cap B_1(0)$ is a narrow region. Then by applying Lemma 11.2.1 (*narrow region principle*) to (11.25), we can obtain (11.26). The proof here is entirely similar to that in Lemma 11.2.1, hence we omit it.

Step 2. Keep moving the plane T_λ until the limiting position

$$\lambda_0 = \sup\{\lambda \leq 0 \mid U_\mu(x), V_\mu(x) \geq 0, x \in \Sigma_\mu \cap B_1(0), \forall \mu < \lambda\}.$$

We claim that

$$\lambda_0 = 0. \tag{11.27}$$

If not, then for $\lambda_0 < 0$ we can show that

$$U_{\lambda_0}(x) \equiv 0, \quad x \in \Sigma_{\lambda_0} \cap B_1(0). \tag{11.28}$$

We postpone the proof of (11.28) for the moment. From (11.28), we know that

$$0 = u_{\lambda_0}(-1, 0') = u(1 + 2\lambda_0, 0') > 0, \quad 0' \in R^{n-1}.$$

This is a contradiction and we obtain (11.27).

We prove (11.28). Suppose (11.28) is not true, then by (11.10)(the strong maximum principle), we have

$$U_{\lambda_0}(x) > 0, \quad x \in \Sigma_{\lambda_0} \cap B_1(0). \tag{11.29}$$

This enables us to keep moving the plane T_{λ_0} to the right. Precisely speaking, there exists some $\varepsilon > 0$ small such that $\lambda_0 + \varepsilon < 0$ and

$$U_\lambda(x) \geq 0, \quad x \in \Sigma_\lambda \cap B_1(0), \quad \lambda \in [\lambda_0, \lambda_0 + \varepsilon). \tag{11.30}$$

The inequality above contradicts the definition of λ_0. Therefore (11.28) must be true.

What's left is to show (11.30). We do it by constructing a narrow region and applying Lemma 11.2.1. The only condition we need to check is the boundary condition of $U_\lambda(x)$. For $\delta > 0$ small, from (11.29) we know there exists a constant C such that

$$U_{\lambda_0}(x) \geq C > 0, \quad x \in \overline{\Sigma_{\lambda-\delta} \cap B_1(0)}.$$

By the continuity of U in λ, we have

$$U_\lambda(x) \geq 0, \quad x \in \overline{\Sigma_{\lambda-\delta} \cap B_1(0)}. \tag{11.31}$$

It's easy to see that

$$U_\lambda(x) \geq 0, \quad x \in \partial((\Sigma_\lambda \backslash \Sigma_{\lambda-\delta}) \cap B_1(0)).$$

Since $\Sigma_\lambda \backslash \Sigma_{\lambda-\delta}$ is a narrow region, by Lemma 11.2.1 we derive that

$$U_\lambda(x) \geq 0, \quad x \in (\Sigma_\lambda \backslash \Sigma_{\lambda-\delta}) \cap B_1(0).$$

Together with (11.31), we arrive at (11.30). This completes the proof of (11.27).

Now we have
$$U_0(x) \geq 0, \quad x \in \Sigma_0 \cap B_1(0).$$

Similarly, one can move T_λ from near 1 to the left and show that
$$U_0(x) \leq 0, \quad x \in \Sigma_0 \cap B_1(0).$$

Thus
$$U_0(x) \equiv 0, \quad x \in \Sigma_0 \cap B_1(0).$$

Since the direction of the x_1-axis is arbitrary, we have actually proved that u is symmetric about the origin.   □

11.3 The Integral Equation Method

Theorem 11.3.1 *Let $s = 1 + \frac{\alpha}{2}$ for $0 < \alpha < 2$. Consider*
$$\begin{cases} (-\triangle)^s u(x) = u^p(x), & x \in \mathbb{R}^n_+, \\ u = 0, & x \notin \mathbb{R}^n_+, \\ \triangle u = 0, & x \in \partial \mathbb{R}^n_+, \end{cases} \tag{11.32}$$

with $p > 1$. If $u \in L^{\frac{n(p-1)}{2s}}(\mathbb{R}^n_+)$ and $-\triangle u > 0$ in $\mathbb{R}^n_+$, then equation (11.32) has no positive solution. Here $(-\triangle)^s = (-\triangle)^{\alpha/2} \circ (-\triangle)$.

We first seek the equivalent integral representation of the pseudo-differential equation.

Lemma 11.3.1 *Suppose u is a positive solution of (11.32). If $-\triangle u > 0$, then u must satisfy the integral equation*
$$u(x) = \int_{\mathbb{R}^n_+} G^+_{2s}(x,y) u^p(y)\, dy, \tag{11.33}$$

where
$$G^+_{2s}(x,y) = \int_{\mathbb{R}^n_+} G^+_2(x,z) G^+_\alpha(z,y)\, dz,$$

and G^+_2 and G^+_α are the Green's functions associated with $-\triangle$ and $(-\triangle)^{\alpha/2}$ in $\mathbb{R}^n_+$ respectively.

Proof. Set $v(x) = -\triangle u(x)$. It's easy to see that (11.32) can be divided into a system of two equations:
$$\begin{cases} (-\triangle)^{\alpha/2} v(x) = u^p(x), & x \in \mathbb{R}^n_+, \\ v(x) = 0, & x \notin \mathbb{R}^n_+, \end{cases} \tag{11.34}$$

and

$$\begin{cases} -\triangle u(x) = v(x), \ x \in \mathbb{R}^n_+, \\ u(x) = 0, \qquad\quad x \notin \mathbb{R}^n_+. \end{cases}$$

We carry out the proof in three steps.

Step 1. Set $P_R = (0, \ldots, 0, R)$ and $B_R(P_R) = \{x \in \mathbb{R}^n \,|\, |x - P_R| < R\}$. Let

$$\bar{w}_R(x) = \int_{B_R(P_R)} G_{R,\alpha}(x, y) u^p(y) dy,$$

where $G_{R,\alpha}(x, y)$ is the Green's function of $(-\triangle)^{\alpha/2}$ on $B_R(P_R)$ (see (5.45) in Chapter 5). Then

$$\begin{cases} (-\triangle)^{\alpha/2} \bar{w}_R(x) = u^p(x), \ x \in B_R(P_R), \\ \bar{w}_R(x) = 0, \qquad\qquad x \notin B_R(P_R). \end{cases}$$

Let

$$\bar{w}(x) = \int_{\mathbb{R}^n_+} G^+_\alpha(x, y) u^p(y) dy.$$

By (11.34), we have

$$\begin{cases} (-\triangle)^{\alpha/2} v(x) = u^p(x), \ x \in B_R(P_R), \\ v(x) \geq 0, \qquad\qquad x \notin B_R(P_R). \end{cases}$$

Set $V_R(x) = v(x) - \bar{w}_R(x)$, and then we have

$$\begin{cases} (-\triangle)^{\alpha/2} V_R(x) = 0, \ x \in B_R(P_R), \\ V_R(x) \geq 0, \qquad\qquad x \notin B_R(P_R). \end{cases}$$

Applying the *maximum principle*, it gives

$$V_R(x) = v(x) - \bar{w}_R(x) \geq 0, \quad x \in B_R(P_R).$$

Therefore, for any fixed $x \in \mathbb{R}^n_+$, there exists an R such that $x \in B_R(P_R)$, and hence

$$\bar{w}_R(x) \leq v(x) < \infty, \quad x \in \mathbb{R}^n_+. \tag{11.35}$$

It's known that

$$G_{R,\alpha}(x, y) \to G^+_\alpha(x, y), \quad as \ R \to \infty. \tag{11.36}$$

Combining (11.35) and (11.36), it yields

$$\bar{w}_R(x) \to \bar{w}(x), \quad as \ R \to \infty. \tag{11.37}$$

Together with (11.35), we deduce that

$$v(x) \geq \bar{w}(x), \quad x \in \mathbb{R}^n_+.$$

Set $V(x) = v(x) - \bar{w}(x)$, then

$$V(x) \geq 0,$$

and

$$\begin{cases} (-\triangle)^{\alpha/2} V(x) = 0, & x \in \mathbb{R}^n_+, \\ V(x) = 0, & x \notin \mathbb{R}^n_+. \end{cases} \tag{11.38}$$

By Theorem 5.2.1 in Chapter 5, we know that if $V(x)$ is a nonnegative solution of (11.38), then

$$V(x) = C_0 \, x_n^{\alpha/2},$$

where $C_0 \geq 0$ is some constant. This implies that

$$v(x) = C_0 \, x_n^{\alpha/2} + \int_{\mathbb{R}^n_+} G_\alpha^+(x,y) u^p(y) dy, \ \ C_0 \geq 0. \tag{11.39}$$

Step 2. Let $B_R^+ = B_R(0) \cap \mathbb{R}^n_+$ and $G_{R,2}^+(x,y)$ be the Green's function of $-\triangle$ in B_R^+. It's known (see [CC]) that

$$G_{R,2}^+(x,y) = \frac{1}{|x-y|^{n-2}} - \frac{1}{(|x-y|^2 + (R - \frac{|x|^2}{R})(R - \frac{|y|^2}{R}))^{\frac{n-2}{2}}}$$
$$- \frac{1}{|x^*-y|^{n-2}} + \frac{1}{(|x^*-y|^2 + (R - \frac{|x|^2}{R})(R - \frac{|y|^2}{R}))^{\frac{n-2}{2}}},$$

and

$$\frac{\partial G_{R,2}^+}{\partial \nu} < 0, \quad x \in B_R^+, \ y \in \partial B_R^+, \tag{11.40}$$

and

$$G_{R,2}^+(x,y) \to G_2^+(x,y), \quad \text{as } R \to \infty. \tag{11.41}$$

Integrating by parts, we have

$$\int_{B_R^+} -\triangle u(y) G_{R,2}^+(x,y) dy$$

$$= \int_{B_R^+} -\triangle G_{R,2}^+(x,y) u(y) dy + \int_{\partial B_R^+} u \frac{\partial G_{R,2}^+}{\partial \nu} ds. \tag{11.42}$$

Combining (11.42), (11.40) with (11.41), it gives

$$\int_{B_R^+} v(y) G_{R,2}^+(x,y) dy \leq u(x), \quad x \in B_R^+. \tag{11.43}$$

Sending $R \to \infty$ in (11.43) yields

$$\int_{\mathbb{R}^n_+} v(y) G_2^+(x,y) dy \leq u(x), \quad x \in \mathbb{R}^n_+.$$

Let

$$H(x) = u(x) - \int_{\mathbb{R}^n_+} G_2^+(x,y) v(y) dy.$$

It's easy to see that

$$\begin{cases} -\triangle H(x) = 0,\ x \in \mathbb{R}^n_+, \\ H(x) = 0, \qquad x \notin \mathbb{R}^n_+. \end{cases}$$

In [CC], the authors proved that

$$H(x) = C_1 x_n, \quad C_1 \geq 0.$$

Therefore,

$$u(x) = C_1 x_n + \int_{\mathbb{R}^n_+} v(y) G_2^+(x,y) dy, \quad C_1 \geq 0. \tag{11.44}$$

Step 3. We claim that

$$C_0 = C_1 = 0.$$

Otherwise, if $C_1 > 0$, then (11.44) implies that

$$u(x) \geq C_1 x_n.$$

For any fixed x, it follows from (11.39) that

$$+\infty > v(x) \geq C_0 x_n^{\alpha/2} + \int_{\mathbb{R}^n_+} G_\alpha^+(x,y)(C_1 y_n)^p dy = \infty.$$

A contradiction. Hence, $C_1 = 0$.

To show that $C_0 > 0$, we notice that (11.39) implies

$$v(x) \geq C_0\, x_n^{\alpha/2}. \tag{11.45}$$

Substituting (11.45) into (11.44), we arrive at

$$+\infty > u(x) \geq \int_{\mathbb{R}^n_+} G_2^+(x,y)\, C_0\, y_n^{\alpha/2} dy = \infty.$$

This, again, is a contradiction. Therefore, $C_0 = 0$.

Thus,

$$v(x) = \int_{\mathbb{R}^n_+} G_\alpha^+(x,y) u^p(y) dy, \tag{11.46}$$

and

$$u(x) = \int_{\mathbb{R}^n_+} G_2^+(x,y) v(y) dy. \tag{11.47}$$

Substituting (11.46) into (11.47), it gives

$$\begin{aligned} u(x) &= \int_{\mathbb{R}^n_+} G_2^+(x,y) \int_{\mathbb{R}^n_+} G_\alpha^+(y,z) u^p(z) dz dy \\ &= \int_{\mathbb{R}^n_+} \left(\int_{\mathbb{R}^n_+} G_2^+(x,y) G_\alpha^+(y,z) dy \right) u^p(z) dz \\ &= \int_{\mathbb{R}^n_+} G_{2s}^+(x,z) u^p(z) dz. \end{aligned}$$

What remains is to show that if u is a solution of the integral equation (11.33), then u satisfies (11.32). Indeed,

$$
\begin{aligned}
(-\triangle)^s u(x) &= \int_{\mathbb{R}^n_+} \left(\int_{\mathbb{R}^n_+} (-\triangle)^s G_2^+(x,y) G_\alpha^+(y,z) dy \right) u^p(z) dz \\
&= \int_{\mathbb{R}^n_+} \left(\int_{\mathbb{R}^n_+} (-\triangle)^{\alpha/2} \circ (-\triangle) G_2^+(x,y) G_\alpha^+(y,z) dy \right) u^p(z) dz \\
&= \int_{\mathbb{R}^n_+} (-\triangle)^{\alpha/2} G_\alpha^+(x,z) u^p(z) dz \\
&= u^p(x). \quad \square
\end{aligned}
$$

Next, we derive some properties of the Green's function $G_{2s}^+(x,y)$, which is essential in the process of moving the planes.

For any real number λ, let

$$
\Sigma_\lambda = \left\{ x = (x', x_n) \in \mathbb{R}^n_+ \mid 0 < x_n < \lambda \right\},
$$

$$
T_\lambda = \left\{ x \in \mathbb{R}^n_+ \mid x_n = \lambda \right\},
$$

and let

$$
x^\lambda = (x_1, x_2, \cdots, 2\lambda - x_n)
$$

be the reflection of the point $x = (x', x_n)$ about the plane T_λ.

It's well-known that for $0 < \alpha \le 2$, the Green's function for $(-\triangle)^{\alpha/2}$ in $\mathbb{R}^n_+$ is

$$
G_\alpha^+(x,y) = \frac{C_{n,\alpha}}{|x-y|^{n-\alpha}} \int_0^{\frac{4x_n y_n}{|x-y|^2}} \frac{b^{\frac{\alpha}{2}-1}}{(1+b)^{n/2}} \, db, \tag{11.48}
$$

and the Green's function for $-\triangle$ in $\mathbb{R}^n_+$ is

$$
G_2^+(x,y) = \frac{1}{|x-y|^{n-2}} - \frac{1}{|x^*-y|^{n-2}}, \tag{11.49}
$$

where x^* is the reflection point of x with respect to $\partial \mathbb{R}^n_+$.

Lemma 11.3.2 *The Green's function $G_{2s}^+(x,y)$ satisfies the following properties:*

1. *For any $x, y \in \Sigma_\lambda$ and $x \ne y$, we have*

$$
G_{2s}^+(x^\lambda, y^\lambda) > \max\{G_{2s}^+(x^\lambda, y), G_{2s}^+(x, y^\lambda)\} \tag{11.50}
$$

and

$$
G_{2s}^+(x^\lambda, y^\lambda) - G_{2s}^+(x, y) > |G_{2s}^+(x^\lambda, y) - G_{2s}^+(x, y^\lambda)|. \tag{11.51}
$$

2. *For any $x \in \Sigma_\lambda$, $y \in \Sigma_\lambda^C$, it holds*

$$G_{2s}^+(x^\lambda, y) > G_{2s}^+(x, y) \quad and \quad G_{2s}^+(y, x^\lambda) > G_{2s}^+(y, x). \qquad (11.52)$$

3. *For x, $y \in \mathbb{R}_+^n$,*

$$G_{2s}^+(x, y) \leq \frac{C}{|x - y|^{n-2s}}. \qquad (11.53)$$

Proof. Let $\tilde{\Sigma}_\lambda$ be the reflection of Σ_λ about T_λ, and $D_\lambda = \mathbb{R}_+^n \setminus (\Sigma_\lambda \cup \tilde{\Sigma}_\lambda)$. To prove the lemma, we use the well-known facts that inequalities (11.50)-(11.52) are also true for $G_2^+(x, y)$ and $G_\alpha^+(x, y)$. Then

1.

$$G_{2s}^+(x^\lambda, y^\lambda) - G_{2s}^+(x^\lambda, y)$$

$$= \int_{\mathbb{R}_+^n} G_2^+(x^\lambda, z)\big(G_\alpha^+(z, y^\lambda) - G_\alpha^+(z, y)\big) dz$$

$$= \int_{\Sigma_\lambda \cup \tilde{\Sigma}_\lambda \cup D_\lambda} G_2^+(x^\lambda, z)\big(G_\alpha^+(z, y^\lambda) - G_\alpha^+(z, y)\big) dz$$

$$= \int_{\Sigma_\lambda} G_2^+(x^\lambda, z)\big(G_\alpha^+(z, y^\lambda) - G_\alpha^+(z, y)\big) dz$$

$$+ \int_{\Sigma_\lambda} G_2^+(x^\lambda, z^\lambda)\big(G_\alpha^+(z^\lambda, y^\lambda) - G_\alpha^+(z^\lambda, y)\big) dz$$

$$+ \int_{D_\lambda} G_2^+(x^\lambda, z)\big(G_\alpha^+(z, y^\lambda) - G_\alpha^+(z, y)\big) dz$$

$$= I_1 + I_2. \qquad (11.54)$$

Since G_α^+ satisfies (11.51), we have

$$G_\alpha^+(z^\lambda, y) - G_\alpha^+(z, y) \geq G_\alpha^+(z, y^\lambda) - G_\alpha^+(z^\lambda, y^\lambda), \quad x, y \in \Sigma_\lambda.$$

Thus

$$I_1 = \int_{\Sigma_\lambda} \bigg(G_2^+(x^\lambda, z)\big[G_\alpha^+(z, y^\lambda) - G_\alpha^+(z, y)\big]$$

$$+ G_2^+(x^\lambda, z^\lambda)\big[G_\alpha^+(z^\lambda, y^\lambda) - G_\alpha^+(z^\lambda, y)\big] \bigg) dz$$

$$\geq \int_{\Sigma_\lambda} \bigg(- G_2^+(x^\lambda, z)[G_\alpha^+(z^\lambda, y^\lambda) - G_\alpha^+(z^\lambda, y)]$$

$$+ G_2^+(x^\lambda, z^\lambda)[G_\alpha^+(z^\lambda, y^\lambda) - G_\alpha^+(z^\lambda, y)] \bigg) dz$$

$$= \int_{\Sigma_\lambda} [G_2^+(x^\lambda, z^\lambda) - G_2^+(x^\lambda, z)][G_\alpha^+(z^\lambda, y^\lambda) - G_\alpha^+(z^\lambda, y)] dz$$

$$> 0. \qquad (11.55)$$

It also follows from (11.52) that for G_α^+ we have

$$I_2 = \int_{D_\lambda} G_2^+(x^\lambda, z)(G_\alpha^+(z, y^\lambda) - G_\alpha^+(z, y)) dz > 0. \qquad (11.56)$$

Combining (11.54), (11.55) and (11.56), it yields

$$G_{2s}^+(x^\lambda, y^\lambda) \geq G_{2s}^+(x^\lambda, y), \ x, y \in \Sigma_\lambda, \ x \neq y.$$

Similarly, one can show that

$$G_{2s}^+(x^\lambda, y^\lambda) \geq G_{2s}^+(x, y^\lambda), \ x, y \in \Sigma_\lambda, \ x \neq y.$$

This verifies (11.50). To prove (11.51), let

$$[G_{2s}^+(x^\lambda, y^\lambda) - G_{2s}^+(x, y)] - [G_{2s}^+(x^\lambda, y) - G_{2s}^+(x, y^\lambda)]$$
$$= \int_{\Sigma_\lambda} K dz + \int_{D_\lambda} Q dz, \qquad (11.57)$$

where

$$K = [G_2^+(x^\lambda, z) + G_2^+(x, z)][G_\alpha^+(z, y^\lambda) - G_\alpha^+(z, y)]$$
$$+ [G_2^+(x^\lambda, z^\lambda) + G_2^+(x, z^\lambda)][G_\alpha^+(z^\lambda, y^\lambda) - G_\alpha^+(z^\lambda, y)]$$
$$> -[G_2^+(x^\lambda, z) + G_2^+(x, z)][G_\alpha^+(z^\lambda, y^\lambda) - G_\alpha^+(z^\lambda, y)]$$
$$+ [G_2^+(x^\lambda, z^\lambda) + G_2^+(x, z^\lambda)][G_\alpha^+(z^\lambda, y^\lambda) - G_\alpha^+(z^\lambda, y)]$$
$$= [G_2^+(x^\lambda, z^\lambda) + G_2^+(x, z^\lambda) - G_2^+(x^\lambda, z) - G_2^+(x, z)]$$
$$\cdot [G_\alpha^+(z^\lambda, y^\lambda) - G_\alpha^+(z^\lambda, y)], \qquad (11.58)$$

and

$$Q = [G_2^+(x^\lambda, z) + G_2^+(x, z)][G_\alpha^+(z, y^\lambda) - G_\alpha^+(z, y)]. \qquad (11.59)$$

Since G_α^+ satisfies both 1 and 2 in the lemma, it's obvious that

$$K, G > 0.$$

This shows that

$$G_{2s}^+(x^\lambda, y^\lambda) - G_{2s}^+(x, y) > G_{2s}^+(x^\lambda, y) - G_{2s}^+(x, y^\lambda).$$

Similarly, one can show that

$$G_{2s}^+(x^\lambda, y^\lambda) - G_{2s}^+(x, y) > G_{2s}^+(x, y^\lambda) - G_{2s}^+(x^\lambda, y).$$

This verifies (11.51).
2. For $x \in \Sigma_\lambda$ and $y \in \Sigma_\lambda^C$, let

$$G_{2s}^+(x^\lambda, y) - G_{2s}^+(x, y) = \int_{\Sigma_\lambda} M dz + \int_{D_\lambda} N dz, \qquad (11.60)$$

where

$$M = [G_2^+(x^\lambda, z) - G_2^+(x, z)]G_\alpha^+(z, y)$$
$$+ [G_2^+(x^\lambda, z^\lambda) - G_2^+(x, z^\lambda)]G_\alpha^+(z^\lambda, y)$$
$$> [G_2^+(x^\lambda, z^\lambda) - G_2^+(x, z^\lambda)][G_\alpha^+(z^\lambda, y) - G_\alpha^+(z, y)] > 0. \quad (11.61)$$

Meanwhile, we have

$$N = [G_2^+(x^\lambda, z) - G_2^+(x, z)]G_\alpha^+(z, y) > 0. \quad (11.62)$$

Combining (11.60), (11.61) and (11.62), we arrive at (11.52).

3. By (11.48) and (11.49), we can see that

$$G_2^+(x, y) \leq \frac{C}{|x - y|^{n-2}}, \quad x, y \in \mathbb{R}_+^n,$$

and

$$G_\alpha^+(x, y) \leq \frac{C}{|x - y|^{n-\alpha}}, \quad x, y \in \mathbb{R}_+^n.$$

Hence

$$G_{2s}^+(x, y)$$
$$= \int_{\mathbb{R}_+^n} G_2^+(x, z)G_\alpha^+(z, y)dz$$
$$\leq \int_{\mathbb{R}_+^n} \frac{C}{|x - z|^{n-2}|z - y|^{n-\alpha}}dz$$
$$= \int_{\bar{z}_n > -y_n} \frac{C}{|x - y - \bar{z}|^{n-2}|\bar{z}|^{n-\alpha}}d\bar{z}, \quad \bar{z} + y = z$$
$$= \frac{C}{|x - y|^{n-2s}} \int_{\tilde{z}_n > -\frac{y_n}{|x-y|}} \frac{d\tilde{z}}{|\tilde{z}|^{n-\alpha}|\frac{x-y}{|x-y|} - \tilde{z}|^{n-2}}, \quad \bar{z} = |x - y|\tilde{z}$$
$$= \frac{C}{|x - y|^{n-2s}}.$$

Now we are ready to prove Theorem 11.3.1.

Proof. Let

$$u_\lambda(x) = u(x^\lambda), \quad w_\lambda(x) = u_\lambda(x) - u(x).$$

Then

Lemma 11.3.3

$$w_\lambda(x) \geq \int_{\Sigma_\lambda} [G_{2s}^+(x^\lambda, y^\lambda) - G_{2s}^+(x, y^\lambda)](u_\lambda^p(y) - u^p(y))\, dy. \quad (11.63)$$

Proof. By (11.33) and Property 2 in Lemma 11.3.2, we have

$$w_\lambda(x) = u_\lambda(x) - u(x)$$

$$= \int_{\mathbb{R}^n_+} G^+_{2s}(x^\lambda, y) u^p(y) dy - \int_{\mathbb{R}^n_+} G^+_{2s}(x, y) u^p(y) dy$$

$$= \int_{\Sigma_\lambda \cup \tilde{\Sigma}_\lambda \cup D_\lambda} G^+_{2s}(x^\lambda, y) u^p(y) dy - \int_{\Sigma_\lambda \cup \tilde{\Sigma}_\lambda \cup D_\lambda} G^+_{2s}(x, y) u^p(y) dy$$

$$= \int_{\Sigma_\lambda} [G^+_{2s}(x^\lambda, y) - G^+_{2s}(x, y)] u^p(y) dy$$

$$+ \int_{\Sigma_\lambda} [G^+_{2s}(x^\lambda, y^\lambda) - G^+_{2s}(x, y^\lambda)] u^p_\lambda(y) dy$$

$$+ \int_{D_\lambda} [G^+_{2s}(x^\lambda, y) - G^+_{2s}(x, y)] u^p(y) dy$$

$$\geq \int_{\Sigma_\lambda} [G^+_{2s}(x^\lambda, y) - G^+_{2s}(x, y)] u^p(y) dy$$

$$+ \int_{\Sigma_\lambda} [G^+_{2s}(x^\lambda, y^\lambda) - G^+_{2s}(x, y^\lambda)] u^p_\lambda(y) dy.$$

By Property 1 in Lemma 11.3.2, one can see that for x, $y \in \Sigma_\lambda$,

$$G^+_{2s}(x^\lambda, y) - G^+_{2s}(x, y) \geq G^+_{2s}(x, y^\lambda) - G^+_{2s}(x^\lambda, y^\lambda).$$

Therefore,

$$w_\lambda(x) \geq \int_{\Sigma_\lambda} [G^+_{2s}(x, y^\lambda) - G^+_{2s}(x^\lambda, y^\lambda)] u^p(y) dy$$

$$+ \int_{\Sigma_\lambda} [G^+_{2s}(x^\lambda, y^\lambda) - G^+_{2s}(x, y^\lambda)] u^p_\lambda(y) dy$$

$$\geq \int_{\Sigma_\lambda} [G^+_{2s}(x^\lambda, y^\lambda) - G^+_{2s}(x, y^\lambda)][u^p_\lambda(y) - u^p(y)] dy.$$

This proves the lemma.

Next we use the method of moving planes to derive a contradiction assuming that (11.32) has positive solutions.

Step 1. Start moving the plane T_λ from near $x_n = 0$ upward along the x_n-axis.

Let $\Sigma^-_\lambda = \{x \in \Sigma_\lambda \mid w_\lambda(x) < 0\}$. We show that

$$w_\lambda(x) \geq 0, \quad a.e. \ \Sigma_\lambda. \tag{11.64}$$

If not, then for any $x^o \in \Sigma^-_\lambda$, by (11.63) and the *mean value theorem*, one have

$$0 < -w_\lambda(x^o) \leq - \int_{\Sigma^-_\lambda} [G^+_{2s}(x^\lambda, y^\lambda) - G^+_{2s}(x, y^\lambda)](u^p_\lambda(y) - u^p(y)) \, dy$$

$$\leq -C \int_{\Sigma^-_\lambda} \frac{1}{|x^\lambda - y^\lambda|^{n-2s}} (u^p_\lambda(y) - u^p(y)) \, dy$$

$$\leq C \int_{\Sigma^-_\lambda} \frac{1}{|x - y|^{n-2s}} p u^{p-1}(y)(-w_\lambda(y)) \, dy.$$

Using an equivalent form of the Hardy-Little-Sobolev inequality (see Lemma 7.2.1) and the Hölder inequality, it follows that

$$\|w_\lambda\|_{L^q(\Sigma_\lambda^-)} \leq C\|u^{p-1}w_\lambda\|_{L^{\frac{nq}{n+2sq}}(\Sigma_\lambda^-)}$$

$$\leq C\|u^{p-1}\|_{L^{\frac{n}{2s}}(\Sigma_\lambda^-)}\|w_\lambda\|_{L^q(\Sigma_\lambda^-)},$$

for any $\frac{n}{n-2s} < q < \infty$. Since $u \in L^{\frac{n(p-1)}{2s}}(\mathbb{R}_+^n)$, for λ sufficiently negative, we have

$$C\|u^{p-1}\|_{L^{\frac{n}{2s}}(\Sigma_\lambda^-)} < \frac{1}{2}. \tag{11.65}$$

Thus

$$\|w_\lambda\|_{L^q(\Sigma_\lambda^-)} < \frac{1}{2}\|w_\lambda\|_{L^q(\Sigma_\lambda^-)},$$

i.e., Σ_λ^- is measure 0.

This proves (11.64) and completes step 1.

Step 2. Keep moving the plane until the limiting position

$$\lambda_0 = \sup\{\lambda \leq \infty \mid w_\mu(x) \geq 0, \ a.e. \ in \ \Sigma_\mu, \ \mu \leq \lambda\}.$$

We assert that

$$\lambda_0 = \infty. \tag{11.66}$$

If not, then for $\lambda_0 < \infty$, we can show that

$$w_{\lambda_0}(x) \equiv 0. \tag{11.67}$$

Thus for any $x^0 \in \partial\mathbb{R}_+^n$, we have

$$0 < u_{\lambda_0}(x^0) = u(x^0) = 0.$$

The contradiction establishes (11.66).

To prove (11.67), we suppose its contrary is true, i.e., there exists a set D of positive measure in Σ_{λ_0} such that

$$w_{\lambda_0}(x) > 0, \quad x \in D. \tag{11.68}$$

Thus, for any $x \in \Sigma_{\lambda_0}$, by (11.63),

$$w_{\lambda_0}(x) \geq \int_{\Sigma_{\lambda_0}} [G_{2s}^+(x^{\lambda_0}, y^{\lambda_0}) - G_{2s}^+(x, y^{\lambda_0})](u_{\lambda_0}^p(y) - u^p(y))\, dy$$

$$\geq \int_D [G_{2s}^+(x^{\lambda_0}, y^{\lambda_0}) - G_{2s}^+(x, y^{\lambda_0})](u_{\lambda_0}^p(y) - u^p(y))\, dy$$

$$> 0.$$

Such strict positivity allows us to keep moving T_{λ_0} to the right while preserving (11.64). In other words, there exists some $\varepsilon > 0$ small such that for any $\lambda \in (\lambda_0, \lambda_0 + \varepsilon)$, it holds that

$$w_\lambda(x) \geq 0, \quad a.e. \ \Sigma_\lambda. \tag{11.69}$$

This is a contradiction with the definition of λ_0. It thus verifies (11.67). The proof of (11.69) is almost the same as that of Theorem 7.2.1 in Chapter 7. Here we omit it.

It follows from $\lambda_0 = \infty$ that $u(x)$ is monotone increasing in the x_n-direction. This violates our assumption that $u \in L^{\frac{n(p-1)}{2s}}(\mathbb{R}^n_+)$. Therefore, (11.32) has no positive solution. $\square$

12

The Regularity for Fractional Equations

12.1 Introduction

In this chapter, we study regularity of solutions to fractional equations. We derive Schauder interior estimates and Hölder estimates up to the boundary for the solutions of the fractional equation

$$(-\triangle)^s u(x) = f(x), \quad x \in \Omega.$$

The Schauder interior estimate is similar to that for the Poisson equation when $s = 1$. It states roughly that if $f \in C^\alpha(D)$ in some open set D, then in any proper subset of D, the regularity of the solution u can be raised by the order of $2s$, the same order as the operator $(-\triangle)^s$. By introducing proper weighted Hölder norms as in the case of Poisson equations, we will be able to control a weighted $C^{2s+\alpha}$ norm of u in Ω in terms of another weighted C^α norm of f in Ω.

However, when consider regularity up to the boundary, the situation in the fractional order equation is quite different from that in the integer order equation (when $s = 1$, the Poisson equation). For integer order equations, a typical example of regularity up to the boundary reads

Proposition 12.1.1 *(See [GT] Theorem 4.13) Let B be a ball in $\mathbb{R}^n$ and u and f functions on B satisfying*

$$u \in C^2(B) \cap C^0(\bar{B}), \; f \in C^\alpha(B), \; \triangle u = f \text{ in } B, \; u = 0 \text{ on } \partial B.$$

Then

$$u \in C^{2+\alpha}(\bar{B}).$$

This proposition tells us that if f is Hölder continuous, and the boundary data is smooth enough, then the regularity up to the boundary of the solution u can still be raised by order 2. This is no longer the case for the fractional equation as illustrated by the following example.

It is well-known that the function

$$\phi(x) = \begin{cases} (1 - |x|^2)^s, & |x| < 1 \\ 0, & |x| \geq 1 \end{cases}$$

satisfies

$$\begin{cases} (-\triangle)^s \phi(x) = C, & x \in B_1(0), \\ \phi = 0, & x \in \mathbb{R}^n \backslash B_1(0). \end{cases}$$

Here $f(x) \equiv C$ is smooth and the Dirichlet data is also smooth. However, one can easily see that $\phi(x)$ is only C^s up to the boundary for $0 < s < 1$.

Most of the results were obtained in [RS1]. We considerably modified the proofs and unfold them in a natural way so that the readers can followed rather easily.

12.2 Schauder Interior Estimates

12.2.1 The Result and Outline of Proof

In this section, unless otherwise stated, we always assume that Ω is a bounded domain in $\mathbb{R}^n$, and $f \in C^\alpha(\Omega)$ for some $\alpha \in (0,1)$.

Consider

$$(-\triangle)^s u(x) = f(x) \qquad x \in \Omega \tag{12.1}$$

with $0 < s < 1$. We will derive a Schauder type interior estimate for the solutions, which is given in terms of weighted Hölder norms, similar to the traditional Schauder estimates for second order elliptic equations.

We first introduce the weighted Hölder norms. Let

$$d_x = dist(x, \partial\Omega), \quad d_{x,y} = \min\{d_x, d_y\}.$$

Write $2s + \alpha = k + \beta$, where $k = 0, 1, 2$ and $\beta \in (0,1)$.

Recall that the usual C^k norm and $C^{k+\beta}$ semi-norm are defined by

$$\|u\|_{C^k(\Omega)} = \sum_{|\gamma| \leq k} \sup_{x \in \Omega} |D^\gamma u(x)|$$

and

$$[u]_{C^{k+\beta}(\Omega)} = \sum_{|\gamma|=k} \sup_{x \in \Omega} \frac{|D^\gamma u(x) - D^\gamma u(y)|}{|x - y|^\beta},$$

while $C^{k+\beta}$ norm is the sum of the two:

$$\|u\|_{C^{k+\beta}(\Omega)} = \|u\|_{C^k(\Omega)} + [u]_{C^{k+\beta}(\Omega)}.$$

Different from the above norms, the following weighted norms contain scaling factors d_x and $d_{x,y}$.

Let

$$\|u\|^*_{C^k(\Omega)} = \sum_{|\gamma| \leq k} \sup_{x \in \Omega} d_x^{|\gamma|} |D^\gamma u(x)|$$

and

$$[u]^*_{C^{k+\beta}(\Omega)} = \sum_{|\gamma|=k} \sup_{x \in \Omega} d^{k+\beta}_{x,y} \frac{|D^\gamma u(x) - D^\gamma u(y)|}{|x-y|^\beta}.$$

Define

$$\|u\|^*_{C^{k+\beta}(\Omega)} = \|u\|^*_{C^k(\Omega)} + [u]^*_{C^{k+\beta}(\Omega)}.$$

When $s = 1$, equation (12.1) becomes

$$-\triangle u(x) = f(x) \qquad x \in \Omega,$$

and the well-known Schauder estimate states that

$$\|u\|^*_{C^{2+\alpha}(\Omega)} \leq C(n,\alpha)(\|u\|_{L^\infty(\Omega)} + \|f\|^{(2)}_{C^\alpha(\Omega)}),$$

where

$$\|f\|^{(a)}_{C^\alpha(\Omega)} = \sup_{x \in \Omega} d^a_x |f(x)| + \sup_{x \in \Omega} d^{a+\alpha}_{x,y} \frac{|f(x) - f(y)|}{|x-y|^\alpha}$$

is another kind of re-scaling norm.

The main purpose of this section is to derive the similar Schauder type estimate for fractional equation (12.1):

Theorem 12.2.1 *Let u be a solution of equation (12.1), then*

$$\|u\|^*_{C^{2s+\alpha}(\Omega)} \leq C(n,\alpha)(\|u\|_{L^\infty(\mathbb{R}^n)} + \|f\|^{(2s)}_{C^\alpha(\Omega)}). \tag{12.2}$$

Notice that the difference between this and the classical Schauder estimate is that here we have the term $\|u\|_{L^\infty(\mathbb{R}^n)}$ instead of $\|u\|_{L^\infty(\Omega)}$, because of the nonlocal nature of the fractional Laplacian:

For any point x in Ω, the value of $(-\triangle)^s u(x)$ depends not only on the values of u in Ω, but also on its values in $\mathbb{R}^n \backslash \Omega$.

Outline of the Proof

i. We first start with ball regions in $\mathbb{R}^n$ by extending f to be 0 outside of $\Omega = B_3(0)$. We estimate the potential

$$w(x) = \int_{\mathbb{R}^n} \frac{C_{n,s}}{|x-y|^{n-2s}} f(y) dy,$$

since $w(x)$ is one of the solution of fractional equation (12.1). We derive the interior estimate

$$\|w\|_{C^{2s+\alpha}(\bar{B}_1(0))} \leq C\|f\|_{C^\alpha(B_2(0))}. \tag{12.3}$$

ii. Then the difference between w and u is an s-harmonic function h, which can be expressed in terms of a Poisson kernel:

$$h(x) = \int_{|y|>2} P_2(y,x)h(y)dy, \ \forall |x| < 2.$$

Using this integral, we obtain

$$\|h\|_{C^{2s+\alpha}(B_1(0))} \le C(\|f\|_{C^\alpha(B_3(0))} + \|u\|_{L^\infty(\mathbb{R}^n)}). \tag{12.4}$$

It follows from (12.3) and (12.4) that

$$\|u\|_{C^{2s+\alpha}(B_1(0))} \le C(\|f\|_{C^\alpha(B_3(0))} + \|u\|_{L^\infty(\mathbb{R}^n)}). \tag{12.5}$$

iii. For any point $x^o \in \Omega$, let $R = \frac{d_{x^o}}{3}$, then $B_{3R}(x^o) \subset \Omega$. Through a rescaling on (12.5), we deduce

$$\sum_{i=0}^{k} R^i \sup_{x \in B_R} |D^i u(x)| + R^{2s+\alpha}[D^k u]_{C^\beta(B_R)}$$

$$\le C(\|u\|_{L^\infty(B_{3R})} + R^{2s} \sup_{x \in B_{3R}} |f(x)| + R^{\alpha+2s}[f]_{C^\alpha(B_{3R(0)})}). \tag{12.6}$$

Based on (12.6), we finally arrive at (12.2).

Here for simplicity we abuse notation for a moment and denote all the k-th order partial derivatives of u by $D^k u$.

12.2.2 The Estimate on the Potential

The fundamental solution of the fractional Laplacian $(-\triangle)^s$ in $\mathbb{R}^n$ is given by

$$N(x,y) = \frac{C_{n,s}}{|x-y|^{n-2s}}, \ n \ge 3. \tag{12.7}$$

Assume that $f \in C^\alpha(\Omega)$. We extend it to be 0 outside of Ω. Let the potential of f be

$$w(x) = \int_{\mathbb{R}^n} N(x,y)f(y)dy. \tag{12.8}$$

Then, as we have seen in Chapter 2,

$$(-\triangle)^s w(x) = f(x), \ \ x \in \Omega.$$

When $s = 1$, in the classical Schauder estimate, there is only one norm, the $C^{2+\alpha}$ norm to be estimated. However, here we need to consider up to three kinds of norms: $C^{k+\beta}$ norms in the cases $k = 0, 1, 2$ respectively, depending on the values of $2s + \alpha \equiv k + \beta$.

For $k = 1$, we need to estimate $\partial_i w$, and for $k = 2$, we also need to estimate $\partial_{ij} w$. The following lemmas provide the expressions for these derivatives.

Lemma 12.2.1 *Assume that Ω_1 is a bounded domain where the divergence theorem holds and $\Omega \subseteq \Omega_1$. Then $w \in C^1(\Omega)$ for $2s + \alpha \in (1, 2)$, and*

$$\partial_i w(x) = \int_{\Omega_1} \partial_i N(x, y)[f(x) - f(y)]dy + f(x) \int_{\partial\Omega_1} N(x, y)\nu_i dS_y, \quad (12.9)$$

where ν_i is the ith component of the unit outward normal vector of $\partial\Omega_1$.

Lemma 12.2.2 *Assume that Ω_1 is a bounded domain where the divergence theorem holds and $\Omega \subseteq \Omega_1$. Then $w \in C^2(\Omega)$ if $2s + \alpha \in (2, 3)$ and*

$$\partial_{ij} w(x) = \int_{\Omega_1} \partial_{ij} N(x, y)[f(y) - f(x)]dy + f(x) \int_{\partial\Omega_1} \partial_i N(x, y)\nu_j dS_y, \quad (12.10)$$

where ν_j is the jth component of the unit outward normal vector of $\partial\Omega_1$.

The proofs of these lemmas are standard calculus arguments and can be found in [GT].

Theorem 12.2.2 *Let $\Omega = B_2(0)$. Assume that $2s + \alpha$ is not an integer. Let $w(x)$ be the potential of f in $B_2(0)$. Then $w \in C^{2s+\alpha}(\bar{B}_1(0))$ and*

$$\|w\|_{C^{2s+\alpha}(\bar{B}_1(0))} \le C\|f\|_{C^\alpha(B_2(0))}. \quad (12.11)$$

Proof. It's easy to see that

$$\|w\|_{L^\infty(B_1)} = C \sup_{B_1(0)} \left| \int_{B_2(0)} \frac{1}{|x - y|^{n-2s}} f(y)dy \right| \le C\|f\|_{L^\infty(B_2(0))}.$$

To estimate the rest of $\|w\|_{C^{2s+\alpha}(\bar{B}_1)}$, we consider three possible cases.

Case i. $2s + \alpha \in (0, 1)$.

To derive the desired bound for

$$[w]_{C^{2s+\alpha}(\bar{B}_1)} = \sup_{x,\bar{x}\in B_1} \frac{|w(x) - w(\bar{x})|}{|x - \bar{x}|^{2s+\alpha}},$$

we show that for any $x, \bar{x} \in B_1(0)$,

$$|w(x) - w(\bar{x})| \le C[f]_{C^\alpha(B_2(0))}|x - \bar{x}|^{2s+\alpha},$$

where C is a constant independent of x and $\bar{x}$.

Let η be the midpoint of x and $\bar{x}$. Then $x, \bar{x} \in B_2(0) \subseteq B_4(\eta)$. Notice that, by symmetry,

$$\int_{B_4(\eta)} \left(\frac{1}{|x - y|^{n-2s}} - \frac{1}{|\bar{x} - y|^{n-2s}} \right) dy = 0,$$

and $suppf \subseteq B_2(0)$, therefore, we have

$$w(x) - w(\bar{x})$$

$$= \int_{B_2(0)} \left(\frac{1}{|x-y|^{n-2s}} - \frac{1}{|\bar{x}-y|^{n-2s}} \right) f(y) dy$$

$$= \int_{B_4(\eta)} \left(\frac{1}{|x-y|^{n-2s}} - \frac{1}{|\bar{x}-y|^{n-2s}} \right) f(y) dy$$

$$= \int_{B_4(\eta)} \left(\frac{1}{|x-y|^{n-2s}} - \frac{1}{|\bar{x}-y|^{n-2s}} \right) (f(y) - f(x)) dy. \qquad (12.12)$$

Let $\delta = |x - \bar{x}|$. We will analyze the above integral on $B_\delta(\eta)$ and $B_4(\eta) \backslash B_\delta(\eta)$ respectively.

Recall that $f \in C^\alpha(B_2(0))$, it thus holds that

$$\left| \int_{B_\delta(\eta)} \frac{1}{|x-y|^{n-2s}} (f(y) - f(x)) dy \right|$$

$$\leq C[f]_{C^\alpha(B_2(0))} \int_{B_\delta(\eta)} \frac{1}{|x-y|^{n-2s-\alpha}} dy$$

$$\leq C[f]_{C^\alpha(B_2(0))} \int_0^\delta \frac{1}{r^{1-2s-\alpha}} dr$$

$$\leq C[f]_{C^\alpha(B_2(0))} |x - \bar{x}|^{2s+\alpha}.$$

Hence

$$\int_{B_\delta(\eta)} \left(\frac{1}{|x-y|^{n-2s}} - \frac{1}{|\bar{x}-y|^{n-2s}} \right) (f(y) - f(x)) dy$$

$$\leq C[f]_{C^\alpha(B_2(0))} |x - \bar{x}|^{2s+\alpha}. \qquad (12.13)$$

By the *mean value theorem*, we know

$$\int_{B_4(\eta) \backslash B_\delta(\eta)} \left(\frac{1}{|x-y|^{n-2s}} - \frac{1}{|\bar{x}-y|^{n-2s}} \right) (f(y) - f(x)) dy$$

$$\leq C \int_{B_4(\eta) \backslash B_\delta(\eta)} \frac{|x - \bar{x}|}{|\xi-y|^{n-2s+1}} [f]_{C^\alpha(B_2(0))} |x - y|^\alpha dy$$

$$\leq C[f]_{C^\alpha(B_2(0))} |x - \bar{x}| \int_{B_4(\eta) \backslash B_\delta(\eta)} \frac{1}{|\xi-y|^{n-2s+1-\alpha}} dy$$

$$\leq C[f]_{C^\alpha(B_2(0))} |x - \bar{x}| \int_\delta^4 \frac{1}{r^{2-2s-\alpha}} dr$$

$$\leq C[f]_{C^\alpha(B_2(0))} (|x - \bar{x}| + |x - \bar{x}|^{2s+\alpha}). \qquad (12.14)$$

Combining (12.12), (12.13) and (12.14), it gives

$$|w(x) - w(\bar{x})| \leq C[f]_{C^\alpha(B_2(0))} |x - \bar{x}|^{2s+\alpha}.$$

This shows that

$$[w]_{C^{2s+\alpha}(\bar{B}_1)} \leq C[f]_{C^\alpha(B_2(0))}.$$

Case ii. $2s + \alpha \in (1, 2)$.

This time we need to estimate $\|\partial_i w(x)\|_{L^\infty(B_1(0))}$ and $[w]_{C^{1+\beta}(B_1(0))}$. By Lemma 12.2.1, for any $x \in B_1(0)$, we have

$$|\partial_i w(x)|$$

$$= \left| \int_{B_2(0)} \partial_i N(x-y)[f(y) - f(x)]\, dy + f(x) \int_{\partial B_2(0)} N(x-y)\frac{y_i}{|y|}\, dS_y \right|$$

$$\leq \left| \int_{B_2(0)} \partial_i N(x-y)[f(y) - f(x)]\, dy \right| + \left| f(x) \int_{\partial B_2(0)} N(x-y)\frac{y_i}{|y|}\, dS_y \right|$$

$$\leq C[f]_{C^\alpha(B_2(0))} \int_{B_2(0)} \frac{1}{|x-y|^{n-2s-\alpha+1}}\, dy + C\|f\|_{L^\infty(B_2(0))}$$

$$\leq C\|f\|_{C^\alpha(B_2(0))}. \tag{12.15}$$

Next we deal with $[w]_{C^{2s+\alpha}(B_1(0))} \equiv [w]_{C^{1+\beta}(B_1(0))}$. Our goal is to show that

$$|\partial_i w(x) - \partial_i w(\bar{x})| \leq C\|f\|_{C^\alpha(B_2(0))}|x - \bar{x}|^{2s+\alpha-1}, \quad \forall x, \bar{x} \in B_1(0).$$

From Lemma 12.2.1, for any $x, \bar{x} \in B_1(0)$, we have

$$\partial_i w(x) - \partial_i w(\bar{x})$$

$$= \int_{B_2(0)} \partial_i N(x-y)[f(y) - f(x)]\, dy - \int_{B_2(0)} \partial_i N(\bar{x}-y)[f(y) - f(\bar{x})]\, dy$$

$$+ f(x) \int_{\partial B_2(0)} N(x-y)\frac{y_i}{|y|}\, dS_y - f(\bar{x}) \int_{\partial B_2(0)} N(\bar{x}-y)\frac{y_i}{|y|}\, dS_y.$$

Let

$$I_1 = \int_{B_2(0)} \partial_i N(x-y)[f(y) - f(x)]\, dy - \int_{B_2(0)} \partial_i N(\bar{x}-y)[f(y) - f(\bar{x})]\, dy,$$

$$I_2 = f(x) \int_{\partial B_2(0)} N(x-y)\frac{y_i}{|y|}\, dS_y - f(\bar{x}) \int_{\partial B_2(0)} N(\bar{x}-y)\frac{y_i}{|y|}\, dS_y.$$

Then

$$\partial_i w(x) - \partial_i w(\bar{x}) = I_1 + I_2. \tag{12.16}$$

We handle I_1 in a way similar to the previous case, i.e., we evaluate the integral in $B_\delta(\eta)$ and $B_2(0)\backslash B_\delta(\eta)$ separately.

$$\int_{B_\delta(\eta)} \partial_i N(x-y)\big(f(y) - f(x)\big)\, dy$$

$$\leq C[f]_{C^\alpha(B_2(0))} \int_{B_\delta(\eta)} \frac{1}{|x-y|^{n-2s-\alpha+1}}\, dy$$

$$\leq C[f]_{C^\alpha(B_2(0))} \int_0^\delta \frac{1}{r^{2-2s-\alpha}}\, dr$$

$$\leq C[f]_{C^\alpha(B_2(0))}|x - \bar{x}|^{2s+\alpha-1}. \tag{12.17}$$

Immediately, one can see that

$$\int_{B_\delta(\eta)} \partial_i N(\bar{x} - y)\big(f(y) - f(\bar{x})\big)dy \leq C[f]_{C^\alpha(B_2(0))} |x - \bar{x}|^{2s+\alpha-1}. \quad (12.18)$$

What remains is

$$\int_{B_2(0)\backslash B_\delta(\eta)} \partial_i N(x - y)\big(f(y) - f(x)\big)dy$$

$$- \int_{B_2(0)\backslash B_\delta(\eta)} \partial_i N(\bar{x} - y)\big(f(y) - f(\bar{x})\big)dy$$

$$= \int_{B_2(0)\backslash B_\delta(\eta)} \big(\partial_i N(x - y) - \partial_i N(\bar{x} - y)\big)\big(f(y) - f(x)\big)dy$$

$$+ \int_{B_2(0)\backslash B_\delta(\eta)} \partial_i N(\bar{x} - y)\big(f(\bar{x}) - f(x)\big)dy$$

$$= I_{11} + I_{12}. \quad (12.19)$$

By the *mean value theorem*, we have

$$|I_{11}| \leq \int_{B_2(0)\backslash B_\delta(\eta)} \frac{|x - \bar{x}|}{|\xi - y|^{n-2s+2}} [f]_{C^\alpha(B_2(0))} |x - y|^\alpha dy$$

$$\leq C[f]_{C^\alpha(B_2(0))} |x - \bar{x}| \int_{B_2(0)\backslash B_\delta(\eta)} \frac{1}{|\xi - y|^{n-2s+2-\alpha}} dy$$

$$\leq C[f]_{C^\alpha(B_2(0))} |x - \bar{x}| \int_\delta^4 \frac{1}{r^{3-2s-\alpha}} dr$$

$$\leq C[f]_{C^\alpha(B_2(0))} \big(|x - \bar{x}| + |x - \bar{x}|^{2s+\alpha-1}\big). \quad (12.20)$$

It follows from the *divergence theorem* that

$$I_{12} = \big(f(\bar{x}) - f(x)\big) \int_{\partial B_2(0)\cup \partial B_\delta(\eta)} N(\bar{x} - y)\nu_i \, dS_y. \quad (12.21)$$

A combination of I_{12} with the remaining boundary terms gives

$$|I_{12} + I_2|$$

$$= \bigg| \big(f(\bar{x}) - f(x)\big) \int_{\partial B_\delta(\eta)} N(\bar{x} - y)\nu_i \, dS_y$$

$$+ f(x) \int_{\partial B_2(0)} \big(N(x - y) - N(\bar{x} - y)\big)\frac{y_i}{|y|} dS_y \bigg|$$

$$\leq C[f]_{C^\alpha(B_2(0))} |x - \bar{x}|^{2s+\alpha-1} + |f(x)| \int_{\partial B_2(0)} \frac{|x - \bar{x}|}{|\xi - y|^{n-2s+1}} dS_y$$

$$\leq C\|f\|_{C^\alpha(B_2(0))} \big(|x - \bar{x}|^{2s+\alpha-1} + |x - \bar{x}|\big). \quad (12.22)$$

By (12.16), (12.17), (12.18), (12.20) and (12.22), we have

$$[w]_{C^{2s+\alpha}(B_1(0))} = \sup_{x,y\in B_1(0)} \frac{|\partial_i w(x) - \partial_i w(\bar{x})|}{|x - \bar{x}|^{2s+\alpha-1}} \leq C\|f\|_{C^\alpha(B_2(0))}.$$

Case iii. $2s + \alpha \in (2, 3)$.

Let i and j be any integers in $[1, n]$. In *Case ii*, we already showed that $\|\partial_i w(x)\|_{L^\infty(B_1(0))}$ can be controlled by $\|f\|_{C^\alpha(B_2(0))}$. What's left is to estimate $\|\partial_{ij} w(x)\|_{L^\infty(B_1(0))}$ and $[w]_{C^{2s+\alpha}(B_1(0))}$.

From Lemma 12.2.2, we know that for any $x \in B_1(0)$,

$$|\partial_{ij} w(x)| \le C[f]_{C^\alpha(B_2(0))} \int_{B_2(0)} \frac{1}{|x-y|^{n-2s-\alpha+2}} dy + C\|f\|_{L^\infty(B_2(0))}$$
$$\le C\|f\|_{C^\alpha(B_2(0))}. \tag{12.23}$$

For any $\bar{x} \in B_1(0)$,

$$\partial_{ij} w(x) - \partial_{ij} w(\bar{x})$$
$$= \int_{B_2(0)} \partial_{ij} N(x, y)[f(y) - f(x)] dy - \int_{B_2(0)} \partial_{ij} N(\bar{x}, y)[f(y) - f(\bar{x})] dy$$
$$+ f(x) \int_{\partial B_2(0)} \partial_i N(x-y) \frac{y_j}{|y|} dS_y - f(\bar{x}) \int_{\partial B_2(0)} \partial_i N(\bar{x} - y) \frac{y_j}{|y|} dS_y.$$

To obtain the same bound for $[w]_{C^{2s+\alpha}(B_1(0))}$, we show that

$$|\partial_{ij} w(x) - \partial_{ij} w(\bar{x})| \le C\|f\|_{C^\alpha(B_2(0))} |x - \bar{x}|^{2s+\alpha-2}, \quad \forall x, \bar{x} \in B_1(0).$$

Let

$$J_1 = \int_{B_2(0)} \partial_{ij} N(x-y)[f(y) - f(x)] \, dy - \int_{B_2(0)} \partial_{ij} N(\bar{x} - y)[f(y) - f(\bar{x})] \, dy,$$

$$J_2 = f(x) \int_{\partial B_2(0)} \partial_i N(x-y) \frac{y_j}{|y|} dS_y - f(\bar{x}) \int_{\partial B_2(0)} \partial_i N(\bar{x} - y) \frac{y_j}{|y|} dS_y.$$

Then

$$\partial_{ij} w(x) - \partial_{ij} w(\bar{x}) = J_1 + J_2.$$

As usual, we integrate J_1 in $B_\delta(\eta)$ and $B_2(0) \backslash B_\delta(\eta)$ respectively. Through a similar argument as in *Case ii*, one can see that

$$\left| \int_{B_\delta(\eta)} \partial_{ij} N(x-y) \big(f(y) - f(x) \big) dy \right| \le C[f]_{C^\alpha(B_2(0))} |x - \bar{x}|^{2s+\alpha-2}. \tag{12.24}$$

Similarly,

$$\left| \int_{B_\delta(\eta)} \partial_{ij} N(\bar{x} - y) \big(f(y) - f(\bar{x}) \big) dy \right| \le C[f]_{C^\alpha(B_2(0))} |x - \bar{x}|^{2s+\alpha-2}. \tag{12.25}$$

At the same time,

$$\int_{B_2(0)\setminus B_\delta(\eta)} \partial_{ij}N(x-y)\big(f(y)-f(x)\big)dy$$

$$-\int_{B_2(0)\setminus B_\delta(\eta)} \partial_{ij}N(\bar{x}-y)\big(f(y)-f(\bar{x})\big)dy$$

$$= \int_{B_2(0)\setminus B_\delta(\eta)} \big(\partial_{ij}N(x-y)-\partial_{ij}N(\bar{x}-y)\big)\big(f(y)-f(x)\big)dy$$

$$+\int_{B_2(0)\setminus B_\delta(\eta)} \partial_{ij}N(\bar{x}-y)\big(f(\bar{x})-f(x)\big)dy$$

$$= J_{11} + J_{12}. \tag{12.26}$$

By the *mean value theorem*, we have

$$|J_{11}| \le \int_{B_2(0)\setminus B_\delta(\eta)} \frac{|x-\bar{x}|}{|\xi-y|^{n-2s+3}} [f]_{C^\alpha(B_2(0))} |x-y|^\alpha dy$$

$$\le C[f]_{C^\alpha(B_2(0))}|x-\bar{x}| \int_{B_2(0)\setminus B_\delta(\eta)} \frac{1}{|\xi-y|^{n-2s+3-\alpha}} dy$$

$$\le C[f]_{C^\alpha(B_2(0))}|x-\bar{x}| \int_\delta^4 \frac{1}{r^{4-2s-\alpha}} dr$$

$$\le C[f]_{C^\alpha(B_2(0))}(|x-\bar{x}| + |x-\bar{x}|^{2s+\alpha-2}). \tag{12.27}$$

Combining J_{12} with J_2, it yields

$$|J_{12} + J_2|$$

$$= \left|\big(f(\bar{x})-f(x)\big) \int_{\partial B_\delta(\eta)} \partial_i N(\bar{x}-y)\,\nu_j\,dS_y\right.$$

$$\left.+f(x) \int_{\partial B_2(0)} \big(\partial_i N(x-y)-\partial_i N(\bar{x}-y)\big)\frac{y_j}{|y|}dS_y\right|$$

$$\le C[f]_{C^\alpha(B_2(0))}|x-\bar{x}|^{2s+\alpha-2} + |f(x)| \int_{\partial B_2(0)} \frac{|x-\bar{x}|}{|\xi-y|^{n-2s+2}} dS_y$$

$$\le C\|f\|_{C^\alpha(B_2(0))}\big(|x-\bar{x}|^{2s+\alpha-2} + |x-\bar{x}|\big). \tag{12.28}$$

From (12.24), (12.25), (12.27) and (12.28), it follows that

$$\sup_{x,\bar{x}\in B_1(0)} \frac{|\partial_{ij}w(x)-\partial_{ij}w(\bar{x})|}{|x-\bar{x}|^{2s+\alpha-2}} \le C\|f\|_{C^\alpha(B_2(0))}. \tag{12.29}$$

Combining the three cases, we arrive at (12.11).
This completes the proof.

12.2.3 The Estimate on s-Harmonic Functions

Previously, we obtained an estimate on $\|w\|_{C^{2s+\alpha}(B_1(0))}$. The difference between the potential w and the solution of (12.1) is an s-harmonic function h, which will be estimated based on a Poisson representation.

Theorem 12.2.3 *Assume that $f \in L^\infty(B_3(0))$, $f = 0$ in $B_3^c(0)$ and $\alpha + 2s$ is not an integer. Let $w(x)$ be the Newton potential of f. Suppose that $u \in L_{2s}(\mathbb{R}^n)$ solves*

$$(-\triangle)^s u(x) = f(x), \quad x \in B_3(0).$$

Let $h(x) = u(x) - w(x)$. Then

$$\|h\|_{C^{2s+\alpha}(B_1(0))} \leq C\left(\|f\|_{L^\infty(B_3(0))} + \int_{|y| \geq 3} \frac{|u(y)|}{(1+|y|)^{n+2s}} dy + \|u\|_{L^\infty(B_3(0))}\right).$$

(12.30)

Proof. In Theorem 4.3.1 Chapter 4, we showed that any $s-$harmonic function $h(x)$ in $B_r(0)$ can be expressed in terms of an integral on $B_r^c(0)$ consisting of some Poisson kernel and h itself:

$$h(x) = \int_{|y|>2} P_2(y,x)h(y)dy, \quad \forall\, |x| < 2.$$

(12.31)

where

$$P_2(y,x) = \begin{cases} \frac{\Gamma(n/2)}{\pi^{\frac{n}{2}+1}} \sin(\pi s)\left(\frac{4-|x|^2}{|y|^2-4}\right)^s \frac{1}{|x-y|^n}, & |y| > 2, \\ 0, & |y| \leq 2. \end{cases}$$

For $x \in B_1(0)$, it's easy to see that

$$|h(x)|$$

$$= \left|\int_{|y|>2} P_2(y,x)h(y)dy\right|$$

$$\leq \int_{2<|y|<3} \frac{C}{(|y|^2-4)^s}|h(y)|dy$$

$$\quad + \int_{|y|>3} \frac{C}{(|y|^2-4)^s}\left(\frac{1}{|x-y|^n}\right)|h(y)|dy$$

$$\leq C\|h\|_{L^\infty(B_3(0))} + \int_{|y|>3} \frac{C}{(1+|y|)^{n+2s}}|h(y)|dy$$

$$\leq C(\|u\|_{L^\infty(B_3(0))} + \|w\|_{L^\infty(B_3(0))})$$

$$\quad + \int_{|y|>3} \frac{C}{(1+|y|)^{n+2s}}|u(y)|dy + \int_{|y|>3} \frac{C}{(1+|y|)^{n+2s}}|w(y)|dy$$

$$\leq C\|u\|_{L^\infty(B_3(0))} + C\|f\|_{L^\infty(B_3(0))} + C\int_{|y|\geq 3} \frac{|u(y)|}{(1+|y|)^{n+2s}} dy. \quad (12.32)$$

With a similar argument, one can show that

$$|D^k h(x)|$$

$$= \left|\int_{|y|>2} D^k P_2(y,x)h(y)dy\right|$$

$$\leq C\|u\|_{L^\infty(B_3(0))} + C\|f\|_{L^\infty(B_3(0))} + C\int_{|y|\geq 3} \frac{|u(y)|}{(1+|y|)^{n+2s}} dy. \quad (12.33)$$

Through a similar process, one can derive expected estimate for the remaining semi-norm $[h]_{C^{2s+\alpha}(B_1(0))}$.

For any $x, \bar{x} \in B_1(0)$, by the *mean value theorem*, we have

$$|D^k h(x) - D^k h(\bar{x})|$$

$$= \left| \int_{|y|>2} [D^k(P_2(y,x)) - D^k(P_2(y,\bar{x}))] h(y) dy \right|$$

$$\leq \left| \int_{|y|>2} D^{k+1}(P_2(y,\zeta)) h(y) dy \right| |x - \bar{x}|$$

$$\leq C \left(\|f\|_{L^\infty(B_3(0))} + \int_{|y|\geq 3} \frac{|u(y)|}{(1+|y|)^{n+2s}} dy + \|u\|_{L^\infty(B_3(0))} \right) |x - \bar{x}|^\beta,$$

$$\tag{12.34}$$

where ζ is some point between x and $\bar{x}$.

A combination of (12.32), (12.33), (12.34) yields

$$\|h\|_{C^{2s+\alpha}(B_1(0))} \leq C \|f\|_{L^\infty(B_3(0))} + \int_{|y|\geq 3} \frac{|u(y)|}{(1+|y|)^{n+2s}} dy + \|u\|_{L^\infty(B_3(0))}.$$

$$\tag{12.35}$$

12.2.4 The Estimate for Solutions on Ball Regions

As an immediate consequence of Theorem 12.2.2 and Theorem 12.2.3, we have

Theorem 12.2.4 *For $\Omega = B_3(0)$. Assume that $\alpha + 2s$ is not an integer. If $u \in L_{2s}(\mathbb{R}^n)$ solves*

$$(-\triangle)^s u(x) = f(x), \quad x \in B_3(0)$$

in the distributional sense, then

$$\|u\|_{C^{2s+\alpha}(B_1(0))} \leq \|f\|_{C^\alpha(B_3(0))} + \int_{|y|\geq 3} \frac{|u(y)|}{(1+|y|)^{n+2s}} dy + \|u\|_{L^\infty(B_3(0))}.$$

From Theorem 12.2.4 it's easy to see that

Theorem 12.2.5 *For $\Omega = B_3(0)$. Assume that $\alpha + 2s$ is not an integer. If $u \in L^\infty(\mathbb{R}^n)$ solves*

$$(-\triangle)^s u(x) = f(x), \quad x \in B_3(0),$$

then

$$\|u\|_{C^{2s+\alpha}(B_1(0))} \leq C(\|f\|_{C^\alpha(B_3(0))} + \|u\|_{L^\infty(\mathbb{R}^n)}). \tag{12.36}$$

12.2.5 The Rescaling

Here we generalize the result in Theorem 12.2.5 from $B_1(0)$ to $B_R(0)$ through a simple rescaling.

Lemma 12.2.3 *Assume that $u_R(x)$ is a solution of*

$$(-\triangle)^s u_R(x) = f_R(x), \quad x \in B_{3R}. \tag{12.37}$$

Then

$$\sum_{i=0}^{k} R^i \sup_{x \in B_R} |D^i u_R(x)| + R^{2s+\alpha}[D^k u_R]_{C^\beta(B_R)}$$

$$\leq C(\|u_R\|_{L^\infty(\mathbb{R}^n)} + R^{2s} \sup_{x \in B_{3R}} |f_R(x)| + R^{\alpha+2s}[f_R]_{C^\alpha(B_{3R})}). \tag{12.38}$$

Proof. Let $u(x) = u_R(Rx)$. Then

$$(-\triangle)^s u(x) = R^{2s}[(-\triangle)^s u_R](Rx) = R^{2s} f_R(Rx) \equiv f(x), \quad x \in B_3(0).$$

Theorem 12.2.5 has provided an estimate on such a u in terms of f:

$$\|u\|_{C^{2s+\alpha}(B_1)} \leq C(\|u\|_{L^\infty(\mathbb{R}^n)} + \|f\|_{C^\alpha(B_3)}). \tag{12.39}$$

Based on this and a rescaling, we will derive (12.38).

In fact,

$$\sup_{x \in B_1} |D^i u(x)| = R^i \sup_{x \in B_R} |D^i u_R(x)|, \quad i = 0, \ldots, k, \tag{12.40}$$

and

$$[D^k u]_{C^\beta(B_1)} = \sup_{x,y \in B_1} \frac{|D^k u(x) - D^k u(y)|}{|x-y|^\beta}$$

$$= \sup_{x,y \in B_1} \frac{|D^k[u_R(Rx)] - D^k[u_R(Ry)]|}{|x-y|^\beta}$$

$$= \sup_{x,y \in B_1} \frac{R^k|[D^k u_R](Rx) - [D^k u_R](Ry)|}{|Rx - Ry|^\beta} R^\beta$$

$$= \sup_{x,y \in B_R} \frac{R^k|D^k u_R(x) - D^k u_R(y)|}{|x-y|^\beta} R^\beta$$

$$= R^{k+\beta}[D^k u_R]_{C^\beta(B_R)}$$

$$= R^{2s+\alpha}[D^k u_R]_{C^\beta(B_R)}. \tag{12.41}$$

By definition, we have

$$\|f\|_{C^\alpha(B_3)} = \sup_{x\in B_3} |f(x)| + \sup_{x,y\in B_3} \frac{|f(x)-f(y)|}{|x-y|^\alpha}$$

$$= \sup_{x\in B_3} |R^{2s}f_R(Rx)| + R^\alpha \sup_{x,y\in B_3} \frac{|R^{2s}f_R(Rx) - R^{2s}f_R(Ry)|}{|Rx-Ry|^\alpha}$$

$$= R^{2s} \sup_{x\in B_{3R}} |f_R(x)| + R^{\alpha+2s} \sup_{x,y\in B_{3R}} \frac{|f_R(x)-f_R(y)|}{|x-y|^\alpha}. \quad (12.42)$$

Combining (12.40), (12.41), and (12.42), it yields

$$R^i \sum_{i=0}^k \sup_{x\in B_R} |D^i u_R(x)| + R^{2s+\alpha} [D^k u_R]_{C^\beta(B_R)}$$

$$\leq C(\|u_R\|_{L^\infty(\mathbb{R}^n)} + R^{2s} \sup_{x\in B_{3R}} |f_R(x)| + R^{\alpha+2s}[f_R(x)]_{C^\alpha(B_{3R})}).$$

This proves (12.38).

In (12.38), rewriting $u_R = u$ and $f_R = f$, we arrive at

Theorem 12.2.6 *Assume that $u(x)$ is a solution of*

$$(-\triangle)^s u(x) = f(x), \quad x \in B_{3R}.$$

Then

$$\sum_{i=0}^k R^i \sup_{x\in B_R} |D^i u(x)| + R^{2s+\alpha}[D^k u]_{C^\beta(B_R)}$$

$$\leq C(\|u\|_{L^\infty(\mathbb{R}^n)} + R^{2s} \sup_{x\in B_{3R}} |f(x)| + R^{\alpha+2s}[f]_{C^\alpha(B_{3R})}). \quad (12.43)$$

12.2.6 The Schauder Estimate on Ω with Weighted Norms

Let

$$d_x = \begin{cases} dist(x, \partial\Omega), & x \in \Omega, \\ 0, & x \notin \Omega. \end{cases}$$

Theorem 12.2.7 *Assume that $\alpha + 2s = k + \beta$ is not an integer. If u is the solution of*

$$\begin{cases} (-\triangle)^s u = f, & x \in \Omega, \\ u = 0, & x \in \mathbb{R}^n \setminus \Omega. \end{cases} \quad (12.44)$$

Then

$$\sum_{i=1}^k \sup_\Omega d_x^i |D^i u(x)| + \sup_\Omega d_{x,y}^{2s+\alpha} \frac{|D^k u(x) - D^k u(y)|}{|x-y|^{\beta'}}$$

$$\leq C\left(\|u\|_{L^\infty(\Omega)} + \sup_\Omega d_x^{2s}|f| + \sup_\Omega d_{x,y}^{2s+\alpha} \frac{|f(x)-f(y)|}{|x-y|^\alpha}\right),$$

$$(12.45)$$

where $d_{x,y} = \min\{d_x, d_y\}$.

Proof. For any $x \in \Omega$, let $R = \frac{1}{3}d_x$ and $B_{3R} = B_{3R}(x)$, then

$$\sum_{i=0}^{k} d_x^i |D^i u(x)|$$

$$\leq C \sum_{i=0}^{k} R^i |D^i u|$$

$$\leq C\big(\|u\|_{L^\infty(\Omega)} + R^{2s}\|f\|_{L^\infty(B_{3R})} + R^{2s+\alpha}[f]_{C^\alpha(B_{3R})}\big)$$

$$= C\big(\|u\|_{L^\infty(\Omega)} + \sup_{B_{3R}} R^{2s}|f| + \sup_{x,y \in B_{3R}} R^{2s+\alpha}\frac{|f(x)-f(y)|}{|x-y|^\alpha}\big)$$

$$\leq C\big(\|u\|_{L^\infty(\Omega)} + \sup_{B_{3R}} d_x^{2s}|f| + \sup_{x,y \in B_{3R}} d_{x,y}^{2s+\alpha}\frac{|f(x)-f(y)|}{|x-y|^\alpha}\big)$$

$$\leq C\big(\|u\|_{L^\infty(\Omega)} + \sup_{x \in \Omega} d_x^{2s}|f| + \sup_{x,y \in \Omega} d_{x,y}^{2s+\alpha}\frac{|f(x)-f(y)|}{|x-y|^\alpha}\big). \qquad (12.46)$$

Next we estimate $\sup_\Omega d_{x,y}^{2s+\alpha}\dfrac{|D^k u(x) - D^k u(y)|}{|x-y|^\beta}$. Given $x, y \in \Omega$, without any loss of generality, we assume that $d_x \leq d_y$.

For $y \in B_R(x)$, we have

$$\sup_{y \in B_R(x)} d_{x,y}^{2s+\alpha}\frac{|D^k u(x) - D^k u(y)|}{|x-y|^\beta}$$

$$\leq C R^{2s+\alpha} \sup_{y \in B_R(x)} \frac{|D^k u(x) - D^k u(y)|}{|x-y|^\beta}$$

$$\leq C\big(\|u\|_{L^\infty(\Omega)} + R^{2s}\|f\|_{C^0(B_{3R})} + R^{2s+\alpha}[f]_{C^\alpha(B_{3R})}\big)$$

$$\leq C\big(\|u\|_{L^\infty(\Omega)} + \sup_\Omega d_x^{2s}|f| + \sup_\Omega d_{x,y}^{2s+\alpha}\frac{|f(x)-f(y)|}{|x-y|^\alpha}\big). \qquad (12.47)$$

For $y \in B_R^c(x)$,

$$\sup_{y \in B_R^c(x)} d_{x,y}^{2s+\alpha} \frac{|D^k u(x) - D^k u(y)|}{|x-y|^\beta}$$

$$= \sup_{y \in B_R^c(x)} d_x^{2s+\alpha} \frac{|D^k u(x) - D^k u(y)|}{|x-y|^\beta}$$

$$\leq CR^{2s+\alpha} \sup_{y \in B_R^c(x)} \frac{|D^k u(x) - D^k u(y)|}{|x-y|^\beta}$$

$$\leq CR^{2s+\alpha-\beta} \left(|D^k u(x)| + \sup_{y \in B_R^c(x)} |D^k u(y)| \right)$$

$$\leq C \left(\sup_{x \in \Omega} R^k |D^k u(x)| + \sup_{y \in B_R^c(x)} R^k |D^k u(y)| \right)$$

$$\leq C \left(\sup_{x \in \Omega} d_x^k |D^k u(x)| + \sup_{y \in B_R^c(x)} d_y^k |D^k u(y)| \right)$$

$$\leq C (\|u\|_{L^\infty(\Omega)} + R^{2s} \|f\|_{L^\infty(B_{3R})} + R^{2s+\alpha} [f]_{C^\alpha(B_{3R})})$$

$$\leq C(\|u\|_{L^\infty(\Omega)} + \sup_\Omega d_x^{2s} |f| + \sup_\Omega d_{x,y}^{2s+\alpha} \frac{|f(x) - f(y)|}{|x-y|^\alpha}). \qquad (12.48)$$

For $d_x \geq d_y$, we can derive the same conclusion for $x \in B_R(y)$ and $x \in B_R^c(y)$.

The conclusion follows naturally from (12.46), (12.47) and (12.48).

12.3 Hölder Regularity up to the Boundary

For a bounded domain $\Omega \subseteq \mathbb{R}^n$, consider

$$\begin{cases} (-\triangle)^s u(x) = f(x), & x \in \Omega, \\ u = 0, & x \in \mathbb{R}^n \backslash \Omega, \end{cases} \qquad (12.49)$$

where $f \in L^\infty(\Omega)$ and $f(x) = 0$ in $\mathbb{R}^n \backslash \Omega$.

12.3.1 Estimates on the Potential

Lemma 12.3.1 (The Potential) *Let*

$$w(x) = \int_{\mathbb{R}^n} \frac{f(y)}{|x-y|^{n-2s}} dy$$

be the potential of f. Then for any $\beta \in (0, 2s)$, it holds that

$$\|w\|_{C^\beta(B_2(0))} \leq C \|f\|_{L^\infty(B_3(0))}. \qquad (12.50)$$

Proof. It is easy to see

$$\|w\|_{L^\infty(B_2(0))} = \sup_{B_2(0)} |\int_{B_2(0)} \frac{f(y)}{|x-y|^{n-2s}} dy| \leq C \|f\|_{L^\infty(B_3(0))}.$$

To estimate the rest of $\|w\|_{C^\beta(B_3(0))}$, we consider two cases.

Case i. $0 < \beta < 2s \leq 1$.

We derive the desired bound for

$$[w]_{C^\beta(B_2(0))} = \sup_{x,\bar{x} \in B_2(0)} \frac{|w(x) - w(\bar{x})|}{|x - \bar{x}|^\beta}.$$

For any $x, \bar{x} \in B_2(0)$, take $\delta = |x - \bar{x}|$, $\xi = \frac{1}{2}(x + \bar{x})$, then we have

$$w(x) - w(\bar{x}) = \int_{B_\delta(\xi)} f(y)\Big(\frac{1}{|x-y|^{n-2s}} - \frac{1}{|\bar{x}-y|^{n-2s}}\Big)dy$$

$$+ \int_{B_3(0)\backslash B_\delta(\xi)} f(y)\Big(\frac{1}{|x-y|^{n-2s}} - \frac{1}{|\bar{x}-y|^{n-2s}}\Big)dy$$

$$= I_1 + I_2. \tag{12.51}$$

To estimate I_1, we notice that

$$\Big|\int_{B_\delta(\xi)} \frac{f(y)}{|x-y|^{n-2s}}dy\Big| \leq C\|f\|_{L^\infty(B_3(0))} \int_{B_\delta(\xi)} \frac{1}{|x-y|^{n-2s}}dy$$

$$\leq C\|f\|_{L^\infty(B_3(0))} \int_{B_{2\delta}(x)} \frac{1}{|x-y|^{n-2s}}dy$$

$$\leq C\|f\|_{L^\infty(B_3(0))}\delta^{2s}.$$

Now it is easy to see that

$$|I_1| \leq \Big|\int_{B_\delta(\xi)} \frac{f(y)}{|x-y|^{n-2s}}dy\Big| + \Big|\int_{B_\delta(\xi)} \frac{f(y)}{|\bar{x}-y|^{n-2s}}dy\Big|$$

$$\leq C\|f\|_{L^\infty(B_3(0))}\delta^{2s}. \tag{12.52}$$

By the *mean value theorem*, we have

$$|I_2| \leq C\|f\|_{L^\infty(B_3(0))}|x - \bar{x}| \int_{B_3(0)\backslash B_\delta(\xi)} \frac{1}{|\hat{x}-y|^{n+1-2s}}dy$$

$$\leq C\|f\|_{L^\infty(B_3(0))}|x - \bar{x}| \int_{B_3(0)\backslash B_\delta(\xi)} \frac{1}{|y-\xi|^{n+1-2s}}dy$$

$$\leq C\|f\|_{L^\infty(B_3(0))}\delta^{2s} \tag{12.53}$$

where $\hat{x}$ is some point between x and $\bar{x}$. By (12.51), (12.52) and (12.53), we derive that

$$\frac{|w(x) - w(\bar{x})|}{|x - \bar{x}|^\beta} \leq C\|f\|_{L^\infty(B_3(0))}, \forall x, \bar{x} \in B_2(0). \tag{12.54}$$

Case ii. $1 \leq \beta < 2s$.

Let $\beta = 1 + \beta'$, similar to (12.54), one can show that

$$\frac{|Dw(x) - Dw(\bar{x})|}{|x - \bar{x}|^{\beta'}} \leq C\|f\|_{L^\infty(B_3(0))}, \forall x, \bar{x} \in B_2(0), \tag{12.55}$$

and for any $x \in B_2(0)$, it holds

$$|Dw(x)| \leq C\|f\|_{L^\infty(B_3(0))} \int_{B_3(0)} \frac{1}{|x - y|^{n+1-2s}} dx \leq C\|f\|_{L^\infty(B_3(0))}.$$

A combination of *Case i* and *Case ii* yields the desired result. This completes the proof.

12.3.2 Estimates of Solutions in Ball Regions

Lemma 12.3.2 *Assume that* $(-\triangle)^s u = f$ *in* $B_3(0)$, *then* $u \in C^\beta(B_1(0))$ *for all* $\beta \in (0, 2s)$ *and*

$$\|u\|_{C^\beta(B_1(0))} \leq C(\|f\|_{L^\infty(B_3(0))} + \int_{|y| \geq 3} \frac{|u(y)|}{(1 + |y|)^{n+2s}} dy + \|u\|_{L^\infty(B_3(0))}).$$

$$(12.56)$$

Proof. Let $h = u - w$, then we have $(-\triangle)^s h = 0$ in $B_3(0)$ and

$$h(x) = \int_{|y| > 2} P_2(y, x) h(y) dy, \ \forall |x| \leq 1. \qquad (12.57)$$

For $x, \bar{x} \in B_1(0)$, by Lemma 12.3.1, through a similar proof as in Theorem 12.2.3, one can show that for $\beta < 1$,

$$\sup_{x,\bar{x} \in B_1(0)} \frac{|h(x) - h(\bar{x})|}{|x - \bar{x}|^\beta}$$

$$\leq C(\|f\|_{L^\infty(B_3(0))} + \int_{|y| \geq 3} \frac{|u(y)|}{(1 + |y|)^{n+2s}} dy + \|u\|_{L^\infty(B_3(0))}),$$

and for $\beta = 1 + \beta'$ with $\beta' \in (0, 2s - 1)$, it holds that

$$\sup_{x,\bar{x} \in B_1(0)} \frac{|Dh(x) - Dh(\bar{x})|}{|x - \bar{x}|^{\beta'}}$$

$$\leq (\|f\|_{L^\infty(B_3(0))} + \int_{|y| \geq 3} \frac{|u(y)|}{(1 + |y|)^{n+2s}} dy + \|u\|_{L^\infty(B_3(0))}),$$

and

$$\sup_{x \in B_1(0)} |Dh(x)| \leq C(\|f\|_{L^\infty(B_3(0))} + \int_{|y| \geq 3} \frac{|u(y)|}{1+|y|^{n+2s}} dy + \|u\|_{L^\infty(B_3(0))}),$$

$$\sup_{x \in B_1(0)} |h(x)| \leq C(\|f\|_{L^\infty(B_3(0))} + \int_{|y| \geq 3} \frac{|u(y)|}{(1+|y|)^{n+2s}} dy + \|u\|_{L^\infty(B_3(0))}).$$

That is

$$\|h\|_{C^\beta(B_1(0))} \leq C(\|f\|_{L^\infty(B_3(0))} + \int_{|y| \geq 3} \frac{|u(y)|}{1 + |y|^{n+2s}} dy + \|u\|_{L^\infty(B_3(0))}).$$

$$(12.58)$$

Combining Lemma 12.3.1 and (12.58), we prove Lemma 12.3.2.

12.3.3 Hölder Estimates up to the Boundary

Lemma 12.3.3 *[RS1] Let Ω be a bounded domain satisfying the exterior ball condition and $f \in L^\infty(\Omega)$. Let $u \in C(\Omega)$ be a solution of (12.49). Then,*

$$|u(x)| \le C\|f\|_{L^\infty(\Omega)}d_x^s(x) \ for \ all \ x \in \Omega,$$

where C is a constant depending only on Ω and s.

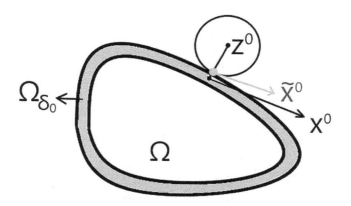

Fig. 12.1

Proof. Since Ω satisfies the exterior ball condition, there exists some $\rho_0 > 0$ such that every point of $\partial\Omega$ can be touched from outside by a ball of radius ρ_0. Let $\Omega_{\delta_0} = \{x \in \Omega \mid dist(x, \partial\Omega) \le \delta_0\}$, then for any $x^0 \in \Omega_{\delta_0}$, there exists an $\tilde{x}^0 \in \partial\Omega$, such that, $dist(x^0, \tilde{x}^0) = dist(x^0, \partial\Omega)$. Hence there exists an exterior ball $B_{\rho_0}(z^0)$ which touches $\tilde{x}^0$ at $\partial\Omega$ (see the Fig. 12.1).

Let $\varphi(x)$ be a solution of

$$\begin{cases} (-\triangle)^s\varphi(x) = 1, & x \in B_{\rho_0}(z^0) \\ \varphi(x) = 0, & x \in B_{\rho_0}^c(z^0). \end{cases} \tag{12.59}$$

It is well-known that

$$\varphi(x) = C_0(\rho_0^2 - |x - z^0|^2)_+^s,$$

is the solution of (12.59) (see [Ge]). Let the Kelvin transform of $\varphi(x)$ be

$$v(x) = \left(\frac{\rho_0}{|x - z^0|}\right)^{n-2s}\varphi\left(\frac{\rho_0^2(x - z^0)}{|x - z^0|^2} + z^0\right),$$

then

$$v(x) = C_0\left(\frac{\rho_0}{|x - z^0|}\right)^n(|x - z^0|^2 - \rho_0^2)^s, \ |x - z^0| \ge \rho_0,$$

and satisfies

$$\begin{cases} (-\triangle)^s v(x) = (\frac{\rho_0}{|x-z^0|})^{n+2s}, \, x \in B^c_{\rho_0}(z^0) \\ v(x) = 0, \qquad\qquad\qquad x \in B_{\rho_0}(z^0). \end{cases}$$

Choose C_Ω such that for all $x \in \Omega \subset B_{\rho_0 + diam\Omega}(z^0) \backslash B_{\rho_0}(z^0)$

$$C_\Omega (\frac{\rho_0}{|x-z^0|})^{n+2s} \geq 1.$$

Obviously, C_Ω depends only on $diam\Omega$. Let $w(x) = C_\Omega v(x) \|f\|_{L^\infty(\Omega)}$, we have

$$\begin{cases} (-\triangle)^s (w(x) \pm u(x)) \geq 0, \, x \in \Omega, \\ w(x) \pm u(x) \geq 0, \qquad\quad x \in \Omega^c. \end{cases}$$

It follows from the *maximum principle* that

$$w(x) \pm u(x) \geq 0, \quad x \in \mathbb{R}^n.$$

Then,

$$|u(x)| \leq w(x) = C_\Omega C_0 (\frac{\rho_0}{|x-z^0|})^n (|x-z^0|^2 - \rho_0^2)^s \|f\|_{L^\infty(\Omega)}. \quad (12.60)$$

Therefore, for any $x \in \Omega$,

$$|u(x)| \leq C_1 \|f\|_{L^\infty(\Omega)}. \quad (12.61)$$

From (12.60), in particular, we obtain

$$|u(x^0)| \leq C_\Omega C_0 (\frac{\rho_0}{|x^0 - z^0|})^n (|x^0 - z^0| + \rho_0)^s (|x^0 - z^0| - \rho_0)^s \|f\|_{L^\infty(\Omega)}$$
$$\leq C(n, s, \Omega, x^0) d^s_{x^0} \|f\|_{L^\infty(\Omega)}.$$

Since Ω_{δ_0} is compact and satisfies the exterior ball condition, $C(n, s, \Omega, x^0)$ has an upper bound for all $x^0 \in \Omega$.

Let $C_M = \sup_{x^0 \in \Omega} C(n, s, \Omega, x^0)$, then for any $x \in \Omega_{\delta_0}$,

$$|u(x)| \leq C_M \|f\|_{L^\infty(\Omega)} d^s_x.$$

Combining this with (12.61), we arrive at

$$|u(x)| \leq C \|f\|_{L^\infty(\Omega)} d^s_x, \forall x \in \Omega.$$

This completes the proof of the lemma.

Lemma 12.3.4 *Let Ω be a bounded domain satisfying the exterior ball condition. Assume that $f \in L^\infty(\Omega)$ and u is a solution to (12.49). Then*

$$u \in C^\beta_{loc}(\Omega) \text{ for all } \beta \in (0, 2s),$$

and for all $x^o \in \Omega$, we have

$$[u]_{C^\beta(\overline{B_R}(x^o))} \leq CR^{s-\beta} \|f\|_{L^\infty(\Omega)}, \quad R = \frac{d_{x^o}}{2}, \quad (12.62)$$

where $C = C(\Omega, s, \beta)$ is a constant.

Proof. For any $x^o \in \Omega$, let $\tilde{u}(x) = u(x^o + Rx)$. Then

$$(-\triangle)^s \tilde{u}(x) = R^{2s} f(x^o + Rx), \quad x \in B_2(0). \tag{12.63}$$

Given $x^o \in \Omega$, there exists an $\eta \in \partial\Omega$ such that $d_{x^o} = |x^o - \eta|$. It's easy to see that for any $x \in \mathbb{R}^n$,

$$d_{x^o + Rx} \leq |x^o + Rx - \eta| \leq |x^o - \eta| + R|x| = R(2 + |x|).$$

Recall that $u = 0$ outside Ω, together with Lemma 12.3.3, it yields that, for any $x \in \mathbb{R}^n$,

$$|\tilde{u}(x)| = |u(x^o + Rx)| \leq C\|f\|_{L^\infty(\Omega)} d^s_{x^o + Rx}$$
$$\leq C\|f\|_{L^\infty(\Omega)} R^s (2 + |x|)^s. \tag{12.64}$$

Applying Theorem 12.2.5 and (12.64) to (12.63), we deduce

$$\|\tilde{u}\|_{C^\beta(\overline{B_1(0)})}$$
$$\leq \int_{\mathbb{R}^n} \frac{|\tilde{u}(x)|}{(1 + |x|)^{n+2s}} dx + R^{2s}\|f(x^o + Rx)\|_{L^\infty(B_2(0))} + \|\tilde{u}(x)\|_{L^\infty(B_2(0))}$$
$$\leq \int_{\mathbb{R}^n} \frac{C\|f\|_{L^\infty(\Omega)} R^s (2 + |x|)^s}{(1 + |x|)^{n+2s}} dx + R^{2s}\|f(x^o + Rx)\|_{L^\infty(B_2(0))}$$
$$\quad + \|\tilde{u}(x)\|_{L^\infty(B_2(0))}$$
$$\leq C\|f\|_{L^\infty(\Omega)} R^s + \|\tilde{u}(x)\|_{L^\infty(B_2(0))}$$
$$\leq C\|f\|_{L^\infty(\Omega)} R^s. \tag{12.65}$$

Notice that

$$[\tilde{u}]_{C^\beta(\overline{B_1(0)})} = R^\beta [u]_{C^\beta(\overline{B_R(x^o)})}. \tag{12.66}$$

By (12.65) and (12.66), we have

$$[u]_{C^\beta(\overline{B_R(x^o)})} \leq C\|f(x)\|_{L^\infty(\Omega)} R^{s-\beta}. \tag{12.67}$$

Theorem 12.3.1 *(see [RS1]) Let Ω be a bounded Lipschitz domain satisfying the interior ball condition, $f \in L^\infty(\Omega)$, and u be a solution of (12.49). Then, $u \in C^s(\mathbb{R}^n)$ and*

$$\|u\|_{C^s(\mathbb{R}^n)} \leq C\|f\|_{L^\infty(\Omega)},$$

where C is a constant depending on Ω and s.

Proof. For any $x, y \in \Omega$, without loss of generality, we assume $d_x \geq d_y$ and $R = \frac{d_x}{2}$.

Case i. $|x - y| \leq R$.

The proof follows from Lemma 12.3.4 by taking $\beta = s$.

Case ii. $|x - y| \geq R$.

From Lemma 12.3.3, we have

$$\frac{|u(x) - u(y)|}{|x - y|^s} \leq \frac{|u(x) - u(y)|}{R^s}$$

$$\leq C(\frac{u(x)}{d_x^s} + \frac{u(y)}{d_x^s})$$

$$\leq C(\frac{u(x)}{d_x^s} + \frac{u(y)}{d_y^s})$$

$$\leq C\|f\|_{L^\infty(\Omega)}.$$

A

Appendix

A.1 For Moving Planes Method

Lemma A.1.1 *For any $\tilde{x} \in \mathbb{R}^n$, let v be the Kelvin transform of u centered at $\tilde{x}$:*

$$v(x) = \frac{1}{|x - \tilde{x}|^{n-\alpha}} u\left(\frac{x - \tilde{x}}{|x - \tilde{x}|^2} + \tilde{x}\right). \tag{A.1}$$

Then

$$(-\triangle)^{\alpha/2} v(x) = \frac{1}{|x - \tilde{x}|^{n+\alpha}} \left((-\triangle)^{\alpha/2} u\right)\left(\frac{x - \tilde{x}}{|x - \tilde{x}|^2} + \tilde{x}\right).$$

Proof. Without loss of generality, let $\tilde{x}$ be the origin. By definition, we know that for any given $x_o \in \mathbb{R}^n$ and $x_o \neq 0$,

$$\left((-\triangle)^{\alpha/2} u\right)\left(\frac{x_o}{|x_o|^2}\right)$$

$$= C_{n,\alpha} PV \int_{\mathbb{R}^n} \frac{u\left(\frac{x_o}{|x_o|^2}\right) - u(y)}{\left|\frac{x_o}{|x_o|^2} - y\right|^{n+\alpha}} dy$$

$$= C_{n,\alpha} PV \int_{\mathbb{R}^n} \frac{u\left(\frac{x_o}{|x_o|^2}\right) - u\left(\frac{z}{|z|^2}\right)}{\left|\frac{x_o}{|x_o|^2} - \frac{z}{|z|^2}\right|^{n+\alpha}} \frac{dz}{|z|^{2n}} \qquad y = \frac{z}{|z|^2}$$

$$= |x_o|^{n+\alpha} C_{n,\alpha} PV \int_{\mathbb{R}^n} \frac{u\left(\frac{x_o}{|x_o|^2}\right) - u\left(\frac{z}{|z|^2}\right)}{|x_o - z|^{n+\alpha}} \frac{dz}{|z|^{n-\alpha}}. \tag{A.2}$$

Let

$$h(z) = -\frac{u\left(\frac{x_o}{|x_o|^2}\right) - u\left(\frac{z}{|z|^2}\right)}{|z|^{n-\alpha}}. \tag{A.3}$$

It's easy to see that

$$h(x_o) = 0.$$

We can rewrite (A.2) as

315

$$((-\triangle)^{\alpha/2}u)(\frac{x_o}{|x_o|^2}) = |x_o|^{n+\alpha}C_{n,\alpha}PV\int_{\mathbb{R}^n}\frac{0-h(z)}{|x_o-z|^{n+\alpha}}dz$$

$$= |x_o|^{n+\alpha}C_{n,\alpha}PV\int_{\mathbb{R}^n}\frac{h(x_o)-h(z)}{|x_o-z|^{n+\alpha}}dz$$

$$= |x_o|^{n+\alpha}(-\triangle)^{\alpha/2}h(x_o). \tag{A.4}$$

On the other hand, by (A.1) and (A.3),

$$(-\triangle)^{\alpha/2}h(z)|_{z=x_o}$$

$$= -(-\triangle)^{\alpha/2}\left(\frac{u(\frac{x_o}{|x_o|^2})-u(\frac{z}{|z|^2})}{|z|^{n-\alpha}}\right)|_{z=x_o}$$

$$= -(-\triangle)^{\alpha/2}\left(\frac{1}{|z|^{n-\alpha}}u(\frac{x_o}{|x_o|^2})-v(z)\right)|_{z=x_o}$$

$$= -(-\triangle)^{\alpha/2}\left(\frac{1}{|z|^{n-\alpha}}\right)\Big|_{z=x_o}u(\frac{x_o}{|x_o|^2})+(-\triangle)^{\alpha/2}v(z)|_{z=x_o}$$

$$= 0+(-\triangle)^{\alpha/2}v(x_o)$$

$$= (-\triangle)^{\alpha/2}v(x_o). \tag{A.5}$$

Combining (A.5) and (A.4) we complete the proof.   □

Next, we prove (6.69) and (6.75). Without loss of generality, we let $x^0 = 0$.

Lemma A.1.2 *Assume that $u \in L_\alpha \cap C^{1,1}_{loc}(\mathbb{R}^n)$ is a nonnegative solution for*

$$(-\triangle)^{\alpha/2}u(x) = u^\tau(x), \quad x \in \mathbb{R}^n,$$

with $\tau > 0$. Let $v(x) = \frac{1}{|x|^{n-\alpha}}u(\frac{x}{|x|^2})$ be the Kelvin transform of u, and

$$w_\lambda(x) = v(x^\lambda) - v(x).$$

Then there exists a constant $C > 0$ such that for any small $\varepsilon > 0$ and λ sufficiently negative,

$$w_\lambda(x) \geq C > 0, \quad x \in B_\varepsilon(0^\lambda)\backslash\{0^\lambda\}.$$

Proof. Let η be a smooth cutoff function such that $\eta(x) \in [0,1]$ in $\mathbb{R}^n$, $supp\,\eta \subset B_2$ and $\eta(x) \equiv 1$ in B_1. Let

$$(-\triangle)^{\alpha/2}\varphi(x) = \eta(x)u^\tau(x).$$

Then

$$\varphi(x) = c_{n,-\alpha}\int_{\mathbb{R}^n}\frac{\eta(y)u^\tau(y)}{|x-y|^{n-\alpha}}dy.$$

It's easy to see that for all $|x|$ sufficiently large, there exists a constant $C > 0$ such that

$$\frac{2C}{|x|^{n-\alpha}} \leq \varphi(x) \leq \frac{3C}{|x|^{n-\alpha}}. \tag{A.6}$$

For $R > 0$ large, let

$$g(x) = u(x) - \varphi(x) + \frac{3C}{R^{n-\alpha}}.$$

Immediately, we have

$$\begin{cases} (-\triangle)^{\alpha/2} g(x) \geq 0, & \text{in } B_R, \\ g(x) \geq 0, & \text{in } B_R^c. \end{cases} \tag{A.7}$$

It follows from the *maximum principle* (see [Si]) that

$$g(x) \geq 0, \quad \text{in } B_R.$$

Thus

$$u(x) - \varphi(x) + \frac{3C}{R^{n-\alpha}} \geq 0, \quad \text{in } B_R.$$

Fix x, as $R \to \infty$,

$$u(x) \geq \varphi(x), \quad \text{in } \mathbb{R}^n. \tag{A.8}$$

Combining (A.6) and (A.8), we arrive at

$$u(x) \geq \frac{2C}{|x|^{n-\alpha}}.$$

It then follows that

$$v_\lambda(x) \geq 2C > 0, \quad x \in B_\varepsilon(0^\lambda) \backslash \{0^\lambda\}.$$

For λ sufficiently negative,

$$v(x) \leq C, \quad x \in \Sigma_\lambda. \tag{A.9}$$

Therefore,

$$w_\lambda(x) \geq C > 0, \quad x \in B_\varepsilon(0^\lambda) \backslash \{0^\lambda\}. \qquad \square$$

Lemma A.1.3 *Let w_λ be defined as before in the previous lemma and let*

$$\lambda_0 = \sup\{\lambda \mid w_\mu(x) \geq 0, \forall x \in \Sigma_\mu \backslash \{0^\mu\}, \mu \leq \lambda\}.$$

If $w_{\lambda_0} \not\equiv 0$, then

$$w_{\lambda_0}(x) \geq c > 0, \quad x \in B_\varepsilon(0^{\lambda_0}) \backslash \{0^{\lambda_0}\}.$$

Proof. By the definition, if

$$w_{\lambda_0}(x) \not\equiv 0, \quad x \in \Sigma_{\lambda_0},$$

then there exists a point x^0 such that

$$w_{\lambda_0}(x^0) > 0.$$

And further, there exists a small positive δ such that

$$w_{\lambda_0}(x) \geq C_0 > 0, \quad x \in B_\delta(x^0).$$

From the integral equation that $w_{\lambda_0}(x)$ satisfies, one can derive that (see the proof of Lemma 3.1 in [CFY])

$$w_{\lambda_0}(x) = C_{n,\alpha} \int_{\Sigma_{\lambda_0}} \left(\frac{1}{|x-y|^{n-\alpha}} - \frac{1}{|x-y^{\lambda_0}|^{n-\alpha}} \right) \left(\frac{v^\tau(y^{\lambda_0})}{|x-y^{\lambda_0}|^\gamma} - \frac{v^\tau(y)}{|x-y|^\gamma} \right) dy,$$

where $\gamma = n + \alpha - \tau(n-\alpha)$.

It follows from the positivity of v that

$$w_{\lambda_0}(x) \geq \int_{B_\delta(x^0)} C \, dy$$
$$> C > 0. \qquad \qquad \square$$

Remark A.1.1 *This lemma can also be derived from Theorem 2 in [LWX].*

Lemma A.1.4 *For λ sufficiently negative, there exists a positive constant C such that*

$$U_\lambda(x), V_\lambda(x) \geq C > 0, x \in B_\varepsilon(0^\lambda) \backslash \{0^\lambda\}. \tag{A.10}$$

Proof. For $x \in \Sigma_\lambda$, as $\lambda \to -\infty$, it's easy to see that

$$\bar{u}(x) \to 0. \tag{A.11}$$

To prove (A.10), it suffices to show that

$$\bar{u}_\lambda(x) \geq C > 0, \quad x \in B_\varepsilon(0^\lambda) \backslash \{0^\lambda\}.$$

Or equivalently,

$$\bar{u}(x) \geq C > 0, \quad x \in B_\varepsilon(0) \backslash \{0\}.$$

Let η be a smooth cutoff function such that $\eta(x) \in [0,1]$ in $\mathbb{R}^n$, $supp\,\eta \subset B_2$ and $\eta(x) \equiv 1$ in B_1. Let

$$(-\triangle)^{\alpha/2}\varphi(x) = \eta(x)f(v(x)).$$

Then

$$\varphi(x) = C_{n,-\alpha} \int_{\mathbb{R}^n} \frac{\eta(y)f(v(y))}{|x-y|^{n-\alpha}} dy = C_{n,-\alpha} \int_{B_2(0)} \frac{\eta(y)f(v(y))}{|x-y|^{n-\alpha}} dy.$$

It's trivial that for $|x|$ sufficiently large,

$$\varphi(x) \sim \frac{1}{|x|^{n-\alpha}}. \tag{A.12}$$

Since for R large,

$$\begin{cases} (-\triangle)^{\alpha/2}(u - \varphi + \frac{C}{R^{n-\alpha}}) \geq 0, \, x \in B_R, \\ (u - \varphi)(x) + \frac{C}{R^{n-\alpha}} \geq 0, \qquad x \in B_R^c, \end{cases} \tag{A.13}$$

by the *maximum principle* (see [Si]), we have

$$(u - \varphi)(x) + \frac{C}{R^{n-\alpha}} \geq 0, \quad x \in B_R.$$

Thus as $R \to \infty$, it yields

$$(u - \varphi)(x) \geq 0, \quad x \in \mathbb{R}^n.$$

For $|x|$ sufficiently large, from (A.12), one can see that for some constant $C > 0$,

$$u(x) \geq \frac{C}{|x|^{n-\alpha}}. \tag{A.14}$$

Hence for $|x|$ small

$$u(\frac{x}{|x|^2}) \geq C|x|^{n-\alpha},$$

and

$$\bar{u}(x) = \frac{1}{|x|^{n-\alpha}} u(\frac{x}{|x|^2}) \geq C.$$

Together with (A.11), it yields that

$$U_\lambda(x) \geq \frac{C}{2} > 0, \quad x \in B_\varepsilon(0^\lambda) \backslash \{0^\lambda\}. \tag{A.15}$$

Through an identical argument, one can show that (A.15) holds for $V_\lambda(x)$ as well. $\quad\square$

Lemma A.1.5 *For $\lambda_o < 0$, if either of $U_{\lambda_o}(x), V_{\lambda_o}(x)$ is not identically 0, then there exists some constant C and $\varepsilon > 0$ small such that*

$$U_{\lambda_o}(x), V_{\lambda_o}(x) \geq C > 0, \quad x \in B_\varepsilon(0^{\lambda_o}) \backslash \{0^{\lambda_o}\}.$$

Proof. From Lemma 2.2 in [CFY], we have the integral expression of U_{λ_o}:

$$U_{\lambda_o}(x) = C_{n,\alpha} \int_{\Sigma_{\lambda_o}} \left(\frac{1}{|x-y|^{n-\alpha}} - \frac{1}{|x-y^{\lambda_o}|^{n-\alpha}} \right)$$
$$\cdot \left(\frac{f(|y^{\lambda_o}|^{n-\beta} \bar{v}_{\lambda_o}(y))}{|y^{\lambda_o}|^{n+\alpha}} - \frac{f(|y|^{n-\beta} \bar{v}(y))}{|y|^{n+\alpha}} \right) dy$$
$$= C_{n,\alpha} \int_{\Sigma_{\lambda_o}} \left(\frac{1}{|x-y|^{n-\alpha}} - \frac{1}{|x-y^{\lambda_o}|^{n-\alpha}} \right)$$
$$\cdot \left(\frac{f(|y^{\lambda_o}|^{n-\beta} \bar{v}_{\lambda_o}(y))}{[|y^{\lambda_o}|^{n-\beta} \bar{v}_{\lambda_o}(y)]^p} \bar{v}_{\lambda_o}^p(y) - \frac{f(|y|^{n-\beta} \bar{v}_{\lambda_o}(y))}{[|y|^{n-\beta} \bar{v}_{\lambda_o}(y)]^p} \bar{v}_{\lambda_o}^p(y) \right.$$
$$\left. + \frac{f(|y|^{n-\beta} \bar{v}_{\lambda_o}(y)) - f(|y|^{n-\beta} \bar{v}(y))}{|y|^{n+\alpha}} \right) dy$$
$$\geq C_{n,\alpha} \int_{\Sigma_{\lambda_o}} \left(\frac{1}{|x-y|^{n-\alpha}} - \frac{1}{|x-y^{\lambda_o}|^{n-\alpha}} \right)$$
$$\cdot \frac{f(|y|^{n-\beta} \bar{v}_{\lambda_o}(y)) - f(|y|^{n-\beta} \bar{v}(y))}{|y|^{n+\alpha}} dy.$$

Since

$$V_{\lambda_0}(x) \not\equiv 0, \quad x \in \Sigma_{\lambda_0},$$

there exists some x^0 such that

$$V_{\lambda_o}(x^0) > 0.$$

Thus, for some $\delta > 0$ small, it holds that

$$f(|y|^{n-\beta}\bar{v}_{\lambda_o}(y)) - f(|y|^{n-\beta}\bar{v}(y)) \geq C > 0, \quad y \in B_\delta(x^0).$$

Therefore,

$$U_{\lambda_o}(x) \geq \int_{B_\delta(x^0)} C \, dy \geq C > 0. \tag{A.16}$$

In a same way, one can show that $V_{\lambda_o}(x)$ also satisfies (A.16).   □

Lemma A.1.6 *For $\lambda_o < 0$, if*

$$V_{\lambda_o}(x) = U_{\lambda_o}(x) \equiv 0, \quad x \in \mathbb{R}^n, \tag{A.17}$$

then

$$\frac{f(t)}{t^{\frac{n+\alpha}{n-\beta}}} = C, \quad t \in (0, \max_{\mathbb{R}^n} u],$$

and

$$\frac{g(t)}{t^{\frac{n+\beta}{n-\alpha}}} = C, \quad t \in (0, \max_{\mathbb{R}^n} v].$$

Proof. We first show $\frac{f(t)}{t^{\frac{n+\alpha}{n-\beta}}} = C, \quad t \in (0, \max_{\mathbb{R}^n} u]$.

When $\lambda_o < 0$, from

$$U_{\lambda_o} \equiv 0, \quad x \in \Sigma_{\lambda_o},$$

we can see that $\bar{u}(x)$ is symmetric about T_{λ_o}. By the arbitrariness of the x_1 direction, it follows that $\bar{u}(x)$ is radially symmetric about some point $Q \neq 0$. Next we show that $(-\triangle)^{\alpha/2}\bar{u}(x)$ shares such symmetry. In other words, for any x and $\tilde{x}$ satisfying $\frac{x+\tilde{x}}{2} = Q$, it holds

$$(-\triangle)^{\alpha/2}\bar{u}(x) = (-\triangle)^{\alpha/2}\bar{u}(\tilde{x}). \tag{A.18}$$

Indeed,

$$(-\triangle)^{\alpha/2}\bar{u}(x) - (-\triangle)^{\alpha/2}\bar{u}(\tilde{x})$$

$$= C P.V. \int_{\mathbb{R}^n} \frac{\bar{u}(x) - \bar{u}(y)}{|x - y|^{n+\alpha}} dy - C P.V. \int_{\mathbb{R}^n} \frac{\bar{u}(\tilde{x}) - \bar{u}(y)}{|\tilde{x} - y|^{n+\alpha}} dy$$

$$= C P.V. \int_{\mathbb{R}^n} [\bar{u}(x) - \bar{u}(y)] \left(\frac{1}{|x - y|^{n+\alpha}} - \frac{1}{|\tilde{x} - y|^{n+\alpha}} \right) dy = 0. \tag{A.19}$$

To prove (A.19), for any given $y \in \mathbb{R}^n$, there exists a corresponding $\tilde{y}$ such that $\frac{y+\tilde{y}}{2} = Q$. By similar triangles (see Fig. A.1) we know that

$$[\bar{u}(x) - \bar{u}(y)]\left(\frac{1}{|x-y|^{n+\alpha}} - \frac{1}{|\tilde{x}-y|^{n+\alpha}}\right)$$

$$= [\bar{u}(x) - \bar{u}(\tilde{y})]\left(\frac{1}{|\tilde{x}-\tilde{y}|^{n+\alpha}} - \frac{1}{|x-\tilde{y}|^{n+\alpha}}\right)$$

$$= -[\bar{u}(x) - \bar{u}(\tilde{y})]\left(\frac{1}{|x-\tilde{y}|^{n+\alpha}} - \frac{1}{|\tilde{x}-\tilde{y}|^{n+\alpha}}\right).$$

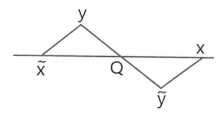

Fig. A.1

Through identical arguments, one can prove that $\bar{v}(x)$ and $(-\triangle)^{\beta/2}\bar{v}(x)$ are also radially symmetric about Q. Further, with (A.17) and (6.168), it is easy to see that $u(x)$ and $v(x)$ are bounded near infinity. In other words, $\max_{\mathbb{R}^n} u$ and $\max_{\mathbb{R}^n} v$ exist.

By (6.169) we have

$$\frac{(-\triangle)^{\alpha/2}\bar{u}(x)}{\bar{v}^{\frac{n+\alpha}{n-\beta}}(x)} = \frac{f(|x|^{n-\beta}\bar{v}(x))}{\left(|x|^{n-\beta}\bar{v}(x)\right)^{\frac{n+\alpha}{n-\beta}}}. \tag{A.20}$$

For convenience, let

$$G(|x-Q|) := \frac{(-\triangle)^{\alpha/2}\bar{u}(x)}{\bar{v}^{\frac{n+\alpha}{n-\beta}}(x)},$$

$$t(x) = |x|^{n-\beta}\bar{v}(x), \qquad F(t) := \frac{f(t)}{t^{\frac{n+\alpha}{n-\beta}}}.$$

Then (A.20) becomes

$$G(|x-Q|) = F(t). \tag{A.21}$$

Next we prove that $\max_{\mathbb{R}^n} t$ and $\min_{\mathbb{R}^n} t$ must be obtained on the line through Q and the origin 0. To see this, consider a $(n-1)$-sphere $\partial B_R(Q)$ for any given $R > 0$. Due to the radial symmetry of $\bar{v}(x)$, there exists some constant $C = C(R)$ such that

$$\bar{v}(x)|_{\partial B_R(Q)} = C.$$

Let O be the origin, P and $\tilde{P}$ be the points of intersection of $\partial B_R(Q)$ and line OQ. Then

$$\bar{v}(P) = \bar{v}(\tilde{P}).$$

Without loss of generality, we assume that P and O are on the same side of Q (see Fig. A.2). Thus $\overline{PO} < \overline{\tilde{P}O}$ and

$$\max_{\partial B_R(Q)} t = t(\tilde{P}) > \min_{\partial B_R(Q)} t = t(P). \tag{A.22}$$

Consider $\mathbb{R}^n$ as the union of a family of concentric spheres with continuously increasing radiuses, and it's easy to see that $t(x)$ must attain on line OQ the global maximum and minimum values.

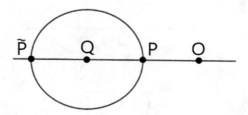

Fig. A.2

We claim that for every pair of such "twin" points $P, \tilde{P} \neq O$ and $P, \tilde{P} \in OQ$, it holds that

$$F(t) \text{ is constant in } [t(P), t(\tilde{P})]. \tag{A.23}$$

By (A.21) we know

$$F(t(P)) = G(|P - Q|) = G(|\tilde{P} - Q|) = F(t(\tilde{P})).$$

From (A.22) and the monotonicity of $F(t)$, (A.23) follows easily.

Let

$$t(\tilde{M}) := \max_{\mathbb{R}^n} t = \max_{\mathbb{R}^n} u$$

and $M = 2Q - \tilde{M}$. Then M and O fall on the same side of Q and

$$\overline{MO} < \overline{\tilde{M}O}, \tag{A.24}$$

since

$$t(x) \to 0 < t(M) < t(\tilde{M}), \text{ as } x \to O.$$

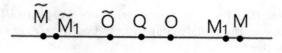

Fig. A.3

Let $P = M$ and $P = O$ respectively in (A.23), it yields

$$F(t) \text{ is constant in } [t(M), t(\tilde{M})], \qquad (A.25)$$

and

$$F(t) \text{ is constant in } (0, t(\tilde{O})]. \qquad (A.26)$$

If $t(\tilde{O}) \geq t(M)$, then from (A.25) and (A.26) we conclude that

$$F(t) \text{ is constant in } (0, \max_{\mathbb{R}^n} u]. \qquad (A.27)$$

Otherwise, we need to bridge the gap between $t(\tilde{O})$ and $t(M)$.

By the continuity of $t(x)$ on $\overline{OM}$, and the fact that

$$t(x) \to 0 < t(\tilde{O}) < t(M), \quad x \to O,$$

one can see that there exists a point $M_1 \in \overline{OM}$ such that $t(M_1) = t(\tilde{O})$. Let $P = M_1$ in (A.23). We have

$$F(t) \text{ is constant in } [t(M_1), t(\tilde{M}_1)],$$

that is,

$$F(t) \text{ is constant in } [t(\tilde{O}), t(\tilde{M}_1)]. \qquad (A.28)$$

Combining (A.26) and (A.28), we have

$$F(t) \text{ is constant in } (0, t(\tilde{M}_1)].$$

By Fig. A.3, it's easy to see that $\tilde{M}_1$ and M_1 are separated by $\overline{O\tilde{O}}$. As a result, it must be true that

$$t(\tilde{M}_1) - t(M_1) \geq C > 0, \text{C independent of } M_1 \text{ and } \tilde{M}_1.$$

In other words, we have made one solid step forward in filling the gap. If $t(\tilde{M}_1) \geq t(M)$, then the proof is done. If not, one can find some $M_2 \in \overline{M_1 M}$ so that $t(M_2) = t(\tilde{M}_1)$ and

$$F(t) \text{ is constant in } (0, t(\tilde{M}_2)].$$

Repeating the process for a finite times by choosing $M_j \in \overline{M_{j-1} M}$ and we arrive at (A.27).

Through a similar argument one can show that

$$\frac{g(t)}{t^{\frac{n+\beta}{n-\alpha}}} = C, \quad t \in (0, \max_{\mathbb{R}^n} v].$$

A.2 For Moving Spheres Method

Lemma A.2.1 *For any $\tilde{x} \in \mathbb{R}^n$, let $\bar{v}$ be the Kelvin transform of u centered at $\tilde{x}$:*

$$\bar{v}(x) = \left(\frac{\lambda}{|x - \tilde{x}|}\right)^{n-\alpha} u\left(\frac{\lambda^2(x - \tilde{x})}{|x - \tilde{x}|^2} + \tilde{x}\right).$$

Then

$$(-\triangle)^{\alpha/2}\bar{v}(x) = \left(\frac{\lambda}{|x - \tilde{x}|}\right)^{n+\alpha} ((-\triangle)^{\alpha/2}u)\left(\frac{\lambda^2(x - \tilde{x})}{|x - \tilde{x}|^2} + \tilde{x}\right).$$

Proof. Without loss of generality, let $\tilde{x}$ be the origin. Let

$$v(x) = \frac{1}{|x|^{n-\alpha}} u\left(\frac{x}{|x|^2}\right). \tag{A.29}$$

Let $x = \frac{y}{\lambda^2}$, then

$$v\left(\frac{y}{\lambda^2}\right) = \left(\frac{\lambda^2}{|y|}\right)^{n-\alpha} u\left(\frac{\lambda^2 y}{|y|^2}\right) = \lambda^{n-\alpha} \bar{v}(y).$$

$$\begin{aligned}
(-\triangle)^{\alpha/2} \bar{v}(y) &= (-\triangle)^{\alpha/2} \left(\frac{1}{\lambda^{n-\alpha}} v\left(\frac{y}{\lambda^2}\right)\right) \\
&= \frac{1}{\lambda^{n-\alpha}} \left((-\triangle)^{\alpha/2} v\right)\left(\frac{y}{\lambda^2}\right) \frac{1}{\lambda^{2\alpha}} \\
&= \frac{1}{\lambda^{n+\alpha}} \frac{1}{|\frac{y}{\lambda^2}|^{n+\alpha}} \left((-\triangle)^{\alpha/2} u\right)\left(\frac{\lambda^2 y}{|y|^2}\right) \\
&= \left(\frac{\lambda}{|y|}\right)^{n+\alpha} \left((-\triangle)^{\alpha/2} u\right)\left(\frac{\lambda^2 y}{|y|^2}\right). \qquad \square
\end{aligned}$$

Here, we prove (8.13).

Lemma A.2.2 *Fix λ small, for $0 < \delta \ll \lambda$, we have*

$$w_\lambda(x) > 1, \quad x \in B_\delta(x_o) \backslash \{x_o\}.$$

Proof. Without loss of generality, let $x_0 = 0$. For $|x|$ large, there exists a positive constant C_o such that

$$\begin{aligned}
u(x) &= \int_{\mathbb{R}^n} \frac{g(u(y))}{|x-y|^{n-\alpha}} dy \\
&\geq \int_{B_1(0)} \frac{g(u(y))}{|x-y|^{n-\alpha}} dy \\
&\geq \int_{B_1(0)} \frac{C_o}{|x-y|^{n-\alpha}} dy \\
&\sim \frac{C}{|x|^{n-\alpha}}.
\end{aligned}$$

For $0 < \delta \ll \lambda$ and $x \in B_\delta(x_o) \backslash \{x_o\}$, it's easy to see that $|x^\lambda|$ takes large values. Hence,

$$u_\lambda(x) = \left(\frac{\lambda}{|x|}\right)^{n-\alpha} u(x^\lambda) \geq \left(\frac{\lambda}{|x|}\right)^{n-\alpha} \frac{C}{|x^\lambda|^{n-\alpha}} = \frac{C}{\lambda^{n-\alpha}}.$$

Choosing λ sufficiently small, it holds that

$$w_\lambda(x) = u_\lambda(x) - u(x) \geq \frac{C}{\lambda^{n-\alpha}} - C > 1. \qquad \square$$

Lemma A.2.3 *Let*

$$\lambda_0 \equiv \sup\{\lambda > 0 \mid w_\mu(x) \geq 0, \ \forall \ x \in B_\mu(x_0)\backslash\{x_o\}, \ \forall 0 < \mu \leq \lambda\}.$$

If $w_{\lambda_0} \not\equiv 0$, then for $\varepsilon > 0$ small we have

$$w_{\lambda_0}(x) \geq c > 0, \quad x \in B_\varepsilon(x_0)\backslash\{x_o\}.$$

Proof. Without any loss of generality, let $x_0 = 0$.

By the definition of λ_0, if

$$w_{\lambda_0}(x) \not\equiv 0, \quad x \in B_{\lambda_0}(0)\backslash\{0\},$$

then there exists a point $\tilde{x} \in B_{\lambda_0}(0)\backslash\{0\}$ such that

$$w_{\lambda_0}(\tilde{x}) > 0.$$

And further, there exists a small positive δ such that

$$w_{\lambda_0}(x) > 0, \quad x \in B_\delta(\tilde{x}).$$

For

$$(-\triangle)^{\alpha/2}u(x) = u^p(x), \quad x \in \mathbb{R}^n,$$

its integral solution takes the form of

$$u(x) = \int_{\mathbb{R}^n} \frac{1}{|x-y|^{n-\alpha}} u^p(y)dy.$$

Let

$$u_{\lambda_0}(x) = (\frac{\lambda_0}{|x|})^{n-\alpha} u(\frac{\lambda_0^2 x}{|x|^2}).$$

Then through elementary calculations, we have

$$(-\triangle)^{\alpha/2}u_{\lambda_0}(x) = (\frac{\lambda_0}{|x|})^\tau u_{\lambda_0}^p(x), \quad \tau = n + \alpha - p(n-\alpha),$$

and

$$u_{\lambda_0}(x) = \int_{\mathbb{R}^n} \frac{1}{|x-y|^{n-\alpha}} (\frac{\lambda_0}{|y|})^\tau u_{\lambda_0}^p(y)dy.$$

It's easy to see that

$$u(x) = \int_{\mathbb{R}^n} \frac{1}{|x-y|^{n-\alpha}} u^p(y)dy$$

$$= \int_{B_{\lambda_0}(0)} \frac{1}{|x-y|^{n-\alpha}} u^p(y)dy + \int_{\mathbb{R}^n \backslash B_{\lambda_0}(0)} \frac{1}{|x-y|^{n-\alpha}} u^p(y)dy$$

$$= \int_{B_{\lambda_0}(0)} \frac{1}{|x-y|^{n-\alpha}} u^p(y)dy + \int_{B_{\lambda_0}(0)} \frac{1}{|\frac{x|y|}{\lambda_0} - \frac{\lambda_0 y}{|y|}|^{n-\alpha}} (\frac{\lambda_0}{|y|})^\tau u_{\lambda_0}^p(y)dy.$$

Similar steps yield

$$u_{\lambda_0}(x) = \int_{B_{\lambda_0}(0)} \frac{1}{|x-y|^{n-\alpha}} \left(\frac{\lambda_0}{|y|}\right)^\tau u_{\lambda_0}^p(y)\, dy + \int_{B_{\lambda_0}(0)} \frac{1}{|\frac{x|y|}{\lambda_0} - \frac{\lambda_0 y}{|y|}|^{n-\alpha}} u^p(y)\, dy.$$

Thus

$$
\begin{aligned}
w_{\lambda_0}(x) &= u_{\lambda_0}(x) - u(x) \\
&= \int_{B_{\lambda_0}(0)} \left(\frac{1}{|x-y|^{n-\alpha}} - \frac{1}{|\frac{x|y|}{\lambda_0} - \frac{\lambda_0 y}{|y|}|^{n-\alpha}} \right) \left(\left(\frac{\lambda_0}{|y|}\right)^\tau u_{\lambda_0}^p(y) - u^p(y) \right) dy \\
&\geq C \int_{B_{\lambda_0}(0)} \left(\frac{1}{|x-y|^{n-\alpha}} - \frac{1}{|\frac{x|y|}{\lambda_0} - \frac{\lambda_0 y}{|y|}|^{n-\alpha}} \right) u^{p-1}(y) w_{\lambda_0}(y)\, dy \\
&\geq C \int_{B_\delta(\tilde{x})} \left(\frac{1}{|x-y|^{n-\alpha}} - \frac{1}{|\frac{x|y|}{\lambda_0} - \frac{\lambda_0 y}{|y|}|^{n-\alpha}} \right) u^{p-1}(y) w_{\lambda_0}(y)\, dy \\
&\geq C \int_{B_\delta(\tilde{x})} C_1 C_2 C_3\, dy > C_4 > 0.
\end{aligned}
$$

Bibliography

[A]   D. Applebaum, Lévy Processes and Stochastic Calculus, 2nd ed, Cambridge Studies in Advanced Mathematics, 116, Cambridge University Press, Cambridge, 2009.

[ABR]   S. Axler, P. Bourdon and W. Ramey, Harmonic Function Theory, 2nd ed, Springer-Verlag New York Inc. 2001.

[B]   J. Bourgain, Global solutions of nonlinear Schrodinger equations, AMS Colloquium Publications, Vol. 46, 1999, AMS Providence, Rhode Island.

[BCPS] C. Brändle, E. Colorado, A. de Pablo and U. Sánchez, A concave−convex elliptic problem involving the fractional Laplacian, Proc. Roy. Soc. Edinburgh, 143(2013) 39-71.

[BDN]   H. Berestycki, C. Dolcetta and L. Nirenberg, Superlinear indefinite elliptic problems and nonlinear Liouville theorems, Topol. Methods Nonlinear Anal. 4(1994) 59-78.

[Be]   J. Bertoin, Lévy Processes, Cambridge Tracts in Mathematics, 121 Cambridge University Press, Cambridge, 1996.

[BGR]   R. M. Blumenthal, R. K. Getoor and D. B. Ray, On the Distribution of First Hits for the Symmetric Stable Processes, 9(1961) 540-554.

[BKN]   K. Bogdan, T. Kulczycki and A. Nowak, Gradient estimates for harmonic and q-harmonic functions of symmetric stable processes, Illinois J. Math. 46(2002) 541-556.

[BL]   A. Bonfiglioli and E. Lanconelli, Liouville-type theorems for real sub-Laplacians, Manuscripta Math. 105(2001) 111-124.

[BNV]   H. Berestycki, L. Nirenberg and S. Varadhan, The principal eigenvalue and maximum principle for second-order elliptic operators in general domains, Comm. Pure Appl. Math. 47(1994) 47-92.

[Bo]   K. Bogdan, The boundary Harnack principle for the fractional Laplacian, Studia Math. 123(1997) 43-80.

[BoG]   J. P. Bouchard and A. Georges, Anomalous diffusion in disordered media, Statistical mechanics, models and physical applications, Physics reports 195 (1990).

[BPGQ] B. Barrios, L. Del Pezzo, J. Garcia-Melian and A. Quaas, A priori bounds and existence of solutions for some nonlocal elliptic problems, Revista Matemática Iberoamericana, 34(2018) 195-220.

[C]   K.-C. Chang, Methods in Nonlinear Analysis, Springer-Verlag Berlin Heidelberg 2005.

[CaS]   L. Caffarelli and L. Silvestre, An extension problem related to the fractional Laplacian, Comm. Partial Differential Equations, 32(2007) 1245-1260.

328   Bibliography

[CaS1]   L. Caffarelli and L. Silvestre, Regularity theory for fully nonlinear integro-differential equations, Comm. Pure. Appl. Math, 62(2009) 597-638.

[CaV]   L. Caffarelli and L. Vasseur, Drift diffusion equations with fractional diffusion and the quasi-geostrophic equation, Ann. Math. 3(2010) 1903-1930.

[CC]   L. Cao and W. Chen, Liouville type theorems for poly-harmonic Navier problems, Disc. Cont. Dyna. Sys. 33(2013) 3937-3955.

[CDL]   W. Chen, L. D'Ambrosio, and Y. Li, Some Liouville theorems for the fractional Laplacian, Nonlinear Analysis, Theory, Methods & Appl, 121(2015) 370-381.

[CDM]   G. Caristi, L. D'Ambrosio and E. Mitidieri, Representation formulae for solutions to some classes of higher order systems and related Liouville theorems, Milan J. Math. 76(2008) 27-67.

[CFY]   W. Chen, Y. Fang, and R. Yang, Liouville theorems involving the fractional Laplacian on a half space, Adv. Math. 274(2014) 167-198.

[CG]   A. Chang, M. González, Fractional Laplacian in conformal geometry, Adv. Math. 2(2011) 1410-1432.

[CGS]   L. Caffarelli, B. Gidas, and J. Spruck, Asymptotic symmetry and local behavior of semilinear elliptic equations with critical Sobolev growth, C.P.A.M. XLII (1989) 271-297.

[CL]   W. Chen and C. Li, Classification of solutions of some nonlinear elliptic equations, Duke Math. J. 63(1991) 615-622.

[CL1]   W. Chen and C. Li, Regularity of solutions for a system of integral equation, Comm. Pure Appl. Anal, 4(2005) 1-8.

[CL2]   W. Chen and C. Li, Methods on Nonlinear Elliptic Equations, AIMS. Ser. Differ. Equ. Dyn. Syst, vol.4 2010.

[CL3]   W. Chen and C. Li, Radial symmetry of solutions for some integral systems of Wolff type, Disc. Cont. Dyn. Sys. 30(2011) 1083-1093.

[CL4]   W. Chen and C. Li, The best constant in some weighted Hardy-Littlewood-Sobolev inequality, Proc. AMS, 136(2008) 955-962.

[CL5]   W. Chen and C. Li, Classification of positive solutions for nonlinear differential and integral systems with critical exponents, Acta Mathematica Scientia, 29(2009) 949-960.

[CL6]   W. Chen and C. Li, A necessary and sufficient condition for the Nirenberg problem, Comm. Pure Appl. Math. 48(1995) 657-667.

[CL7]   W. Chen and C. Li, Super polyharmonic property of solutions for PDE systems and its applications, C.P.A.A. 6(2013) 2497-2514.

[CL8]   W. Chen and C. Li, A priori estimates for prescribing scalar curvature equations, Annals of Math. 145(1997) 547-564.

[CL9]   W. Chen and C. Li, Maximum principle for the fractional p-Laplacian and symmetry of solutions, Adv. Math. 335(2018) 735-758.

[CLL]   W. Chen, C. Li, and Y. Li, A direct method of moving planes for the fractional Laplacian, Adv. in Math. 308 (2017) 404-437.

[CLL1]   W. Chen, C. Li, and Y. Li, A direct blow-up and rescaling argument on nonlocal elliptic equations, International J. Math. 27(2016) 1-20.

[CLO]   W. Chen, C. Li, and B. Ou, Classification of solutions for an integral equation, Comm. Pure Appl. Math. 59(2006) 330-343.

[CLO1]   W. Chen, C. Li, and B. Ou, Qualitative properties of solutions for an integral equation, Disc. Cont. Dyn. Sys. 12(2005) 347-354.

[CLO2]   W. Chen, C. Li, and B. Ou, Classification of solutions for a system of integral equations, Commun. Partial Differ. Equ. 30(2005) 59-65.

[Co]   P. Constantin, Euler equations, Navier-Stokes equations and turbulence, in Mathematical Foundation of Turbulent Viscous Flows, Vol. 1871 of Lecture Notes in Math. Springer, Berlin, 2006.

[CoT]   R. Cont and P. Tankov, Financial Modelling with Jump Processes, Chapman & Hall/CRC Financial Mathematics Series, Boca Raton, Fl, 2004.

[CRS]   L. Caffarelli, J. Roquejoffre and Y. Sire, Variational problems with free boundaries for the fractional Laplacian, J. Eur. Math. Soc. 12(2010) 1151-1179.

[CS]   X. Cabré and Y. Sire, Nonlinear equations for fractional Laplacians, I: Regularity, maximum principles, and Hamiltonian estimates, Annales de l'Institut Henri Poincare (C) Non Linear Analysis. Elsevier Masson, 31(2014) 23-53.

[CS1]   L. Caffarelli and L. Silvestre, Regularity theory for fully nonlinear integro-differential equations, Comm. Pure Appl. Math. 62(2009)597-638.

[CSS]   L. Caffarelli, S. Salsa and L. Silverstre, Regularity estimates for the solution and the free boundary of the obstacle problem for the fractional Laplacian, Invent. Math. 171(2008) 425-461.

[CT]   X. Cabré and J. Tan, Positive solutions of nonlinear problems involving the square root of the Laplacian, Adv. Math. 224(2010) 2052-2093.

[CY]   A. Chang and P. Yang, On uniqueness of an n-th order differential equation in conformal geometry, Math. Res. Letters, 4(1997) 1-12.

[CZ]   W. Chen and J. Zhu, Radial symmetry and regularity of solutions for poly-harmonic Dirichlet problems, J. Math. Anal. Appl. 2(2011) 744-753.

[DC]   I. C. Dolcetta and A. Cutri, On the Liouville property for sublaplacians, Ann. Scuola Norm. Sup. Pisa Cl. Sci. 25(1997) 239-256.

[DK]   H. Dong and D. Kim, Schauder estimates for a class of non-local elliptic equations, Disc. Cont. Dyn. Sys. 33(2013) 2319-2347.

[DS]   L. Dupaigne and Y. Sire, A Liouville theorem for non local elliptic equations, Symmetry for elliptic PDEs, Contemp. Math. 528(2010) 105-114.

[F]   L. Fraenkel, An Introduction to Maximum Principles and Symmetry in Elliptic Problems, Cambridge Unversity Press, New York, 2000.

[Fa]   M. Fall, Entire s-harmonic functions are affine, Proc. Amer. Math. Soc. 144 (2016) 2587-2592.

[FC]   Y. Fang and W. Chen, A Liouville type theorem for poly-harmonic Dirichlet problem in a half space, Advances in Math. 229(2012) 2835-2867.

[FW]   M. Fall and T. Weth, Nonexistence results for a class of fractional elliptic boundary values problems, J. Funct. Anal. 263(2012) 2205-2227.

[FZ]   Y. Fang and J. Zhang, Nonexistence of positive solution for an integral equation on a half-space R_+^n, Comm. Pure and Applied Analysis, 12(2013) 663-678.

[G]   G. Grubb, Distributions and Operators, Springer, New York, 2009.

[Ge]   R. K. Getoor, First passenge times for symmetric stable processes in space, Trans. Amer. Math. Soc. 101(1961) 75-90.

[GNN]   B. Gidas, W. Ni, and L. Nirenberg, Symmetry of positive solutions of nonlinear elliptic equations in R^n, Mathematical Analysis and Applications, vol. 7a of the book series *Advances in Mathematics*, Academic Press, New York, 1981.

[GT]   D. Gilbarg and N. S. Trudinger, Elliptic partial differential equations of second order. Reprint of the 1998 edition. Classics in Mathematics. Springer, Berlin, 2001.

[GS]   B. Gidas and J. Spruck, A priori bounds for positive solutions of nonlinear elliptic equations, Comm. Partial Differential Equations, 6(1981) 883-901.

[Gu]   Q. Guan, Integration by parts formula for regional fractional Laplacian, Comm. Math. Phys. 266(2006) 289-329.

[GZ]   R. Graham and M. Zworski, Scattering matrix in conformal geometry, Invent. Math. 1(2003) 89-118.

[JLX1]   T. Jin, Y. Li, and J. Xiong, On a fractional Nirenberg problem, part I: blow up analysis and compactness of solutions, J. Eur. Math. Soc. 6(2014) 1111-1171.

[JLX2]   T. Jin, Y. Li, and J. Xiong, On a fractional Nirenberg problem, part II: existence of solutions, Int. Math. Res. Not. 6(2015) 1555-1589.

[JW]   S. Jarohs and Tobias Weth, Symmetry via antisymmetric maximum principles in nonlocal problems of variable order, Annali di Matematica Pura ed Applicata (1923-), doi: 10.1007/s10231-014-0462-y.

[Ku]   T. Kulczycki, Properties of Green function of symmetric stable processes, Probability and Mathematical Statistics, 17(1997) 339-364.

[L]   N. S. Landkof, Foundations of modern potential theory, Springer-Verlag Berlin Heidelberg, New York, 1972. Translated from the Russian by A. P. Doohovskoy, Die Grundlehren der mathematischen Wissenschaften, Band 180.

[Li]   C. Li, Local asymptotic symmetry of singular solutions to nonlinear elliptic equations, Invent. Math. 123(1996) 221-231.

[Lie1]   E. Lieb, Existence and uniqueness of the minimizing solution of Choquard nonlinear equation, Stud. Appl. Math, 57(1977) 93-105.

[Lin]   C. Lin, A classification of solutions of a conformally invariant fourth order equation in R^n, Comment. Math. Helv. 73(1998) 206-231.

[LiZ]   Y. Li and M. Zhu, Uniqueness theorems through the method of moving spheres, Duke Math. J. 80(1995) 383-417.

[LL]   E. Lieb and M. Loss, Analysis, American Mathmatical Society, Volume 14, 2001.

[LM]   C. Li and L. Ma, Uniqueness of positive bound states to Shrodinger systems with critical exponents, SIAM J. of Appl. Anal. 40(2008) 1049-1057.

[LS]   M. Lazzo and P. Schmidt, Nonexistence criteria for polyharmonic boundary-value problems, Analysis International mathematical journal of analysis and its applications, 28(2008) 449-460.

[LS1]   M. Lazzo and P. Schmidt, Oscillatory radial solutions for subcritical biharmonic equations, J. Differential Equations, 247(2009) 1479-1504.

[LWX]   C. Li, Z. Wu and H. Xu, Maximum principles and Bocher type theorems, Proceedings of the National Academy of Sciences, June 20, 2018.

[LZ]   G. Lu and J. Zhu, An overdetermined problem in Riesz-potential and fractional Laplacian, Nonlinear Analysis, 75(2012) 3036-3048.

[LZ1]   G. Lu and J. Zhu, Axial symmetry and regularity of solutions to an integral equation in a half space, Pacific J. Math. 253(2012) 455-473.

[M]   L. Modica, A gradient bound and a Liouville theorem for nonlinear Poisson equations, Comm. Pure Appl. Math. 38(1985) 679-684.

[MC]   L. Ma and D. Chen, A Liouville type theorem for an integral system, Comm. Pure Appl. Anal. 5(2006) 855-859.

[MC1]   L. Ma and D. Chen, Radial symmetry and monotonicity results for an integral equation, J. Math. Anal. Appl. 2(2008) 943-949.

[MCL]   C. Ma, W. Chen, and C. Li, Regularity of solutions for an integral system of Wolff type, Adv. Math. 3(2011) 2676-2699.

[Mi]   E. Mitidieri, Nonexistence of positive solutions of semilinear elliptic systems in R^N, Differential & Integral Equations, 9(1996) 465-479.

[ML]   L. Ma and B. Liu, Symmetry results for decay solutions of elliptic systems in the whole space, Advances in Math. 225(2010) 3052-3063.

[MLL]   P. Ma, F. Li and Y. Li, A Pohozaev Identity for the Fractional Hénon System, Acta Mathematica Sinica, English Series 33(2017) 1382-1396.

[MZ]   L. Ma and L. Zhao, Classification of positive solitary solutions of the nonlinear Choquard equation, Arch. Rat. Mech. Anal. 195(2010) 455-467.

[MZ1]   L. Ma and L. Zhao, Sharp thresholds of blow up and global existence for the coupled nonlinear Schrodinger system, J. Math. Phys. 49(2008); doi:10.1063/1.2939238.

[NPV]   E. Nezza, G. Palatucci, and E. Valdinoci, Hitchhikers guide to the fractional Sobolev spaces, Bull. Sci. math. 5(2011) 521-573.

[QR]   J. Qing and D. Raske, On positive solutions to semilinear conformally invariant equations on locally conformally flat manifolds, International Mathematics Research Notices, Vol. 2006, Article ID 94172, 1-20.

[R]   P. H. Rabinowitz, Differential Equations, Springer Berlin Heidelberg, 1982.

[RS]   X. Ros-Oton and J. Serra, The Pohozaev identity for the fractional Laplacian, J. Arch. Rational Mech. Anal. 213(2014) 587-628. doi:10.1007/s00205-014-0740-2

[RS1]   X. Ros-Oton and J. Serra, The Dirichlet problem for the fractional Laplacian: Regularity up to the boundary, Journal de Mathématiques Pures et Appliquées, 101(2014) 275-302.

[S]   Y. Shen, A Liouville theorem for harmonic maps, American Journal of Mathematics, 117(1995) 773-785.

[Sc]   R. Schoen, On the number of constant scalar curvature metrics in a conformal class, Differential Geometry, 311-320, Pitman Monogr. Surveys Pure Appl. Math., 52, Longman Sci. Tech., Harlow, 1991.

[Se]   J. Serrin, A symmetry problem in potential theory, Arch. Rat. Mech. Anal. 43(1971) 304-318.

[Si]   L. Silvestre, Regularity of the obstacle problem for a fractional power of the Laplace operator, Comm. Pure Appl. Math. 60(2007) 67-112.

[SV]   R. Servadei and E. Valdinoci, The Brezis-Nirenberg result for the fractional Laplacian, Trans. Amer. Math. Soc. 367(2015) 67-102.

[SZ]   P. R. Stinga and C. Zhang, Harnack's inequality for fractional nonlocal equations, Disc. Cont. Dyna. Sys. 33(2013) 3153-3170.

[TZ]   V. Tarasov and G. Zaslasvky, Fractional dynamics of systems with long-range interaction, Comm. Nonl. Sci. Numer. Simul. 11(2006) 885-889.

[WX]   J. Wei and X. Xu, Classification of solutions of higher order conformally invariant equations, Math. Ann. 313(1999) 207-228.

[Z]   M. Zhu, Liouville theorems on some indefinite equations, Proc. Roy. Soc. Edinburgh Sect. A Math. 129(1999) 649-661.

[ZCLC]   L. Zhang, W. Chen, C. Li, and T. Cheng, A Liouville theorem for α-harmonic functions in R_+^n, Disc. Cont. Dyn. Sys. 36(2016) 1721-1736.

[ZF]   C. Zhong and X. Fan, Introduction to Nonlinear Analysis, Lanzhou University press, 2011.

[ZCCY]   R. Zhuo, W. Chen, X. Cui, and Z. Yuan, Symmetry and non-existence of solutions for a nonlinear system involving the fractional Laplacian, Disc. Cont. Dyn. Sys. 2(2015) 1125-1141.